Holography MarketPlace

6th Edition

*The reference text
and sourcebook for
holography worldwide.*

The original version of this book contained holograms from the vendors and many of them
are no longer in business. Therefore this version of the book contains everything that was
in the original version but it has no holograms.

Edited by Franz Ross and Alan Rhody

Copyright h 1997 Ross Books

Library Catalogue Information
Holography MarketPlace - Sixth Edition
Bibliography:
1. Includes Index.
2. Holography
3. Directories - Holography Industry
4. Photography
5. Physics
ISBN 978-0-89496-099-4
Printed in Canada

About The Front Cover Hologram

The unique hologram featured on this year's cover of the HOLOGRAPHY MARKETPLACE is an illustration of the current advancement in the use of computer generated art and video in holograms. The hologram is unique in that it is a combination of three different types of images all contained in a single hologram.

The image of St. Basil Cathedral, bordering Red Square in Moscow, Russia, was created from a photograph which was separated into a multiplane image consisting of 14 levels. In the past, conventional holography has been limited to less than six levels.

The image of the Mobius is a computer generated image rendered from a three dimensional model. The stereogram of the gentleman comes from video footage. Note that at a specific angle the hologram achieves full color in all three images on all levels.

The hologram was a joint effort between Linda Law Holographics, Simian Company Inc., and STI. Linda Law is an artist who specializes in rendering two and three dimensional artwork to be used in the development of holograms. She created the artwork for the Mobius image and the St. Basil Cathedral. In addition, she designed the printed graphics on the cover to compliment the hologram.

The Simian Company originated the hologram on their unique origination system. A number of computer files containing the various images and backgrounds were combined into a single file. Tone, color and lighting were adjusted and the hologram was output onto a photo-resist plate.

The plate was sent to Simian's sister company, STI ,who then created the nickel master and matrix production shims and embossed the image into polyester film. The film was then metallized, adhesive coated, a liner was added and then it was die cut into finished labels. These labels were sent to the publisher for application.

See the advertising insert in this book and the business directory listings for additional information about these companies.

Correction: The Mobius image on the cover is owned exclusively by A D 2000 INC's HoloBank. The image was designed by Peter Scheir of that company.

Foreword

The sixth edition of the Holography MarketPlace has a number of new items that will be of interest to professional holographers, designers and sales/marketing managers, as well as more basic information that will appeal to the interested businessperson, student and casual reader.

New Articles: For holographers, we have updated and reorganized the section on silver halide recording materials. We have compiled the previously hard-to-get information about the Russian emulsions from Siavich, along with information about Kodak's and Agfa's current product lines. This information, along with the names of companies that still sell discontinued Ilford materials, should greatly assist holographers searching for basic recording supplies. For beginners, we have included articles about where to find affordable lasers and how to make holograms "on the kitchen counter". For experienced holographers, we have presented useful information about the newest laser systems and color reflection holography techniques.

For art directors and designers, we have updated and out- lined the procedures used to generate "camera-ready" artwork for commercial holographic applications. Of special interest to computer graphic artists is the section that describes the digital hardware and software tools that are currently being employed in the industry. We have also included an interview with several professional modelmakers who create the highly detailed 3D miniatures frequently used as subject matter in holograms.

The Database: For marketing managers, salespeople, and researchers, we have updated our international corporate directory of businesses involved in the holography industry. Over the past months, we have painstakingly called and talked to nearly every one of the hundreds of companies listed in North America, Australia and most of Europe. We notice and encourage an increase in participation from Eastern European and Russian businesses. We are absolutely certain that there is no database even remotely close to the one you are holding in tenns of accuracy and completeness. We would appreciate if you would notify us (in writing) of any updates, mistakes, omissions and pertinent new listings.

Other Reference Material: Please notice that we have increased our coverage of the tools and equipment used by holographers. In future publications we plan to report more extensively about the replication and finishing industries. We will also be expanding our coverage of industrial holography, including Non Destructive Testing, optical storage devices and the fabrication of CGHs, DOEs and HOEs.

Holograms: Once again, we are proud to include a wonderful and useful "sample kit" of holograms that represents the state of the industry, today. When you contact our advertisers, please let them know you appreciate seeing actual samples of their work.

Thanks to everyone that assisted with this edition.

Franz Ross & Alan Rhody
Berkeley, CA - Winter 1977

Table of Contents

Index to Advertisers

RARE & COLLECTABLE EDITIONS

The following titles are now available in limited quantities. Each volume contains pertinent reference material, historically significant information and collectable holograms!

Now only **$19.95** each or **$50.00** for entire set of **2 - 5** !

HOLOGRAPHY MARKETPLACE 2nd EDITION (1990)

Includes the hologram: *Statue of Liberty (American Bank Note) embossed (2"x3")* Plus: Introduction to Holography; Emulsions and Recording Materials; Holographic Optical; Elements; Color Holography; Business of Holography; Holographic Distribution Process; Embossed Holograms; Computer Generated Holograms; Holographic Non Destructive Testing; Database of Businesses; Database of Individuals; Bibliography; Glossary

HOLOGRAPHY MARKETPLACE 3rd EDITION (1991)

Includes these holograms: *Brain Skull (Polaroid) two channel image photopolymer (3"x3"); Earth/Space/Grid (American Bank Note) embossed (3"x3"); Floating Alphabet (American Bank Note) embossed (3"x4"); Magic Wizard (American Bank Note) embossed (1"x1"); Woman/Fruit/Flowers (Light Impressions) embossed true color still life (4"x4"); Prehistoric Man (Bridgestone Graphic Technologies) embossed (3"dia.); Space Shuttle in Orbit (Archeozoic/Polaroid) photopolymer (1"x1")* Plus: Introduction to Holography; Varieties of Holograms; Recording Materials; Lasers; Holographic Optical Elements ; Non Destructive Testing; Computer Generated Holograms; Holography in Education; Embossed Holograms; Business of Holography; Businesses by Category; Database of Individuals; Bibliography; Glossary

HOLOGRAPHY MARKETPLACE 4th EDITION (1993)

Includes these holograms: *Transamerica Pyramid (Polaroid) photopolymer (2.5"x3"); Earth/Grid (AD2000 Inc./ABN) embossed (3"x3"); Disney Characters (Holograms De Mexico) color embossed (2"x3"); Ghostbusters (American Bank Note) 2 channel embossed (4"x5"); Butterfly (Holography Presses On) embossed diffraction (.5"); Egyptian King (Holopress) embossed (2"x2"); 4 Image Montage (The Diffraction Company) embossed (4"x4"); Inaugural Invitations (CFC Applied Holographics) embossed (1.5"x1.5"); Bouquet (CFC Applied Holographics) full color embossed (2.5"x3.5"); Folding Package (Transfer Print Foils/Light Impressions) embossed (3"x5"); Earth/Lab (Global Images/Chromagem) embossed (5"x5")* Plus: Sales and Distribution; Direct Mail Marketing; Model Making; Holography Basics; Advanced Principles; Holographic Optical Elements; Heads Up Display; Computer Generated Holograms; Holographic Non Destructive Testing; Embossed Holograms; Lasers; Recording Materials; Main Business Listings; Bibliography; Glossary

HOLOGRAPHY MARKETPLACE 5th EDITION (1995)

This edition features *TV- a limited edition photopolymer hologram which was produced using a newly developed process that incorporates digital, optical, and holographic technologies to create a truly amazing image!* Plus: *Harry 4"x5", Chinese Lion Dancers 3.5"x5" (The Lasersmith); Mount Rushmore 6"x9", Rock Solid (TPF); Space Shuttle, Flag, Eagle, Fireworks 5.5"x11" (Crown Roll Leaf); Map, (Hologramas de Mexico); Initials (HPO); Wizard 2"x3" polymer (Lazer Wizardry); Butterfly 3"x3", Valid (CFC/Applied); Tiger (Krystal Holographics); Matrix (Dimensional Arts)* Plus: Sales and Distribution; Holographic Stereograms; Embossed Holograms; Lasers; Recording Materials; Business Listings, Appendix and more.

ORDERING INFORMATION

Continental USA	Single $25	Entire Set 2-5 $65	All prices **include** Shipping and Handling
Alaska, Hawaii, Canada, Mexico	Single $30	Entire Set 2-5 $75	VISA, MC, AMEX, check or money
All Other Overseas Countries	Single $40	Entire Set 2-5 $85	order (US$) Payable to: ROSS BOOKS

P.O. Box 4340 Berkeley, CA 94704
Toll Free in USA 800-367-0930
ph. 510-841-2474, fax 510-841-2695
email sales@rossbooks.com

1

Introduction to Holography

This chapter assumes that you have no prior knowledge of holography. It discusses the difference between holography and photography, it details the unique attributes of the medium, and it explains the process by which basic holograms are created.

For holography beginners, the chapter includes a step-by-step explanation of how you can easily produce your own holograms. For more advanced practitioners, it concludes with a look at one of the techniques involved in producing color reflection holograms.

Introduction

You wake up in the morning and go out to breakfast. The restaurant is giving away hologram trading cards with each meal. You grab one for your son because you know he collects them. You finish your breakfast and head off to your job. Not paying close attention to the road at this early hour, you exceed the posted speed limit. The friendly Highway Patrol Officer who pulls you over takes your driver's license and examines it closely in the sunlight. He is looking for a hologram laminate that indicates your license is authentic.

At work, you open a new package of computer software. The box is sealed with a hologram label. On the way home from work you stop at the music store to buy a CD for your daughter. A 3D holographic portrait of her favorite group is on the CD's cover and, even more amazing, the band members appear to dance a bit as you tilt the box back and forth! You pick out a greeting card to go with the gift. It has colorful holographic foil decorating the front. Then you pay for your purchases with a credit card that has a hologram on it, too.

This scenario is an example of the frequency with which we see holograms in our everyday life. Unlike a decade ago, holographic images and holographic materials have entered the mainstream and are commonly used for a variety of commercial applications. This is partly due to the standardization of manufacturing methods and the inevitable maturation of the entire industry. More important

is the fact that the clients who commission holographic originations are achieving the results they desire.

Holography has proven that it is a viable technology that can be successfully integrated into a wide range of commercial endeavors including advertising campaigns. marketing promotions, security programs and retail sales. The medium has established a track record of deterring counterfeiters, attracting shoppers, and adding value to products. As the technology evolves further and potential users become more informed about the holography marketplace, a host of other applications will certainly arise.

Holography compared to Photography

Although often compared with photography. holography is really a completely different medium. Holograph: is based on different optical principles than photograph: and holograms have different physical attributes than photographs. They are the comparable only because they both are ways of recording an image onto a piece of photosensitive material (film), and, at times, similar equipment and materials are used in making both items.

The most apparent difference between a photograph and a hologram is image dimension and image depth. For instance, when we look at a photograph and move it from side to side we are unable to see "around" the scene or perceive any depth. We only see a flat (two dimensional) picture displayed on the surface of the film. This picture is actually a collection of light and dark shapes that we recognize as a particular subject. OUT memory might remind us that the subject really has dimension and depth, but this visual information is not recorded in the photograph.

A hologram is also flat, but the picture "on it" is not. When we look at a hologram and move it from side to side we can see many different views of the scene. We can also look behind foreground elements to see things in the background. This property is called parallax and it is closely tied to the process of visual perception. The Random House Dictionary defines parallax as, "the apparent displacement of an observed object due to the difference between two points of view". Our brain automatically uses these multiple views to create image dimension and image depth.

The relevant point here is that if we see at least two views of the same object, We can perceive a three dimensional image. A hologram inherently records and displays many views of the same object -- a photograph is limited to only one. (In fact, the word "hologram", coined by physicist Dennis Gabor in 1948, is commonly defined as "whole picture" based on its Greek roots. The terms "holography" and 'holographic' are typically used to discuss anything related to holograms, though the word "holograph" actually has another, unrelated meaning.)

To better illustrate the difference between the two media, imagine looking out through a small window at a particular view. If the window represented a photograph, you would be frozen in front of it in one position. Consequently, you would only see one perspective of that scene and would have no way of perceiving depth. However, if the window were a hologram you could move around the window and see different parts of the scene from many different viewing angles. The scene would look three dimensional and display depth. To further elaborate on this analogy, if this window could "remember" a scene, and you could carry it around and show it to someone else later -- they would be seeing a hologram of that scene.

It is possible to create an image with depth using two similar, but slightly different flat photographs if these "stereo pairs" are presented to our eyes in just the right way. Perhaps you are familiar with stereogram posters, "Viewmaster™" binocular viewers or virtual reality helmets. These artificial methods work, but they require people to either look at things unnaturally or to employ special viewing equipment. To recreate even the most basic real world renditions requires expensive computer driven display devices that are capable of generating thousands and thousands of stereo pairs. For many commercial applications it is more practical to use holography when dimensional imaging is required.

More Unique Holographic Attributes

Another difference between holograms and photographs is that holographers can position their images to "project off" or "float over" the surface of the film. One popular holographic image is of a water faucet that projects a foot or so out of the picture frame. A viewer can reach right out and put his hand right through this apparently solid image. Other holographic images can be positioned to appear some distance "behind" the picture frame. Still others

straddle the surface of the film (called the image plane). No photograph can do that!

These fascinating attributes are often exaggerated in futuristic movies and television shows. Many potential clients ask for holograms that can float in mid air or want miniature holographic people to be projected into the middle of a room. They assume that two laser beams can cross in space and a hologram will appear. Although these volumetric three dimensional displays are certainly being researched, they have little to do with holography. It is important to note that all holographic images are created by the interaction of a beam of light with a specially recorded piece of material, typically a sheet of film. You can hang this film on a wall, you can lay it on a table, you can even attach it to a piece of paper ---- but the image is always "attached" to the holographic recording material.

Some Display Limitations

Holograms do have two significant limitations which should be addressed, as they definitely affect commercial applications. One is viewing angle. Unlike a normal picture hanging on the wall or in a magazine, holographic images can only be seen within certain viewing parameters. If a viewer moves to far off-center, the holographic image will disappear. Another common problem is lighting. All holograms need to be properly illuminated in order to be seen. The lighting conditions that exist in many display environments are typically not the best for viewing holograms. Potential users must anticipate how their holograms are going to be displayed before starting production. There are ways to solve these problems using good design practices, as well as technological approaches which are currently being developed by researchers worldwide. Both these issues are covered in greater detail throughout this book.

Another Major Difference

The degree to which you can look around an object and the distance the image forms in front of, or behind the film depends on how the hologram was made. This brings up another major difference between the two mediums -- the procedures used to record imagery. We are all familiar with photography; we need a camera, film and an adequate amount of light. The light reflects off our subject, passes through the camera lens and exposes the film. Bright subjects expose the film a lot, darker subjects expose the film less. The picture we see is composed of varying tones.

Holography is different. It records an image in an entirely different way. Film is still employed, however, conventional cameras and ordinary lighting are not. Instead, to make a hologram the film needs to be exposed by a beam of coherent light -- that is, light which is composed of lightwaves that have identical frequencies and which are vibrating in phase. Light from the sun and from lightbulbs will not work. These sources emit light of varying frequencies with randomly varying phase.

Lasers do emit such light, and are therefore utilized. As certain lasers have become more affordable, making holo-

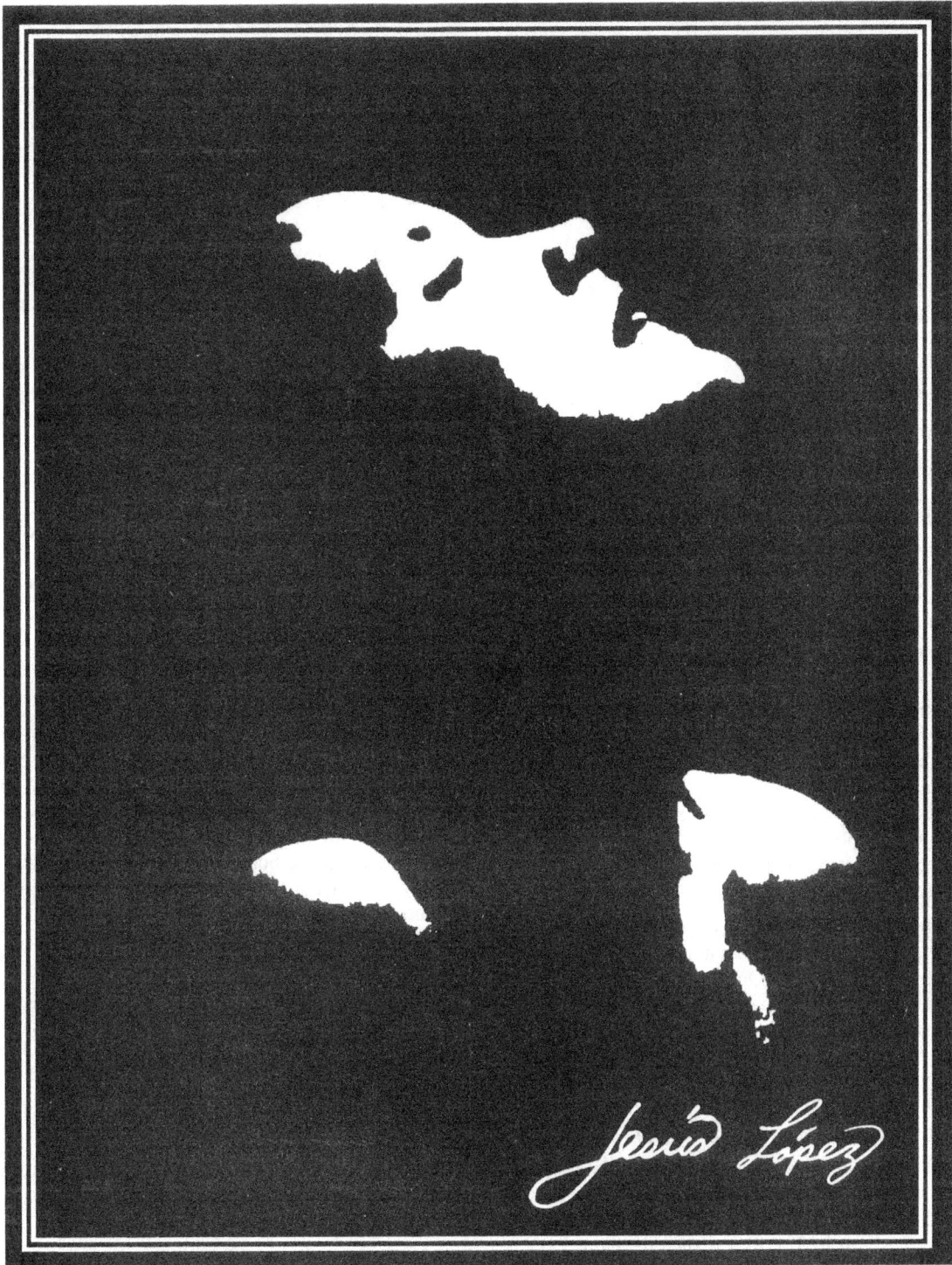

"*The Muse*" (8x10) in.,Multiplex Stereogram Achromatic Transmission Transfer

Creating the World's Most Collectible Holographic Fine Art.
Specialists in 4-Dimensional Female Figure Studies and
Holographic Sculptures.

LOPEZ'S GALLERY INTERNATIONAL

CORPORATE OFFICE: 500 North Michigan Ave., Suite 1920,Chicago, IL 60611-3704, U.S.A.
GALLERY STUDIO: 258 Territorial Road, Benton Harbor, MI 49022, U.S.A.
Telephone: (800) 4-D FINE ART • Fax: (773) 248-9527• Internet/Email: fineart4D@aol.com

grams has become more practical. Since most of these laser-are not portable, and many delicate optical components are used in the process to further manipulate the laser beam, most holographic recordings are made in a darkened holographic studio, where conditions can be precisely controlled.

The Holographic Process

The process used to produce commercial holograms consists of three main parts --

1) recording the image,

2) regenerating the image so that we can see it, and

3) replicating it, if it looks good.

This process is somewhat analogous to the process used in making an audio recording. Obviously, holography deals with visual images rather than sound, but the basic production steps can be compared. We begin by recording the performer during a session in a recording studio using specialized equipment. Next, we "play back" the original recording in order to listen to it. If we like what we hear, we can duplicate the original recording (stored on a "master tape") onto cassettes, CDs, etc. for sale to the public.

Let's consider the first step in the holographic process -- recording the image. This step is discussed in greater detail elsewhere in this book, but for beginners the following explanation should suffice. As mentioned, holographers must use the coherent light from a laser to illuminate the subject, or object being recorded. The recording is captured on a photosensitive material, which is typically a sheet of special, high resolution black and white film.

In most holographic recording set ups, the single beam of light leaving the laser is immediately divided into two smaller beams of equal length. One of these smaller beams travels directly to the film. The other beam reflects off the object and back onto the same film. As the two beams of light converge on the film they interact and combine to form a complex pattern. This pattern is called an **interference pattern.** An interference pattern created in this way and recorded by any means is called a **hologram**.

The interference pattern is recorded onto the film during an exposure that lasts from fractions of a second to minutes, depending on the film's sensitivity and the amount of laser light reaching it. The film is then developed and processed in order to permanently store the interference pattern, i.e. a hologram is produced. But, if we looked at it now, we would only see an indecipherable microscopic pattern of closely spaced overlapping lines. No recognizable image, just a recording of the interference pattern.

To see the image which was recorded, we must perform second procedure -- image "playback" . By properly illuminating the film (in this case, with the laser light that originally exposed it), a viewable image will appear. The image that results is also commonly referred to as a hologram, though the phrase "holographic image" is more correct.

The image that is produced is a replica of the unique set of light waves which bounced off our subject and exposed the film during the recording process. How does this work? The interference pattern stored on the film interacts with the incoming laser light to reconstruct an image of our subject. Whatever the film "saw" during the recording process, a viewer looking at the finished hologram would see. If we made a hologram of a physical object, we would see a three dimensional image of that object. Since it is impractical to use lasers to playback all holograms, methods were developed that allowed holograms to be seen and enjoyed under ordinary lighting conditions.

The third step of the process is replication. If the holographic recording process was successful, the original hologram can then be mass replicated in a variety of ways for commercial applications.

This simplifies the process quite a bit, but the basic facts to learn are that: a laser is needed to illuminate our subject and thereby expose the film; an interference pattern is recorded onto the film (not a visible picture); once this film is developed and illuminated correctly, a three dimensional image of the subject is regenerated.

Lighting Holograms Properly

To fully grasp what holograms look like, and to understand how much their look depends upon proper illumination and viewing angle, one should examine the holograms in this book under various light sources. (All the holograms in this book are designed to be lit with a light source positioned above, and on the reader's side of the hologram. Hold the book in front of you and tilt it back and forth until the holographic images appear brightest and most distinct.)

Use a Single, Point Source

An overhead, single beam of bright light (such as direct sunlight) works best. However, sunlight is not the only source of light that works with these holograms. There are a whole range of other light sources with which to view a hologram; some are better than others. Just remember that holograms require a light source that mimics the laser light that originally made the hologram. Ideally, it should contain the original exposure wavelength, should have enough intensity to replay the hologram, and should emanate from a single point source (such as a small light bulb). Good light sources cast sharp shadows. Florescent lights with tube bulbs found in most offices do not work well.

Thus, whenever you go into a shop that specializes in holograms, one usually finds that the shop has subdued, overhead lighting with a single spotlight focused on each hologram, or group of holograms. This serves the dual purpose of creating a pleasant lighting environment, as well as providing a proper illumination for holograms. People who display holograms in their homes find that an inexpensive way to illuminate them is by using a clear (unfrosted) light bulb with a single, small filament inside. Bulbs with vertical filaments often work better than bulbs with horizontal filaments. Halogen spotlights work very well, too. These bulbs are available at any shop with a large selection of lamps and lighting supplies.

Avoid Multiple Lights and Diffuse Lighting

Whether or not a hologram can be seen well in ordinary room lighting also depends on how the hologram was designed. Well designed holograms that are intended to be viewed in typical commercial environments (rooms with multiple light sources, rooms with lots of fluorescent lights) depict subjects with very little depth. This is due to the fact that when a hologram is illuminated with light coming from different sources at different angles, the hologram forms images at all the different angles dictated by the various sources. This mixture of images "blurs out" the image the holographer is attempting to recreate. However, if there is little or no depth to the object, all the images being created by the different sources appear to be focusing in the same place and the image will be viewable. Diffuse light sources, such as fluorescent lights, also create images which are usually too blurry to see. Thus, shallow images are more recognizable under diffuse lighting.

Two Types of Lasers Used in Holography

We will discuss lasers in depth later in this publication, but it is important that we touch on the subject in this introductory chapter, too. The type of laser you use affects what subjects you can record and we want to discuss that next.

There are two kinds of lasers used by holographers; the Continuous Wave (CW) laser and the Pulsed Laser. The CW laser emits a steady wave of laser light, whereas the pulsed laser emits laser light in bursts. The CW laser is by far the most common laser used in holography. The power of a CW laser is typically measured in watts (w). In holography labs, most of these lasers fall in the 5 to 50 mw (milliwatt) range.

Remember that what we are recording on the plate are two laser beams converging (or interfering) with each other at the plate. If the object moves even a microscopic amount' (on the order of a fraction of a wavelength) from one moment to the next, we will record two different interference patterns and the holographic image will look blurry or won't even appear.

An exposure with a CW laser can take less than a second to several minutes. Because there cannot be any motion at all during the exposure, we need eliminate any vibration coming from the ground. To do this we make or buy a vibration isolation table on which to put our laser, optics, and objects. Since it is absolutely critical that we have no motion at all, the subjects that we holograph with CW lasers have to be "dead" or immobile, objects.

Pulsed lasers, quite the opposite of CW, emit extremely quick bursts of very powerful laser light. The output is measured in joules. Consequently, the exposure time is much shorter than a CW laser. Exposures can be made in nanoseconds (one nanosecond is one billionth of a second). You do not need a vibration isolation table for the pulsed laser. What can you shoot? Anything you want. You can shoot people, splashing water, animals. Why such freedom? Because your subject cannot move significantly in a nanosecond .

What are the drawbacks of pulsed lasers? Why doesn't everyone buy one? The answer is money. They typically cost tens of thousands of dollars or more. And require a lot of extra overhead and care. Lasers don't last forever and when a pulsed laser bums out it is expensive to fix. Holographers are anxiously awaiting, with cash in hand. A low-cost, easily maintained pulsed laser.

Recording Materials

Although we have discussed how holograms are made. we have not discussed the photosensitive materials that they are recorded on. In photography, the most common item used to capture images is a silver halide emulsion coated onto a film base. In holography, there are a number of materials used to record your image. The most common recording media are:

I) silver halide

2) photoresist

3) photopolymer

4) dichromated gelatin

Note that we use the phrase "recording materials" instead of emulsions. That is because not all the items used to capture holographic images are emulsions. In an embossed hologram, for example, the holographic image is literally stamped into clear plastic or foil using a mechanical, rather than an optical process. Holograms have even been recording on chocolate candies and lollipops. A discussion of the different recording materials requires a chapter of its own, which you will find later in this book.

Artwork Origination

One important topic that we should cover in this section is what kinds of subject matter you can record in a hologram. As with any creative art form there are many choices available, and much depends on what you want to accomplish. To simplify matters, we will list some of the most common things that are used for holographic subjects.

1) *3D models* (sculptures, miniature models. or actual objects).

2) *2D models* (flat graphics, illustrations. photos. etc.)

3) *Stereographic composites* (specially shot motion pictures, video, or computer graphic files that are arranged to produce 3D imagery).

Although some of the above topics are covered in more detail later in this book, we will give an overview of each now.

3D models: These are the most common subjects recorded in holograms. What your model is. of course. depends on the type of laser you are going to use for your exposure. With a pulsed laser, as we have already mentioned. you can shoot live subjects and just about anything you wish. Most holograms, however, are made with a CW laser and immobile objects are required as subjects. Holographers will frequently commission sculptors to create highly detailed miniatures of things which can't fit in the holography studio.

2D models: You will see this type of artwork being used with great abundance in embossed holograms. Camera-ready art for 2D holograms is created much the same way you would create artwork for a printer, except that the graphics can be positioned to appear on several different levels.

2D/3D models: You can also have a combination of photos, line art and 3D objects in your final hologram, although the depth of the 3D object is often limited due to practical considerations.

Stereographic composites: The holographic stereogram is one of the most exciting compositions in holography today. It allows artists to incorporate a wide range of visual effects in their images - especially animation and dimension. Today's computer technology is making the production process more accessible and the finished products more refined.

We should point out that several techniques are used to make holographic stereograms. Sorting out the jargon can become confusing. Some of the names you will hear that refer to holographic stereograms are:

Holographic Stereograms - Probably the most common, safest, and most inclusive name. It is used to name any hologram that belongs to the group of holograms that are designed to achieve their effect by utilizing a human's capacity for stereo vision.

Integral Holograms - In general, this is a term that refers to a finished image that is constructed from many discrete units.

Multiplex Hologram - Describes a hologram produced using a system developed by Lloyd Cross and refined by the Multiplex Company that utilizes stereographic and integral techniques. This is probably the first commercial holographic stereogram process developed and it involves filming a subject on a rotating stage.

Most historians credit Lloyd Cross and his cohorts in San Francisco with the development of a process that resulted in the first reliable method for producing a holographic stereogram -- it resulted in a three-dimensional cinematic image that appeared to "float in space". Their method, developed in the early 1970's, allowed live subjects, life-size models and special visual effects to be incorporated into their holograms in a practical and afford- able way, as expensive pulsed lasers were unobtainable.

In order to commercialize the endeavor, Cross and his colleagues manufactured a motorized display unit for their freestanding 360 degree version. They also developed a stationary wall-mounted unit that displayed 120 degrees of viewing angle as the person moved around it. The idea of creating a self-contained holographic display device was quite revolutionary and very admirable. The complete units, which incorporated an inexpensive light source (an unfrosted light bulb with a vertical filament) along with the hologram, sold for several hundred dollars. The Multiplex Company has been producing units based on this process for over twenty years.

Here is a simplified description of the process:

1) Make a rotating stage.

2) Place an object or a scene with live actors on the stage.

3) Set up a stationary movie camera in front of the stage.

4) Film the subject as the stage rotates 360 degrees, making sure to shoot at least three frames for each degree of rotation. In addition to the stage moving, the subject is allowed to move slightly in a manner that will result in a smooth animated sequence. Rapid or uneven motion, however, will create undesired "blurring" effects.

5) Develop the movie footage in a normal manner.

6) HOP Transfer: We now want to make a hologram of each frame of the movie footage. These holograms will be sequentially exposed onto a sheet of film using a holography setup whose elements are collectively referred to as the Holographic Optical Printer. The HOP setup illumi-

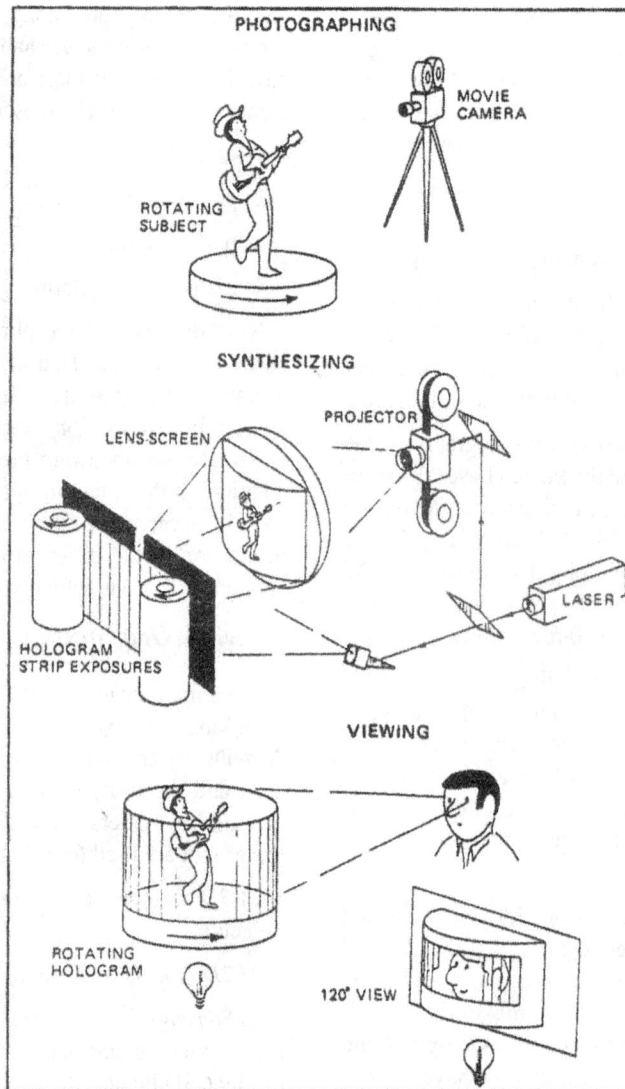
Figure 1.1

nates each individual frame of movie footage with laser light. Another laser beam meets the beam that went through our movie frame at the emulsion by another path to create the hologram. Each frame is optically "condensed" into a narrow strip on the film using lenses and a mechanical slit aperture that restricts the image to one, narrow, vertical slit. The film is advanced and the process is repeated. A series of vertical slit holograms, running the length of the film, results. [See Figure 1.1]

7) After the process is complete, you will have a length of film with hundreds of thin vertical holograms on it. Once processed, you can take the film and wrap it into a cylinder shape. When the film is illuminated from inside the cylinder (behind the film) with an appropriate light source, the viewer will see an apparently solid image floating in space inside the cylinder! As the cylinder rotates, or the viewer walks around it, the image looks fully dimensional and appears to move!

These dramatic effects result from the fact that each of the viewer's eyes sees a slightly different image at the same time. Our brain then combines these images to give us a "stereogram" effect. One limitation to Cross's approach is that this technique creates images that display horizontal parallax only (i.e., you cannot see above and below the image). This is very adequate in most situations because in life we generally inspect images by looking side to side and not over and under the image.

Subsequently, holographers produced variations of Cross's concept. Some made stereograms with different degrees of view, commonly 60 or 90 degrees. Others began shooting the cinema by moving the camera along a track (instead of moving the stage). They went on to flattening out the cylinder, which allowed the holographic stereogram to be produced and handled more easily. Eventually, researchers embossed these holograms onto mirror backed plastics or produced copies which allowed front lighting (which is more practical in most situations).

A major advance in display technology was the introduction of the LCD (Liquid Crystal Display). It did not take long for holographers making stereograms to see the benefits of using the LCD as a source for the image being recorded. LCD origination substitutes graphics displayed on a Liquid Crystal Display screen for cinematic footage. This allows digitized images (with all their advantages) to be easily incorporated into a hologram. A wide variety of cinematic, video and still images can be scanned in a computer, manipulated, and displayed electronically using LCD technology. Computer assisted design and origination of artwork will be discussed further in a later chapter.

Categorizing Holograms

Perhaps the best way to proceed from here is to explain, step-by-step, the making of two of the most fundamental types of holograms -- **transmission** holograms and **reflection** holograms. Very simply, the terms "transmission" and "reflection" refer to how the hologram is illuminated during the viewing process. Transmission holograms require that the illuminating beam of light pass through the hologram in order for an image to be to be seen. Therefore, these holograms must be backlit. The Multiplex holograms just described are transmission holograms.

Conversely, reflection holograms require that the illuminating beam emanates from a source on the viewer's side of the hologram. The light reflects off of the hologram back to the viewer's eyes. Most of the embossed holograms in this book seem to be reflection holograms, however they are actually transmission holograms with a mirror attached to the back. These terms are one of the main ways used to categorize all holograms.

In The Recording Studio

Suppose one enters a studio where a simple hologram is about to be made. There is a special vibration-free table in the room. On the table is a laser, some mirrors and a piece of photosensitive material positioned in a "plate holder". This photosensitive material is typically a silver halide emulsion coated onto a glass plate, and thus will be referred to as the "recording plate". Everything on the table is arranged in a carefully measured manner. The object to be holographically recorded is positioned in front of the plate holder in the middle of the table.

As mentioned, to record a hologram we use a laser, which emits a single beam of light at one wavelength. We cannot use just any light as our source because the light from common light bulbs or daylight contains many constantly- changing wavelengths - it is not coherent. If we make the exposure using incoherent light, the changing wavelengths would create a multitude of interference patterns and the resulting holographic image would be completely blurred and useless.

Making Transmission Holograms

To make a transmission hologram, first we turn on the laser and aim the beam at Mirror 1. Due to the fact that Mirror 1 is only partially reflective, part of the beam is reflected toward Mirror 3, and the other part passes through Mirror 1 to Mirror 2. Because the beam is split, Mirror I is referred to as a "beamsplitter". [See Figure 1.2]

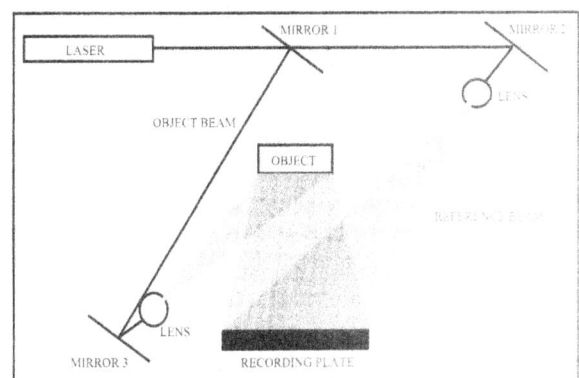

Figure 1.2 Basic Set up - Transmission Hologram

The beam that passes through Mirror I to Mirror 2 two is called the reference beam. After the reference beam strikes Mirror 2, it is reflected through a lens toward the recording plate. The lens' function is to spread the beam so that it will cover the entire plate (in some cases, the lens is placed in front of Mirror 2; in either case its function is the same--to spread the beam). The reference beam's path always ends at the recording plate without ever illuminating the object.

At the same time, the other beam, which we call the object beam, reflects off Mirror 3 and also passes through a lens. This lens spreads the beam out so that it illuminates the entire object. The laser light reflects off the object (hence the name object beam) and strikes the photographic recording plate. The two beams must travel exactly the same distance so that when they recombine at the recording plate they will be in sync with each other, and an interference pattern will be formed.

After exposure, the photosensitive plate is developed, and the resulting developed plate is the hologram. Holding the developed plate up to light, we see that the plate is semitransparent. On closer inspection we see that the dark- ness of the plate is caused by developed emulsion. The plate seems to have countless swirls of threadlike developed emulsion which are called fringes. The fringe patterns look like the swirls that make up your fingerprints or the boundaries on topographic maps. There appears to be no order to the swirls.

Viewing a Hologram in the Studio

To see the image, we put the recording plate back in the plate holder on the table in exactly the same place it was for the exposure. Then we remove the object and Mirrors 1 and 3 from the table. Now, when the laser is turned on again only the reference beam illuminates the plate. When you look through the plate, an image is seen of the origirral object, in its original place and at its original depth. [See Figure 1.3J This reconstructed image is indistinguishable from what you would see if the object was not removed! The first time holography students see this happen they are quite amazed!

A detailed explanation of why this happens would occupy many pages. A simple explanation might go like this: the two beams beam strike the photosensitive recording material at the same time. Since they both originated from the same laser beam, and traveled equal distances, they are in precisely in sync with each other. When two such waves of light recombine, their interaction produces an interference pattern. This pattern is recorded on the photosensitive material during the exposure step. When we develop and process the recording plate, the interference pattern is stored on the plate as the fringes that we see. Because we are recording the interaction of lightwaves which are quite small, the fringe patterns are microscopic (on the order of 1,000 or more fringes per mm).

After development, if we aim only the reference beam at the plate (at exactly the same angle that originally exposed the plate) the interference pattern which was recorded naturally causes the light waves passing by to change direction. This phenomenon is called diffraction. It occurs whenever light passes through small apertures, like the space between fringes. This diffracted light has exactly the same form as the beam that was originally reflected from the object. In other words, a hologram regenerates or recreates the way light reflects from an object without the object being there. To a properly positioned viewer, it looks just like the original object!

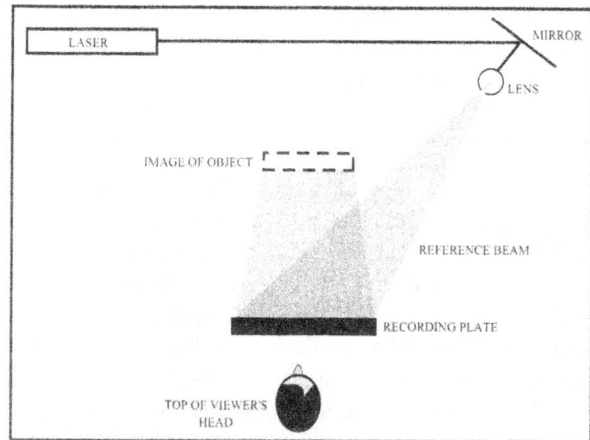

Figure 1.3 Viewing the image using a laser

Making Reflection Holograms

How are reflection holograms made? If we start with the basic setup previously depicted, but transfer the reference beam around with mirrors so it illuminates the recording plate from the back instead of from the same side as the object beam, we create a reflection hologram, It is that simple. [See Figure 1.4J

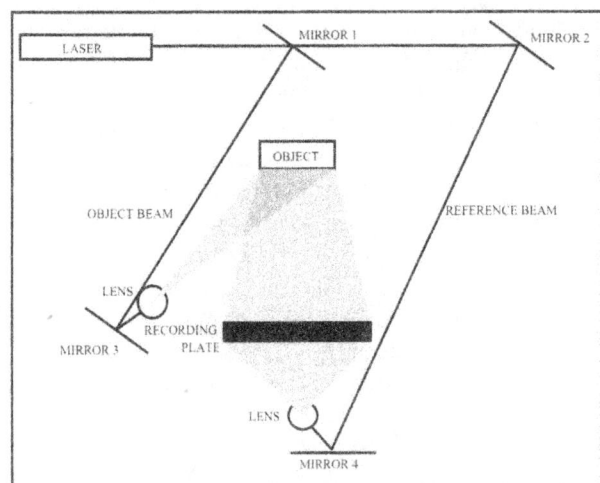

Figure 1.4 Basic Set up - Reflection Hologram

Within the two major divisions of holograms (reflection and transmission), there are many variations. Like any other specialized field, holography has its own lingo, and in some cases the same hologram can be described using more than one name.

The Master Hologram, the H-1, the H-2

It is important that we cover the topic of the H-1 and master hologram in this introduction because it is a fundamental procedure in the making of almost every commercial hologram. H-1 stands for "hologram one", which simply means it is the first hologram you make on the path to your desired final hologram. Sometimes the H-1 is the master hologram from which you make multiple copies. Frequently, though, there is more than one hologram that needs to be made before you get the finished hologram from which you will make copies. If this is the case, the next hologram in the sequence is called the H-2, and then H-3, and so forth.

A question that immediately comes to mind is, "Why would anyone want to make an H-2 ?" Well, historically one of the big problems that holographers had was placing the subject exactly where they wanted it. Suppose, for ex- ample, you want the object in the final hologram to appear half in front and half behind the recording plate. How would you do it? You obviously can't do it on your first shot because the object would have to be going right through your photographic plate.

This problem was solved by the following procedure:

1) Make an H-1 transmission hologram.

Since the H-1 hologram creates an image of the object, why not use the image made by our H-1 as our subject and make a hologram (H-2) of the image made by the H-1 ?

2) In other words, make a hologram of our hologram. This H-2 hologram can be a transmission or reflection hologram, depending on your need. It sounds strange, because you are making a hologram of an image and not an object. But it works. [See Figure J.5]

3) Now, since you can make a hologram of the H-1's image, take time to move the image around to wherever you want it positioned. In this case, adjust the H-2 recording plate so that the image of the object is half in front and half behind the plate and then make your H-2. The problem of getting half of the object in front of the plate, and half behind, is solved.

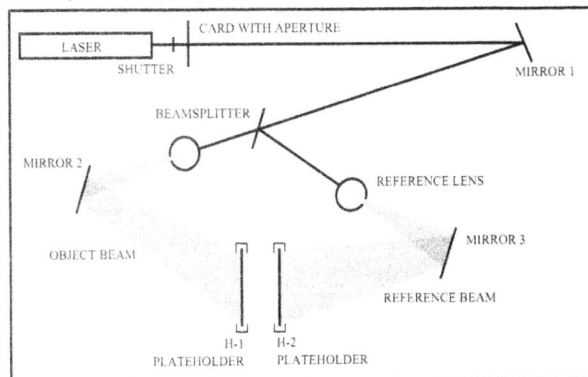

Figure J.5 Reflection H-2 being made from Transmission H-1

In short, there are at least three good reasons why an H-2 should be made:

I) The H-2 allows you to reposition the image of your subject. When you reposition your image from the H-1, you may make your subject focus out in front of the recording plate, behind the plate, or anywhere within the limits of your equipment (you are usually limited by the laser's ability and the quality of the optics). The creative potential here is enormous because you are able to move solid objects around as if they are ghosts. You can have two objects occupying the same space, etc. The process of moving the image around to make the H-2 is called image planing.

2) It gives the holographer a chance to brighten up the image. Since you may move your image anywhere, you can focus the image right at the recording plate. This concentrates the light directly on the recording material and brightens up the image considerably. This is commonly done in silver halide reflection holograms.

3) It saves time on remakes. If you develop the H-2 and decide you don't like the position of your subject astride the recording plate, you don't have to find the original subject and set it up again. This can be important if there are large costs in arranging the H-1 shot.

Going through the pains of making H-1, H-2, etc. to produce a master for commercial replication is usually required. It is technically possible to get results from the first shot-but most professional holographers shoot a series of holograms in order to end up with a suitable production tool.

Viewing Holograms Without Using a Laser

We mentioned earlier that although it is necessary to use a laser to make a transmission hologram, it is not always necessary to use a laser to see a transmission hologram. In fact, most transmission holograms can be seen in sunlight. This may seem confusing, because we have said that in order to see a holographic image you have to shine the laser reference beam that made the hologram on the plate. This is true, but sunlight contains a multitude of wavelengths, including the one we used to make our exposure. The sun is such a great distance from earth that it appears to be a single beam of light shining on our plate. It would seem that we have only to position the plate at the proper angle, and we should see our image.

This is logical, but it also stands to reason that if sunlight passed through a transmission hologram we would also get images being formed by all of the other wavelengths that are somewhat close to the wavelength of your reference beam. These other frequencies of light would diffract at a somewhat different angle than the original reference beam. The result would be a multitude of images forming right next to each other, creating a blur instead of a clear, crisp image.

That's exactly what does happen and it took a while for a solution to be developed. Around 1969, Dr. Steve Benton came up with a solution. The resulting hologram is sometimes referred to as a Benton hologram, or more frequently, a rainbow hologram.

Rainbow Holograms

Benton reasoned that since our problem is too much imagery at the point of reconstruction for our object, why not block off some of it? In other words, suppose we put up an opaque mask against the transmission hologram, with a long, narrow horizontal slit through which we view our transmission hologram. This would certainly clean out a lot of the annoying secondary images that are blurring the primary image's reconstruction. [See Figure 1.6]

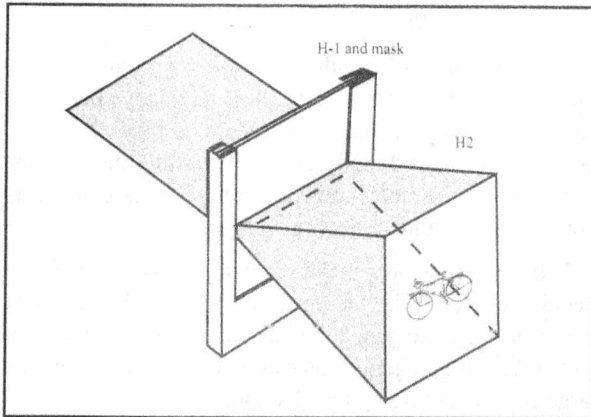

Figure 1.6 Creating a daylight viewable Transmission Hologram

Figure 1.6 Creating a daylight viewable Transmission Hologram

This "cleaning" comes at a price, however, because the mask causes loss of vertical parallax (the ability to be able to see over and under our object). We would, however, still have our horizontal parallax (ability to see side-to-side around the object). Humans, with feet fixed on the ground and eyes on a horizontal plane, are actually more accustomed to horizontal parallax than vertical.

The procedure to produce this masked hologram is as follows:

1) First a normal transmission hologram is made.

2) Next, a transfer copy of the transmission hologram is made, but an opaque card with a horizontal slit is placed between H-1 and H-2.

If the copy hologram is viewed in the frequency of our laser light, the eyes must be positioned at the real image of the slit to see the holographic image.

Now, imagine viewing this H-2 hologram in two different colors of light. A hologram of the image made through the slit will be played back, but each of the two wavelengths of light will diffract through the hologram fringes at a slightly different angle. There will be two different images of the object, each a different color and each at a slightly different vertical position.

Next, think of the image in white light or sunlight. All of the wavelengths will reconstruct their own image, all slightly displaced vertically with respect to one another. We are faced with the same problem we had with the original transmission

hologram, except the images being recreated have no vertical parallax. As you move up and down in front of the plate the color of the image will shift through all of the colors of the rainbow (hence the name "rainbow hologram"). As you move from side to side you will have horizontal parallax because nothing has been done to destroy it. By careful planning, the image may be made any desired color, or even a combination of colors (a multi- color rainbow hologram).

In effect, the hologram is filtering the white light, while all that is sacrificed is vertical parallax, which, as we mentioned, our two horizontally-positioned eyes usually don't miss anyway. Rainbow images are often extremely bright, because all of the frequencies in white light are being used to form the image. So the rainbow hologram technique is a way of making a transmission hologram sunlight-viewable. Other names for this are daylight-viewable or white light viewable. They all mean the same--a hologram you can see without the need of a laser.

Image Projection of Holograms

Although transmission holograms seem to be naturally designed to create a hologram with considerable projection, one can also make reflection holograms that have a great deal of projection. In fact, reflection holograms with considerable projection are a favorite among artistic holographers and the buying public. They are favored be- cause they can be hung on the wall and illuminated just like a painting, whereas transmission holograms need to be lit from behind, often requiring a much larger viewing area.

Laser-viewable transmission holograms can demonstrate amazing depth and projection when the correct equipment is used to make and display them. It should be noted that the depth of the holographic image is not so much a function of the power of the laser as it is the coherence length of laser light (you can read more about coherence length in the "laser chapter"). Theoretically, the maximum image projection in front of the hologram plate can be as great as the projection in back of the plate (depth of the image). Unfortunately, it is difficult for our brains to make sense of greatly projected images. Because of this, and the fact that there usually are optical distortions created in the image planing process, projected distances in transmission holography are usually kept under four feet.

Laser transmission holograms have the widest parallax and display deep images best. There are laser transmission holograms, for example, of people and objects in a 4000 cubic foot room, made by pulsed lasers. Not surprisingly, projected hologram images like this generate one of the highest shock and thrill responses from viewers.

Pseudoscopic and Orthoscopic

Both laser-viewable and white light transmission holograms share a fascinating property. As you recall, after the developing stage, the hologram can be put it back in the plateholder for viewing. If the original subject has been removed from the table, the viewer can look through the holographic plate and see the image of the object appearing behind the plate at its original position on the table.

Now comes the interesting feature. Take the transmission hologram out of its plateholder, flip it over, and put it back in the plateholder. Step back and look at the plate. You see the image forming out in front of the plateholder (between you and the plateholder). It focuses in air the same distance in front of the plateholder as it originally sat behind the plateholder. You also see that the image is a pseudoscopic image.

What is pseudoscopic? An image as normally seen in everyday life is an orthoscopic image. A pseudoscopic image is the opposite of this. For example, if one's viewpoint moves to the right, you do not see more of the right view of the image but the left view, and when the viewer raises his viewpoint, the lower part of the subject comes into view instead of the upper part. A pseudoscopic image yields an exciting effect, but it can be confusing to the viewer. Some artists, however, have produced exciting pseudoscopic work using geometric shapes like wide spirals, pyramids and cones.

The aforementioned "introduction to holography" might make the holographic production process seem difficult and esoteric. However, many people with very little knowledge about lasers, optics, and physics (including children) have been taught to make holograms in their classrooms and even in their own homes. The following article is included for those of you who wish to create your own holograms. There is no better way to gain a more complete understanding of the process than by doing it your-self. We encourage you to try.

(Editor's Note - Items identical or similar to those mentioned in the following articles may be obtained from other sources. We also recommend that beginners read the newly revised "HOLOGRAPHY HANDBOOK -- Making Holo-grams the Easy Way", available from Ross Books, Berkeley, CA.)

Low Budget Holography for Beginners

by Tung H. Jeong

Like photography, holography can be very simple or extremely complex. This article is written for the absolute beginner with a limited budget who wishes to get into the exciting field of holography. Besides being "fun", making and understanding holograms is also highly educational. No less than three Nobel Prizes in physics have been awarded to discoverers (Lippmann - 1908, Bragg - 1915, Gabor - 1971) whose contributions made holography possible. Yet, like using an automatic camera, one can learn to make holograms easily before understanding them.

A budget of about $200 to $400 is required to make holograms that can be shown with ordinary white spot lights (or the sun). The cost is largely determined by the power output of the Helium-Neon (HeNe) laser, which can cost from about $SO (O.S milliwatt) to $2S0 (S milliwatt). Thus if you already have a laser, the total,material, including film or plates and processing chemicals, cost less than $200.

In this article, I will describe a complete procedure for making small "white light reflection" holograms using the Holokit™ (sold by the Integraf company) along with "household" items like plywood, trays, and readily avail- able common articles. The Holokit includes the following items: a box of 30 2.Sx2.S inch holographic plates, a 1D-3 chemical processing kit (complete with instructions), a beamspreader, educational materials, and a set of four "lazy" balls along with one "busy" ball. These lazy balls are made with a material called Norsorex and have no "bounce" at all at room temperature, thus serve as ideal vibration isolators.

The experimenter will need to obtain a HeNe laser (we recommend S.Omilliwatt output), a piece of 2x4 feet 3/4 inch plywood, one can of Krylon (or similar brand) flat black spray paint, four 1-liter (or quart) plastic bottles, a squeegee (or windshield wiper), three plastic trays (big enough to submerge a 2.Sx2.S inch glass plate), a wooden clothes pin, a night light (S Watts), a six inch piece o

"2x4" wood, one white and one black paper card (about 4xS inch), some "play dough" (or a putty that will not harden), a measuring cup, three steel paper clamps, three round-headed screws (small), and four washers (for 114" screws). Unless the tap water in your area is exceptionally pure, buy three gallons of distilled water. Finally, to look at the holograms you make, a spot light is needed. Generally, any light that originates from a spot will do: the sun, a slide projector, a track light, or a pin light.

Begin by painting the plywood and the 6 inch section of "2x4" black. Fasten the three round-headed screws on the end of the "2x4" so that it can stand without rocking. Glue the back of the beamspreader near the top on one side of the "2x4". Never touch the silvered concave surface of the mirror. When not in use, put a flat piece of tape across the mirror so that no dusk can enter.

First mix the 1D-3 processing chemical according to the instructions that accompany it. All ingredients are proportionately packaged. Mix the chemicals with distilled water and put them in the three bottles and label them accordingly. Mixing is quicker if the water is heated. Let them cool to room temperature when used. If the stoppers on these bottles are tight, the liquid chemical will stay good for several months at room temperature; longer if refrigerated (but not frozen). Make sure they are out of reach of children, and clearly distinguish them from food!

Now it is time to set up the holography system. On a sturdy table (kitchen counters are perfect), arrange the equipment as shown in the following diagram. [See Figure 1.7]

Put the "lazy balls" on top of the washers, so that they won't roll, and place them [or other vibration isolation devices - ed.] under the board. Tilt the laser slightly upwards using a v-block (cut from wood) so that the beam reaches the curved surface of the beamspreader and is spread across the board as shown. Always avoid looking directly into an undiverged laser beam!

The beam can be steered by sliding the "2x4" on the board. Arrange for the spreading beam to cross the board diagonally so that it illuminates an object of your choice. Use a solid object that will not deform or rock. Soft plastic or fabric will moves perpetually and should be avoided. It should look bright under the red laser light. Put the white card in the position of the holographic plate to mark the best location for uniform exposure. If crater-like fringes inside the red spot are seen, it means there is dirt or dust on the mirror where the beam is reflected. Do not touch the mirror surface! Blow the dust away with bottled dry air from a photographic store (if you want to spend the money), or just move the laser and/or the beamspreader until a clean spot is seen on the card. At this time, put some play-dough or putty around all points of the laser, v-block, "2x4", and any object that touches the board. This is done to prevent movement.

Now turn out all the lights except a night light, and cover it with obstacles until you can barely see after dark adaptation (a few minutes). No light should shine on top of the

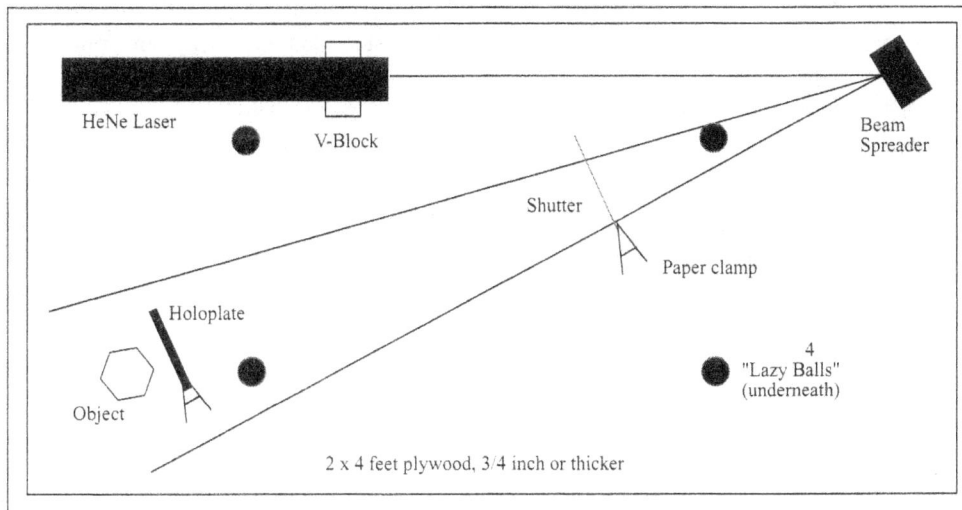

Figure 1.7 A simple set up

board. Now turn the room lights back on and prepare the chemicals.

Layout the three processing trays, preferably one on each side of a sink with running water (at room temperature). The third tray is for post-development and can be placed anywhere where it can be exposed to bright light (sun light is best). If a sink is not available (for example, I do this demonstration in front of audiences in lecture rooms or conference rooms), use a bucket or large tray of water. It is also advisable to wear goggles to protect the eyes from accidental splashes.

Now mix equal parts of solutions A and B (enough to submerge a holographic plate held on the comer with a clothe pin). Now you are ready for the experiment. Turn off the light and get dark adapted. Block the spreading beam with the black card. Remove the white card from the steel paper clamp. Take the box of holographic plate- (Agfa 8E75) and snap it open (it takes two hands and two tries!), take out one plate, put it down on the board, and immediately close the box (you must hear two snaps)

Pick up the plate and check which side has the emulsion on it (the sticky side; if you can't tell, skip the test, it's not crucial). Hold the plate vertically by clamping a quarter inch of its edge, with the emulsion side facing the object. Move the plate-clamp assembly until the emulsion (or the glass) touches the object. Now make sure the plate-clamp- object-board assembly does not move by putting more putty around the clamp and between the lower exposed comer of the plate and the board. Push the plate down until its lower edge is in solid contact with the board. Now wait for 30 seconds for things to settle down.

Lift the shutter off the board, but continue to block the beam. Wait two seconds (this is important, as you moved the board while lifting the shutter!). Now quickly pull the shutter up and exposed the entire holographic plate with laser light. (The shutter is actually better if located be- tween the laser and the beamspreader where the beam is narrow. But beware that no scattered light from it gets to the plates before the exposure.) Now the direct light arriving at the plate acts as a

reference beam, interfering with transmitted light that scatters off the object and returned to the plate, forming a white light reflection hologram.

What is the-exposure time? You can calculate it approximately if you are given the fact that it takes one microjoule of light energy to correctly expose one square centimeter of-theplate. But here are some practical numbers: assuming that the size of the red spot on the white card (previously) is just bigger than 2.5 inches in diameter, and your laser has an output of 1.0 milliwatt (mW), the correct exposure is about one second. Thus, if the laser is 2.0 mW, the exposure is 1/2 sec.; 3.0 mW - 1/3 sec; 5 mW - 115 sec; etc. It is useless to give you more precise number than the above because many factors can cause variations on the exposure time. Experimental calibration is the only way.

Now clip the plate on one comer with the clothe pin and completely submerge the plate in the developing tray (equal mixture of A and B solutions). Slowly agitate. Pull it out and look through the plate against some lit area (a white wall or a piece of white paper, and watch out for the dripping!) A correctly exposed plate should become al- most opaque in about one minute. If it turns black to soon, cut down the exposure; if it takes over two minutes, in- crease the exposure. Note that the developer stains badly, so avoid accidental drippings.

When the plate is almost completely opaque, take it out of the developer and rinse it for two minutes in room temperature water (running water is better), then submerge it in the bleaching solution. In about one minute, the plate will look completely clear. Bleach it another 10 seconds and rinse for several minutes. Now, remove the clothe pin and hold the plate by the edge and squeegee both sides. In areas with very hard water, a rinse in distilled water is helpful. The hologram cannot be view until it is completely dry. You can just wait, or use warm (not hot) air from a hair drier.

To view the hologram, shine a spot light at it from the same direction as the original laser light that exposed it. The image will look much blighter if the emulsion side of the hologram facing the object is sprayed with a flat black paint. If a good three dimensional image is seen, you are ready to try other experiments.

****

For more advanced students or professional holographers with access to higher priced lasers and additional supplies, we've included an article that explains one of the process used to create color reflection holograms.

Color Reflection Holograms with One Laser

by Hans I. Bjelkhagen and Tung H. Jeong

The subject of making color holograms has been taught as one of six tutorials in the Holography Workshops since July 1994 at Lake Forest College. Throughout the academic year, students perform independent studies on the subject, one of which involves using minimum equipment for making full color, full parallax, Denisyuk type of reflection holograms. This article specifically addresses the latter method.

The details of color holography has been published in a review paper] with 126 references. It describes a general method using the combined outputs from several different lasers . In this article, we will discuss only the essential steps in making the hologram using the Spectrum 70 Argon-Krypton ion laser from Coherent, Inc. We assume that the reader is already familiar with the concepts and procedures for making single color Denisyuk holograms. [Fi ure 1.8 shows the layout of equipment.]

On top of a vibration isolated table, a secondary vibration isolation system is necessary for supporting the water-cooled mixed gas laser due to the movement of water. We use four 1.5 inch diameter "lazy balls "2 under a 2-inch thick board to support the laser. These balls are made with a material called Norsorex and have no "bounce" at all at room temperature, thus serve as ideal vibration isolators.

The laser can be tuned to a variety of color from red to blue. Our choices are the 647 nm, 531 nm, and 476 nm lines. To identify these lines, the extra beam (due to the internal reflections of the exit mirror) of several milliwatts from the laser is directed by a min "Of through a holographic diffraction grating. The first diffracted order of each line can be readily identified. The main output is directed by two mitTors through a shutter and then a spatial filter which expands the beam to the Denisyuk setup.

The procedure is complicated by the fact that whenever a, new line is tuned, the output is different and there is no assurance that the line is in a single axial mode. Thus, before the shutter, a thick piece of uncoated glass with a small wedge is place to reflect two beams, one to a power meter and the other to a Michelson interferometer with unequal optical paths. The path difference is approximately I m. The interference pattern is spread out by a concave mitTor onto a screen. The image from the screen can be relayed by a video camera to a monitor located near the location of the Denisyuk setup. Thus the holographer receives simultaneous information on the wavelength, power, and coherence of the light before each sequential exposure.

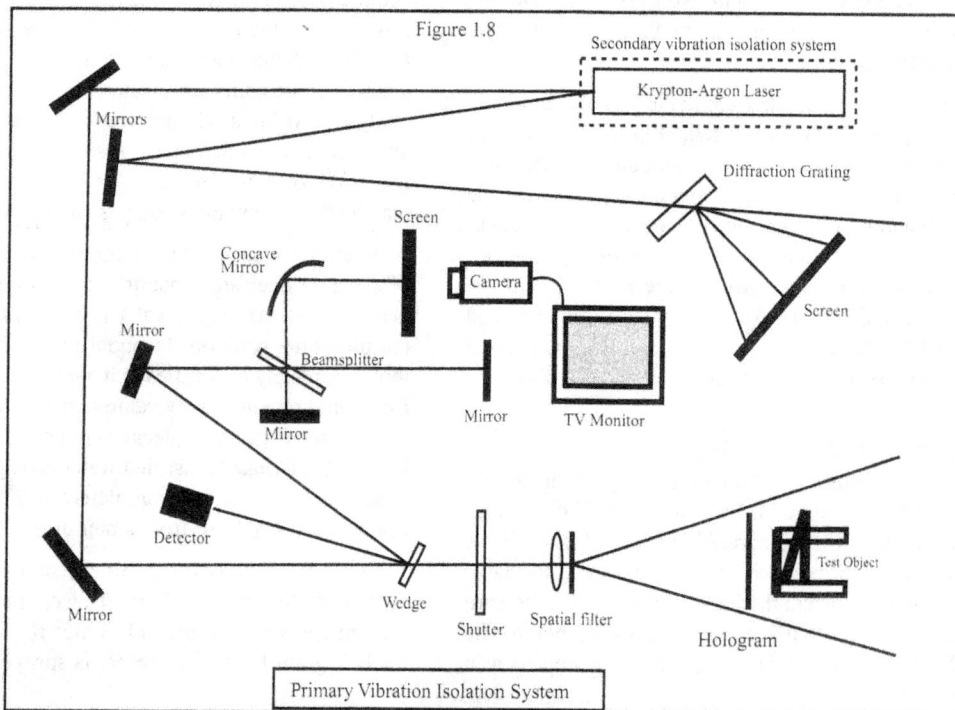

Figure 1.8

The best silver halide emulsion that we have tested is the PFG-03C coated on glass, manufactured by Slavich, outside Moscow[3]. Table I is a summary of characteristics of this emulsion.

Table I: Characteristics of the Slavich color emulsion.

Silver Halide Material.................. PFG-03C

Emulsion thickness......................7mrn

Grain size....................................12-20 run

Resolution...................................-10,000 lp/mm

Blue sensitivity............................-1.0 - 1.5 x 10.3 J/cm^2

Green sensitivity..........................-1/2 - 1.6 x 10-3 J/cm^2

Red sensitivity..............................-0.8 - 1.2 x 10-3J/cm^2

Color sensitivity peaks at 633 nm and 530 nm

Using the data from the table, with the understanding that there is variation from batch to batch, one can determine the correct exposure time experimentally. We normally expose red first, followed by green and then blue.

The processing of the plates is critical -- for this reason, the process is presented here in great detail. The emulsion is rather soft, and it is important to harden the emulsion before the development and bleaching takes place. Emul-

sion shrinkage and other emulsion distortions caused by the active solutions used for the processing must be avoided. The following bath is used for this first processing step:

Formaldehyde 37% (Formalin)..........10 ml (l0.2g)

Potassium bromide..............................2 g

Sodium carbonate (anhydrous)...........5 g

Distilled water.. .... 1 liter

The time in this solution is 6 minutes. The developer used is the holographic "CWC2 developer":

Catechol...10 g

Ascorbic acid.....................................5 g

Sodium sulfite (anhydrous)................5 g

Urea...50 g

Sodium carbonate (anhydrous)..........30 g

Distilled water................................. 1 liter

The developing time is 3 minutes at 20°C.

'The catechol based CW-C2 developer has become one of

---

1 9 9 7

"Hands-on" Holography Workshop
Advanced Holography Tutorials
International Symposium with Exhibition

• July 7-11 Holography Workshop: no previous experience needed. "Hands-on" learning in making transmission, reflection, cylindrical, focused image, rainbow, HOE, and other holograms. (Twenty-sixth season)

• July 14-19 One-day tutorials by world experts on: Photo-chemistry on silver halide, commercial holographic imaging, embossing technology, non-silver and photopolymer material, color holography. and pulsed protraiture. (Also planned, computer HOE design)

• July 21-25 Sixth International Symposium on Display Holography, with accompanying art forum for the exhibition of fine art pieces and display of commercial work. Abstract are requested on the artistic. scientific, and cultural aspects of display holography. A business and career session is included. Proceeding to be published by SPIE.

Submit abstracts to: Prof. Tung H. Jeong, Director, Center for Photonics Studies, Lake Forest College. 555 N. Sheridan Road, Lake Forest, IL 60045 USA. Send inquiries to Mrs. Virginia Crist. Administrative Director. Telephone: 847-735-5160; Fax: 847-735-6291; e-mail: CRIST@LFC.EDU

the most successful developers for the processing of the CW-laser-exposed monochrome reflection holograms. The use of urea serves to increase developer's penetration into the emulsion which is important for uniform development of the recorded layers within the emulsion depth. Catechol has also a tanning effect on the emulsion but with less staining effect as compared to pyrogallol. Therefore, this developer can be considered suitable for processing color holograms.

The bleach bath used to convert the developed silver hologram into a phase hologram is very critical. The bleach must create an almost stain-free clear emulsion in order not to affect the color image. In addition, no emulsion shrinkage can be accepted which would change the colors of the image. A special rehalogenating bleach for holography was used here. It is based on the idea of mixing a bleach by using an oxidation process between persulfate and com- mon developing agent, e.g., ascorbic acid, amidol, metol, and hydroquinone. A set of new rehalogenating bleach baths for holography was previously introduced by Bjelkhagen[4] et al. These baths have very good performance concerning both high efficiency and low noise, and some of them introduce no emulsion shrinkage.

These bleaches have been named PBU (Phillips- Bjelkhagen Ultimate) bleaches followed by the name of the developing agent on which they are based. The ami- dol-based rehalogenating bleach, "PBU-amidol bleach" was selected for the color processing and is mixed the following way:

Cupric bromide......................... 1g

Potassium persulfate..................10 g

Citric acid..................................50 g

Potassium bromide....................20 g

Distilled water............................1 liter

After the above mentioned chemicals have been mixed, add 1 g amidol [$(NH_2)_2C_6H_3OH.2HCl$, 2,4-diaminophenol dihydrocloride]. The bleach can be used after a few minutes of being mixed. Enough oxidation of the developing agent amidol must take place. Dilute one part stock solution with two parts distilled water for use. The bleaching time is normally about S minutes. The process must continue until the plate is completely clear. After the bleach is finished the plate is washed for at least 10 minutes and then soaked in water to which 20 mill acetic acid (glacial) have been added. This is done in order to prevent print-out of the finished hologram. Washing and drying must be done so that no shrinkage occurs. The best way to dry the plates is to let them slowly dry in air of room temperature. Warm air will introduce an emulsion shrinkage.

The processing steps are summarized in Table II.

Table II: Color Holography Processing Steps

l. Tanning in a Formaldehyde solution........6 min

2. Short rinse.

3. Develop in the CWC2 developer..............3 min

4. Wash

5. Bleach in the PBU-amidol bleach .... 5 min

6. Wash

7. Soak in acetic acid bath.................................1min

8. Short rinse.

9. Wash in distilled water with wetting agent added.

10. Air dry the holograms.

After the hologram is completely dried, spray paint the emulsion side (which should face the object during recording) with a flat black paint. Our choice has been a product from Krylon, available in most hardware stores.

A photograph of a typical color hologram made in our laboratory is included in the color pages of this book.

For further information regarding this article, classes and workshops, contact:

Lake Forest College 555 N. Sheridan Road Lake Forest, IL 60045 or e-mail: Jeong@LFC.edu

References:

1. Hans, I. Bjelkhagen, Tung H. Jeong, and Dalibor Vukievi , "Color Reflection Holograms Recorded in a Panchromatic Ultra-high-Resolution Single-Layer Silver Halide Emulsion", Journal of Imaging Science and Technology, Volume 40, No.2, pp. 134-146 (March/April, 1996)

2. A package of 4-1.5 inch diameter "lazy balls" and 1 "busy ball" is available from Integraf for $15.00.

3. Tntegraf is a United States distributor of this emulsion.

4. Hans T. Bjelkhagen, Silver Halide Recording Materials for Holography and Their Processing, Springer-Verlag (1993)

2

Artwork Origination

The first step in producing a hologram is designing and preparing suitable artwork This is a crucial step, as the type and quality of the subject matter recorded in the hologram will determine how the final image will look. This chapter discusses the newest ways of generating artwork (using digital tools), as well as more traditional techniques. An interview with professional modelmakers is included.

Digital and optical based imaging systems are merg- ing in the holography studio to produce visual displays that cannot be reproduced by either technology alone. The first section of this chapter will explain how computer graphic artists can use readily available computer hardware and software to generate "camera-ready" art suitable for holographic reproduction, and thereby expand their visual repertoire. It will also explore some related developments in the field of "computer-aided holography".

Using Computers to Originate and Fabricate Holographic Imagery

It is now possible for computer-graphic artists using readily accessible hardware and software programs to electronically generate "camera-ready" artwork that holographers can assemble into images that display dimension, depth, projection and motion. Many holographers are substituting this digitally originated artwork in place of the time-consuming drawings, hard-to-record physical objects and expensive cinematic shoots that they traditionally utilized. They are using the computer to increase flexibility and versatility in the design and production processes, as well as to cut production costs. This merger between electronic imaging systems and optical-based ones is resulting in new and profitable opportunities for all those involved, especially the artists and designers that are able to best utilize both media to achieve their client's goals.

In most instances, the "camera ready" artwork prepared by the design team is output as a series of computer graphics files which must be sent to a hologram origination facility. These computer files correspond to the various graphic elements of the holographic image being produced. In brief, the holographer uses these graphics files to generate a "master" hologram which is recorded on a high resolution photosensitive material using a CW laser and specialized optics. The master hologram can then be mass-replicated in a manner suitable for commercial applications.

In the future, we may see affordable desktop "holo-printers" that connect to any graphics workstation and produce finished holograms as easily as today's computer printers output pages of paper. Several variations of these "instant" hologram machines are currently being developed by a handful of companies around the world. (The Business Directory in this book lists companies that manufacture and market these devices.)

The first part of this section will deal with image design and production techniques relating to the simplest, and most commonly requested type of hologram - a 2D/3D embossed hologram. The second part of this section will discuss more complex holograms that are created using LCD screens, digitized imagery, computer generated animation programs and other computer graphic techniques.

Before we discuss the digital origination of holographic artwork in detail, we recommend that new readers review the basics of holography by reading the introductory chapter in this book.

the basics of holography by reading the introductory chapter in this book.

Why Use a Hologram?

As illustrated by the advertisements in this book, holography's unique attributes make the medium a very useful and attractive one to commercial users. Advertisers like holograms because they have great visual appeal. People are generally fascinated by holographic images, and will stop and pay attention for a moment. Manufacturers frequently incorporate holograms into their packaging because it attracts shoppers to a product located on a crowded shelf. Security-conscious companies use holograms because they're very difficult to duplicate. Publishers use holograms to illustrate their books and magazines in new ways. Since most of these users need to combine holographic imagery with conventional printing techniques, holograms were not widely used until a way was developed to reproduce holograms in a cost effective manner that could be easily incorporated with existing manufacturing methods.

Making Holography Commercially Feasible

To this end, a technique called embossed holography was invented, whereby a hologram could be mass pro- duced by mechanical, rather than optical methods (This technique is used a lot in commercial applications). In brief, the original "master" hologram is copied onto a metal plated stamping die, which is then used to repeatedly press (emboss) the holographic interference pattern into rolls of very thin sheets of plastic or foil, using conventional methods and machinery. Using this method, great quantities of holograms can be run off at very high speed, bringing the unit price of the hologram to cents per square inch, or less. Most importantly, these holograms can be easily attached to paper or plastic surfaces with commonly-used adhesives or hot stamping presses. Tens of millions of holograms have been produced using this method world-wide, especially by companies in the print and packaging industries.

Using Your Computer to Design and Generate Artwork for Simple 2D/3D Embossed Holograms.

2D/3D Holograms

Since the most widely-used commercial holograms are multi-level "2D/3D" embossed holograms, it would be beneficial to discuss design considerations for this type of hologram in detail. For the purposes of this article, imagine breaking a conventional print ad into three levels of related graphics ~ a foreground, a middle ground, and a background. Picture each element to be a separate flat graphic (a 2D) or better yet, a photographic transparency. (For example, the foreground image might be a corporate logo, the middle image a picture of a product, and the background image a landscape.) Arrange the three elements front to back, yet separated from each other by a 1/4 inch or so of space (3D). This array represents the multi-level imagery associated with a typical 2D/3D hologram.

Digital Design Tools

These "levels" of artwork can be easily generated using digital tools familiar to most computer graphics artists. Adobe's Illustrator, Corel Draw and MacroMedia's Freehand can be used to create original drawings. Adobe's Photoshop can be used to import and touch up a client's existing artwork to make it suitable for holographic reproduction. Adobe's Postscript or Microsoft's Truetype font collections are often utilized to create logos. Kai's Power Tools and Bryce programs can be used to further enhance imagery. In short, a variety of software programs are capable of doing the job.

Once the artwork is finished, the digital files are sent to the hologram production studio either electronically or physically. There, an in-house designer will typically use Adobe's Separator, Quark Express or Photoshop to break the image into the appropriate component layers, if the original artist has not already done so. These files will be sent to an image-setter (a machine that outputs film) which will generate film transparencies corresponding to the different levels of imagery. These transparencies will be copied onto rigid glass plates. These glass plates will then be stacked in a sequential array. This array will constitute the physical object that will be recorded using the holo- graphic mastering techniques described in other parts of this book.

Design Guidelines - Dimension and Depth

Designing an image for a 2D/3D hologram is similar, but not identical, to designing an image for print. Since the artist is designing a multi-level image, subject matter should obviously be positioned to take advantage of these unique dimensional properties. For clarity, the most important elements of the scene are usually placed directly on the image plane. Foreground elements intended to float "above" the image plane should be easily recognizable, as they will blur out under less than ideal lighting. The same applies to background images with great depth. Drop shadows, textures and shadings are often incorporated to exaggerate dimensional effects.

Designers commonly arrange the different image levels in one of two ways:

- So that in the finished hologram the foreground image appears to "float" slightly above the surface of the embossed material, the primary image is on the hologram's surface (called the "image plane"), and the background is behind the other two.

- Or, the foreground image might be positioned directly on the image plane, with the mid-ground and background images underneath, which further exaggerates the apparent depth of the image. (See the Crown Roll Leaf advertisement in this book for good examples of the latter arrangement.)

Design Guidelines - Parallax

Designers must also consider, and should definitely take advantage of, parallax - the ability to look around the sides of an image. It is important to note that due to standardized holographic production methods, most embossed holograms only display horizontal parallax i.e.. a side to side view, rather than an over and under one. This is usually adequate as it mimics the way we ordinarily look at the world.

To use parallax effectively, new graphic design parameters must be addressed. Foreground images positioned to float off the hologram's surface need to be sized correctly so edges do not "cut-off" prematurely if a viewer moves off center. And since background imagery (which is unseen and therefore not included in ordinary pictures) will be in sight when a viewer looks "behind" the foreground elements, the graphics for every underlying layer should extend completely from one side of the hologram to the other. For these reasons, and for finishing purposes, foreground and background artwork should be oversized in relation to the final size of the hologram. Only the imagery that is planned to appear on the hologram's surface should be actual size.

Although parallax considerably expands the viewing zone of the image, this attribute does have certain restrictions. If the viewer moves too far off center, the image will disappear. It is wise to consult with the holographer to determine the viewing parameters of a particular manufacturing process before starting design work.

Design Guidelines - Size

Although embossed holograms can be produced in a variety of sizes, cost considerations, manufacturing equipment, and marketing requirements favor making holograms of 6"x 6" or less. It can be quite a challenge. to achieve the visual impact required by a client while working on such a small canvas. To save production time and expense, it is common to gang a number of smaller images on one "master" hologram. "4-ups", "9-ups" and "36-ups" are common arrangements. The entire set of images is replicated together, and then each separate image is die cut out during the finishing process.

Design Guidelines - Color

The designer should also always consult with the holographer beforehand to determine what colors are best utilized in a particular type of hologram. Colored inks and pigments are not used at all. Rather, embossed holograms create different colors by bending light to varying degrees. It is like creating a customized prism that will direct the light according to the designer's wishes. Therefore, the graphic artist needs only to assign colors to specific image areas -the holographer "colors it in" during the exposure process by adjusting his optical set up.

Another unique property of embossed holograms is

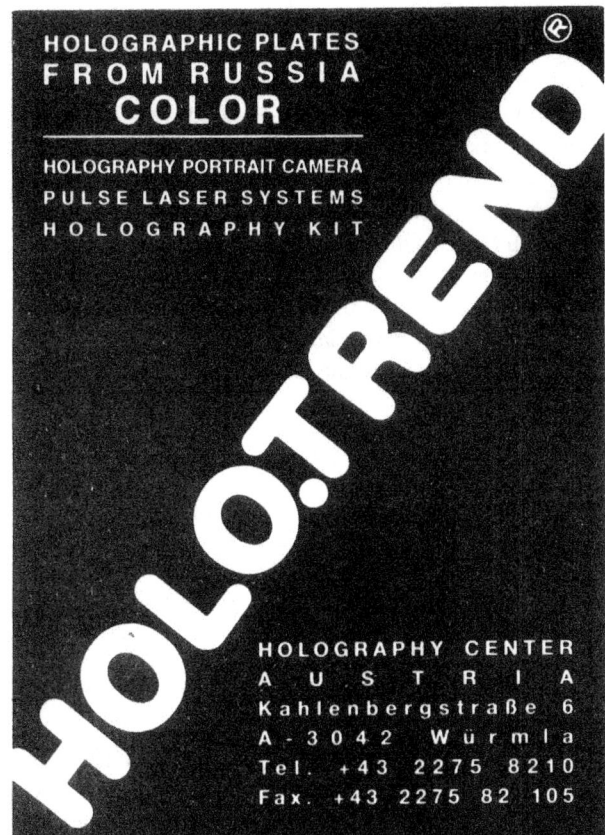

that they do not display permanent colors - , the viewer moves up and down in relation to the hologram, the colors will shift through the entire rainbow. In practice, it is common to take advantage of these unnatural color shifts to emulate movement and increase visual impact. Clients usually desire the brilliant, dynamic effects which result.

The designer only needs to specify the colors that are intended to play back when the finished hologram is viewed directly at eye level under proper illumination. Unlike standard four-color printing that utilizes combinations of cyan, magenta, yellow, and black (CMYK), the primary colors utilized in embossed holography are red, blue, and green (RGB). These three colors can be combined holographically to create yellow, white and a range of secondary colors, (complex colors are a result of halftone screen mixing of RGB; different densities result in different colors.) In holography, black results from unexposed areas of the film.

Some hologram mastering facilities are able to recreate "true color" using proprietary techniques based on pixilated renditions of the artwork. In some cases the studio will provide a palate of colors that designers can work from. Other studios claim that any color can be holographically duplicated. Again, we advise that designers consult with the production house to determine the most appropriate way to proceed. Ask to see samples that have been produced for other clients.

Experience dictates that lots of black and white colored image areas are visually unappealing. (Matte white is commonly reserved for registration marks and trademark information.) Drab colors should be avoided in favor of the bright colors inherent to this process - red, yellow, blue and green. Colors should be arranged to best contrast imagery and accent dimensional effects. Each area of the image should be assigned a color separate from its neighbor, and different from the areas that may overlie or underlie it. Colors in the center of the color spectrum are usually recommended for major image components, as they are the brightest and will be the last to disappear as the viewer shifts position.

At The Holography Studio

Once the graphics work has been done, files should be organized, labeled and shipped to the holography studio for review. Most studios are capable of accepting a variety of file formats including diskette, DAT, SyQuest disks, and optical disks. At this point the designer's work is finished.

Once the holographer records the image the resulting master hologram can be proofed by the designer and/or client. If it is acceptable, it is sent to a production facility to be replicated. After replication, the holograms are finished in accordance to the client's wishes. Most embossed holograms are backed with an adhesive for hot stamping to paper stock or delivered on rolls for "peel and stick" applications. They work best when attached to a rigid material that has a flat and smooth surface, such as magazine covers, bank cards and cardboard packaging.

Production Steps - Creating A Simple Embossed "2D/3D" Multi-level Hologram

Image Design

1. Designer consults with holographer regarding job-specific design requirements.

2. Designer prepares client's artwork for holographic reproduction. Creates a multi-level image "on paper" consisting of b&w line art drawings and/or photographic images.

3. Drawings, photos and graphics that comprise image are scanned into the computer using Photoshop. If the desired imagery must be copied from existing corporate artwork, Photoshop tools can be used to extract the desired graphics.

Digital Image Assembly

4 Digitized image is assembled and checked. Line art is cleaned up (Black outlines should separate image components and all lines must be unbroken).

5. Bit map images converted to postscript using Streamline.

6. Postscript images imported into Illustrator. Image is broken in into multiple levels - one file created per image plane (i.e. an image with a foreground, a midground and a background requires three files).

7. Designer assigns appropriate colors to image components on each level.

8. Completed files sent to holography studio by diskette, SyQuest, DAT, optical disk, e-mail, etc.

Creating a production tool

9. These files are reviewed and imported into new Illustrator "master" file standardized for that particular holography studio. Images are ganged if necessary; sized to fit production equipment; pre-designed cut guides, registration marks, and TM symbols are added.

10. Adobe Separator is used to generate color separation "sub-files" for imagery levels that are multi-color and composite "sub-files" for imagery levels of single color. These "sub-files" will be used to generate masks that the holographer will use when exposing the recording material. Photoshop in its ROB mode generates necessary separations if non primary colors are being copied from existing artwork.

11. Output "sub-file" separations on paper to check color assignment.

12. Take Illustrator "sub-files" (on SyQuest drive) to image setter for output.

13. Image setter outputs film positive transparencies - one per "sub file". Colored areas are now solid black.

14. Holographer uses these film positives make glass negatives on Kodak HRP (color areas are now clear to allow laser light to pass through and expose plate at specified angle per selected color).

Holographic recording

15. Holographer shoots one layer at a time, one color at a time. Each time masks are positioned to block off potions of the recording material that should not be exposed. Every time another color is required, the holographer must adjust the optical setup to change the reference beam's angle.

16. Holographer repeats the exposure process for each level of imagery until the master hologram is complete.

Holographic replication

17. Finished hologram is checked for flaws and prepared for electroplating.

18. The hologram is metalized and stamping dies are created that reproduce the microscopic patterns on the original hologram.

19. These stamping dies are used to emboss the hologram on rolls of foil or plastic.

20. Holograms are die cut, finished, and sent off for application (hot stamping, packaging, etc.).

(Editor's note - Special thanks to computer artist Mike Grogan ofHolographic Dimensions Inc., Florida for the information he provided.)

The aforementioned information describes how computer graphic artists can use relatively simple holographic techniques to incorporate image depth, parallax and new colors into their work. More complex holographic techniques are currently being perfected that allow designers to output visual effects "on paper" that most people believe are forever confined to the monitor screen - motion and dimension.

In brief, these fully dimensional, animated images (called holographic stereograms) are assembled in the computer, output as holograms, and then "printed" by embossing them, or copying them onto photopolymer film. Since the added complexity is in the origination and mastering steps, these enhanced holograms remain affordable; as mass replicating a complex holographic image is no more expensive than mass replicating a simple 2D/3D. However, the ability to publish 3D animated images has even greater commercial potential!

Computer Aided Production of More Complex Holograms

This next section will discuss the evolution of holographic stereograms and explain why and how the computer has been incorporated into the stereogram pro- duction process.

Holographic Stereograms

The unique origination and mastering techniques which result in "holographic stereograms" have been developed and used by a select group of holographers and cinematographers over the past twenty five years, but the advent of new optical devices, affordable computers and graphics software is allowing traditional illustrators and graphics designers to participate in the creation of these animated 3D holographic images.

Without delving into extremely technical explanations, " it is important to note an important property of the holographic medium - many different images can be recorded on a single piece of holographic film. As one of these "multi exposure" hologram moves past a viewer (or the viewer tilts the hologram), each successive image is revealed. If these successive images are arranged properly, the viewer might see an entire animated scene (a mini movie comprised of related frames), or a three dimensional scene, or both! These effects are created by taking advantage of a human's capacity for stereo vision - two slightly different views of the same object are simultaneously displayed by one hologram, and the viewer "sees" a three dimensional composite image. A similar effect is achieved when viewing a stereo pair of photographs or by staring at stereographic posters.

Steve Larson, president of Laser Images, a hologram research and manufacturing facility, makes this distinction between holographic stereograms and conventional stereoscopic devices, "Similar in some respects to the old Viewmaster™ concept (a binocular-like 3D slide viewer

that displayed two near identical pictures taken from slightly different angles), a holographic stereogram differs in that the number of stereographic pairs [can be] in the hundreds, instead of the single pair that was provided with the Viewmaster. What that means to the hoiographer/ artist, is that now, motion and time can be captured and displayed holographically."

Early Holographic Stereograms and Cinematography

Holographic stereograms were traditionally created by placing a subject in the center of a rotating stage and filming the scene as it turned 360 degrees. A few seconds of subject motion (such as a wink or a kiss) might also be captured on film as the stage rotated. Each frame of movie footage was then projected and re-recorded, one by one, on a thin vertical slice of holographic film, using a special device called a 'holographic optical printer'. When the holographic film was developed, it held hundreds of views of the subject (one view per slice) - each from a slightly different perspective. This length of film was typically rolled into a cylinder or an arc and mounted in a holder. When the hologram was illuminated and displayed properly, it appeared that the subject was "inside" the cylinder and the viewer could - see-entirely around the subject. [Refer back to Figure 1.1] If the original footage incorporated motion, the viewer could see a few seconds of animation, as well. Eventually, the cylinder was flattened out into a strip of film, which was easier to replicate and display.

Substituting photographic film footage for the miniature 3D models and/or simple multi-level graphics used in tabletop holography also allowed holographers to -leave the confines of the production studio and capture scenes which previously could not be recorded easily: landscapes, people, and large objects. This drastically increased the opportunities for advertisers to incorporate holograms into their marketing campaigns - many could supply the appropriate film footage or could contract holographers to shoot it for them.

High-resolution video footage and computer-generated imagery were gradually integrated into these stereogram recording sessions for the same reasons they are used in other visual arts - speed, versatility, and to lower production expenses. These electronic devices also allow a wider range of special visual effects to be incorporated into the design process. However, many holographers still prefer using a series of photos or cinematic footage, due to quality concerns. The photographic footage can easily be scanned in and digitized, if required.

Source Material - Film, Video, or Computer Graphics?

Clients that want to produce holographic stereograms using existing corporate artwork can supply holographers with photos, film, video or computer files. Glen Gustafson, a cinematographer who has worked on holography projects, offers his opinion regarding the use of film and video as source material.

"There are instances when high quality video, such as Beta SP, is the best choice for shooting a stereogram. If the printed image is going to be small, the superior resolution of film will be lost in the lenticular screen used in creating the H-1 (though film still retains its advantage with regard to contrast). Digitizing video for manipulation in Photoshop, etc. is cheaper than scanning film, but the resolution is 525 lines compared to film's 4,000. Each frame of 35mm negative contains approximately 50 megabytes of image data, allowing film to record a much more complete range of tones. This produces a smoother and more pleasing image, with detail in the lightest and darkest areas. Much of the work done in Photoshop was made necessary by shooting video in the first place, resulting in harsh contrast and blazing reds, in addition to the low resolution," says Gustafson.

Steve Larson relates his own experience," My initial work with holographic stereograms was done using 35 mm film creating monochromatic images. This worked quite well, but when I progressed to creating color stereograms, the process became quite cumbersome as each frame had to be color-separated, registered and sequenced. I quickly looked toward computer controlled production methods that would let the software do the monotonous jobs. In addition, I needed to find a film-capture/modeling pack- age that would provide a source of modeling, animation, and rendering in high resolutions, while keeping my bud- get in mind. My search quickly led to NewTek, another Kansas company such as ourselves. I got an Amiga and Video Toaster and started playing, and I haven't turned back since. Today, we use Video Toaster on the Amiga and LightWave 4.0 on the PCs. They are networked together so that we can transfer files back and forth between the platforms."

Walter Spierings, president of Dutch Holographic Laboratory B.Y., a firm involved in the development of digital holographic systems, speaks about his experiences with his MPGH (multiple photo-generated hologram) system which uses a computer driven camera snapping picture after picture as it moves along a 10 foot rail.

"Our first computer generated holograms were recorded...using 36 frames projected through a converted photographic camera, as compared to the current range of 150 to 250 frames. Then I purchased a Nikon camera that could load and shoot 250 frames, and began automating the stereogram printer. ..all of the control equipment has since been replaced by state-of-the-art IBM computers and interface cards. The use of additional frames and the automation of the recording process resulted in an enormous improvement in quality. The result, along with improvements in registration, was a final image that appeared more solid...The 35mm slide film produces high quality holograms, but the new LCD technology suggested itself as a possible electronic interface for computer data and our Holoprinter [an automated stereogram production machine], says Spierings.

Benefits of Using The Computer

Besides automating control and registration processes, the computer is used for several purposes:

- to digitize graphics from a variety of sources (photographs, video, cinema, drawings, etc.);
- to manipulate and edit these images (paint surfaces, add lighting effects, render many different perspectives, ad or subtract frames, etc.);
- to generate a series of sequential graphic cells ; and
- to output the appropriate computer files. Once the graphics work has been done, the files are organized, labeled and shipped to the holography studio. At this point the designer's work is finished.

Steve Smith, president of The Lasersmith. a U.S. based origination and production studio, provides an example of how his company created a holographic stereogram called 'Chinese Lion Dancers' which appeared on the back cover of the Holography Marketplace 5th Edition. "First, the images of the lion dancers were recorded with the Lasersmith 36-camera imaging system. Next, the chosen animation set of 36 frames were digitized with Kodak's Photo CD. These images were then brought into a 8100 PowerMac, balanced, and an alpha channel of each lion dancer outline was created in Photoshop. Next, a computer-modeled scene was created in the Byte by Byte software's Sculpt 4D. 36 frames were rendered to match the perspective and parallax of the filming setup; all of these files were composited in Photoshop utilizing the previously-saved alpha channels. The resulting film strip was imaged onto an RGB Transmission H-1 master hologram and transferred to the final photoresist master", relates Smith .

Modeling and Rendering

As mentioned, digital devices can either originate imagery or add a wide range of special visual effects to existing artwork. Most recently, affordable modeling and layout programs such as LightWave have been used to create realistic looking images using no physical objects as subjectmatter. Hundreds of different perspectives of a "cyber- scene" can be rendered, combined and output as a fully dimensional hologram.

Steve Smith states in last year's Holograph" Marketplace, "Early on, pricy workstations such as the first series of Silicon Graphics were required to run expensive image modeling and rendering programs; these cost upwards of $20,000, generally out of the range of professional holographers. With the release of powerful yet lower priced graphics workstations, such as the Pentium and the PowerMac, and with more capable yet affordable image modeling and rendering software, a new realm of imag- ing techniques is setting the stage to present the computer aided holographic stereogram as the way to create holo- graphic stereograms."

In brief, the computer graphics altist first models a real or an imaginary object or scene "on screen". NewTek's

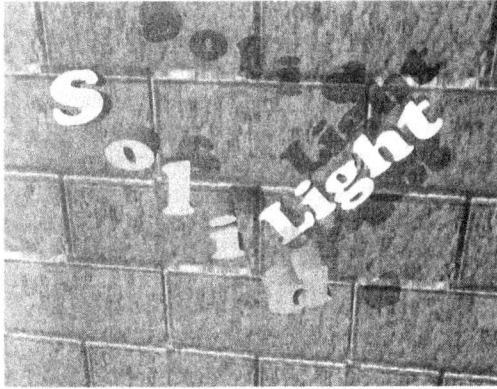

Lightwave 3D, Byte by Byte's **Sculpt 4D** or Alias/Wavefront's *Power Animator* are good modeling programs: to use. The computer is then used to render an appropriate sequence of graphics files which correspond to the various visual perspectives which the designer wants to appear in the finished hologram. Pentiums and PowerMacs will be used to render simple rotations. DeskStation Technologies' Raptor rendering engine or SGI hardware have been used to render more complex imagery, i.e. hundreds of different perspectives of one image.

George Sivy, a world renowned modelmaker who specializes in holographic design, elaborates on the afore-mentioned process, "I use an accelerated Amiga 3000 Tower computer running VideoToaster to import and mix existing imagery. I use LightWave to model, light, surface and manipulate single objects or even entire scenes. In layout mode, I animate the object, and animate a virtual camera in accordance to the holographer's requirements. (I typically either rotate the object in relation to a stationary camera or pan the scene in a smooth horizontal path.) I use a Raptor rendering engine provided by DeskStation Technologies to generate 180 frames. This translates to 6 seconds of real time animation which may be incorporated into the final hologram. Next, the rendered frames are edited further using the VideoFlyer and copied onto S-VHS tape for proofing. Once approved, the files are downloaded onto an appropriate storage medium and shipped to the holographer."

At The Holography Studio

Clients that want to generate their own imagery are ad- ised to consult closely with the holographer before originating and delivering any artwork. Most holography studios are capable of accepting a variety of file formats which you can ship to them electronically or by diskettes, DAT, SyQuest disks, or optical disks. Since there are currently only a few computer graphics artists familiar with the procedures required to create animated stereograms, the hographer's in-house artists will probably need to review your files, clean them up and re-sequence them.

In brief, this digitized artwork is then output to an LCD, frame by frame. Each individual picture displayed on the LCD is recorded sequentially from the display screen onto a single holographic plate by using a specially engineered transfer device and assorted optical components. The holographer ends up with an array of matched images on the "master" hologram that combine to achieve animated and/or dimensional effects. Once the holographer records this image the resulting "master" hologram can be proofed again by the designer and/or client. If it is acceptable, it is typically sent to a replication facility.

Hologram Printers

Several companies are currently developing holographic "printers" that can be attached to a computer graphics workstation to output "instant" dimensional and/or animated images. Here are comments from a few prominent holographers regarding their work in this area.

DHL

Walter Spierings explains about the DHL "one-step" system, "A major hardware development for us was the design of the Digital Holoprinter for the production of full color embossing masters. The Digital Holoprinter is similar to a regular Holoprinter with LCD technology as the interface device, which allows for a direct connection between the Holoprinter and the computer. Images designed in the 3D Holosoftware TRACES (a software package tailored to SGI workstations) can be automatically transferred into photoresist masters from which the embossed holographic material, e:g. stickers are produced. With this new LCD technique, pixels are invisible, comparable to 35mm slide film quality... The office Holoprinter will [also] use a high resolution LCD screen as the interface between the computer and the final hologram. The development process could be similar to ordinary copiers, using a dry process with DuPont's polymer, requiring only ultraviolet light and heat to enhance the development of the image."

Spierings continues, "Bringing holography into the hands of the public through links with the world of computers offers as yet unimagined possibilities for creativity. Like a fax machine, computer or photocopier, the Desktop Holoprinter will be a 'black box user tool': nobody will need to know exactly how it works, but everybody will know how to work with it. Designers and engineers in industrial, architectural, and packaging firms can routinely use 3D visualization techniques frequently, and easily. Dimensional prototyping can be readily communicated, analyzed, modified and printed again from the computer workstation."

Spatial Imaging

Rob Munday, President of Spatial Imaging Ltd., a U.K. based company and an acknowledged leader in the field, concurs, " Digital and optical holographic technologies have been married using LCD stereogram systems such as DIHO (Digital input/Holographic output). My belief is that in the future we will see a reduction in the size of digital hologram systems... to provide truly desktop hologram printers. Now that holograms can be made from any digital image data, more efficiently and more cheaply, their application in the graphic arts will increase enormously".

Astor

Francis Tufty, of Advanced Holographic Laboratories/ Astor Universal/another prominent U.K. based holo- graphic manufacturer, adds that, "The use of computer...graphics in the field of holography has speeded up the preparation of artwork, streamlined the proofing process and given creative control over the imagery back to the commissioning designers and artists. There is also a more fundamental marriage between the two disciplines. Newly developing digital graphics/holographic hybrid technologies are offering visual displays and effects that until now have been impossible for either to deliver singularly. The ultimate hybrid will be near real time video acquisition. digital manipulation, and Spatial Light Modular holographic recordings of full color, animated and true to life surfaces, textures and shades."

Imagination Plantation

Noah Hurwitz of Imagination Plantation, a San Francisco based digital design company that has worked on holography projects, summarizes, "The advent of powerful, computer based 3D content-creation tools has changed the landscape of holography...The capabilities are growing exponentially and the consumer's appetite is growing even larger...The computer allows for imagery that would have been previously impossible due to time, budget, or technical constraints, to be a reality."

Clearly, these experts agree that digital designers should learn about holographic technologies if they want to provide their clients with the best possible imagery. What is new and unique today will be commonplace tomorrow. The best designers will explore this fascinating medium thoroughly and utilize it when appropriate.

(Editor's Note - More information regarding the topic of holographic stereograms can be found in the Holography Marketplace 5th Edition, which can be ordered from ROSS BOOKS. There is a unique and noteworthy example of a fully dimensional, animated holographic image --that integrates video footage and that was produced using digital tools -- on the cover of this edition.)

Although some three-dimensional holographic images are being produced using compilations of a series of digitized graphics, most holograms are still created by recording the laser light reflected off of a (3D) physical object, such as a highly-detailed miniature model or a specially- sculpted scene. Therefore, the process o f creating a custom designed holographic image usually requires the use of an expert sculptor or modelmaker who can best translate a client's ideas into a three-dimensional artwork appropriate for holographic reproduction.

3D Model Making

To gain an basic understanding of the elements involved in producing a successful model for holograms, we consulted George Sivy, of Richmond Development Group (formerly Gray Scale Studios). Sivy has been a holography model maker for eleven years. He worked for Polaroid during their first years in holography, making models utilized for custom and stock images (such as the popular "Brain/Skull" which appeared on the cover of the Holography Marketplace'S 3rd Edition). In addition, he has collaborated on numerous commercial projects, ranging from embossed security holograms to photopolymer holograms designed for the giftware market.

Sivy is convinced that well-designed and well executed artwork is the basic element from which a successful hologram is created. Of the three components Sivy lists as necessary to the creation of a holographic image: the artist; the holographer; and the manufacturer; "Quality work from all three is important. Artwork, however, is key, in that good. well-conceived and carefully executed artwork will carry mediocre holography and/or manufacturing. Even the very best holography and the highest quality manufacturing can't make up for poorly-done artwork."

He has three basic suggestions for those who are working with modelmakers to create a hologram that "works": plan in advance; remain in direct contact with the artist from start to finish; and be willing to pay for the artist's time and expertise.

Communication

Since the client is often unfamiliar with the technological processes that are used to create a hologram and salespeople frequently gloss over the medium's limitations, the modelmaker often assumes the role of educator as well as craftsperson. This requires that modelmakers be effective communicators who can work with both conceptual and technical issues. "A hologram, under the best of circumstances, is still an interpretation of the client's 2D artwork," states Sivy.

"Nine out of ten of these people do not know what to expect in terms of time and other factors, which go into creating a piece," he states. Because of this, he says, there can be "communication gaps, misunderstandings, and deadlines which are too tight..Expectations should be more appropriately formed."

Sivy advises that the client be brought into the loop even

when art directors or ad agencies are involved. Ideally, the client, the client's art director, the salesperson, the modelmaker, the holographer, the replicator, and the finisher will all agree on the various elements that will ensure a top quality result. All parties should agree on a realistic production schedule and should communicate regularly. Since the modelmaker is usually responsible for the first tangible output seen by the client, the credibility of the entire process is affected by the professionalism displayed by this participant.

Design

Sivy divides the design process into two major parts - conception and execution. To best utilize the holographic medium, designers must remember that "a hologram is not merely a representation of an object in 3D, but rather it is a 'window' into a three dimensional space. The more of this space which is utilized, the greater the visual impact of the finished image."

Standing directly in front of the window, however, permits us only a limited view of the contents. Imagine yourself moving from side to side - which allows you to see around a given object and into areas of the room which would otherwise be hidden. This illustrates the concept of parallax view - a very unique and basic characteristic of holography which "represents considerable potential for innovative design and composition". Designers should predetermine viewing angles with the holographer and communicate these measurements to the model maker in order to maximize the usable image area. Designers also need to take into account that a holographic image has "volume" - it can have depth (an object can be behind the window) and/or projection (an object can appear in front of the window). Images should be designed to best utilize this front-to-back dimension, also.

It is important to note that different replication materials have different capabilities regarding the amount of depth and projection that an image can display. This "depth of field" is closely tied to how the hologram will be illuminated. Embossed and dichromate holograms can only reconstruct fairly shallow images (approximately one inch), while photopolymer films can replay images of several inches or more. Holograms produced on silver halide glass plates can reconstruct images several feet deep under proper illumination.

A modelmaker who is familiar with the physical limitations imposed by the recording materials and anticipates the conditions under which the finished hologram will be viewed can create an image that replays clearly in most situations. Therefore, many modelmakers place the most important visual components of their images on the image plane, which stays in focus under less-than-ideal lighting conditions.

Execution

After the design process comes execution, a step which utilizes a modelmaker's craftsmanship and technical proficiency. In terms of materials, Sivy uses "what will solve a given problem." Although some materials and combina-

tions that he uses are proprietary, he did mention the readily available Sculpey, a synthetic clay that can be baked. This ensures a stable model, as most holographic mastering uses continuous wave lasers which require that absolutely no motion is present during exposure periods. Long exposure times might require even more stable materials, and Sivy may use Sculpey first, then make a mold which is used to generate an even more stable scene.

After the model is created, it is often painted to create contrasting areas of light and dark. Since most, but not all, holograms are intended to be reproduced as monochromatic images, the "coloring" process is quite different than in ordinary modelmaking. Less experienced model makers will often work under a safelight which duplicates the laser light which will illuminate the -model during exposure .

Good model makers will also utilize textures, shading and special effects to maximize the hologram's visual impact. Again, this requires communication between the design team, the model maker and the holographer.

Some Additional Opinions

In order to further elaborate on Sivy's statements for our many readers outside the holography industry, especially those contemplating integrating holography into their sales and marketing programs, we interviewed several other professional modelmakers to get a "behind the scenes" viewpoint of this crucial step in the hologram production process.

The participants were: Joseph Farina o f The Laser Holography Workshop; David Merritt of Louis Paul Jonas Studios, Inc.; and Amy Medford and Leonid Siveriver of Avant-Garde Studio. (See directory listings for additional corporate information.) Although each business brings different experiences to the industry, all have learned how to apply their various skills to create successful models for holographic applications.

1. How does the model making process for holography differ from conventional model making?

JF: A model must be, made as a component of the holographic process, rather than a model in its own right. The more the modelmaker knows about this process, the better. For instance, each holographic recording material has its own characteristics which will influence the depth of the sculpted model. Also, the surface finish of the model can have a critical effect on the final hologram and its reflectance should be optimized for the wavelength of the laser being used. Color should not be considered, only reflectance. (For this reason, models are often painted in shades of gray.) Since the model must remain completely stable during exposure we must use materials which keep the risk of vibration to an absolute minimum.

HOLOGRAMAS S.A. DE C.V.

With the highest technology
We manufacture Holograms for:

- Security Seals, Self-Destructive Labels.
- Promotional items.
- Hot Stamping Materials.
- Flexible Packaging Materials.
- Machinery for Holograms Application.

SECURITY IN OUR HANDS HAS A NEW PERSPECTIVE...

retail VISION™

HOLOGRAMS

A
CROWNING
INNOVATION
THAT
DRAMATICALLY
BOOSTS
SALES

Atlanta 1996

CROWN
ROLL LEAF, INC.® &
HOLO-GRAFX™

TURN OVER ▶

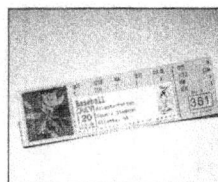

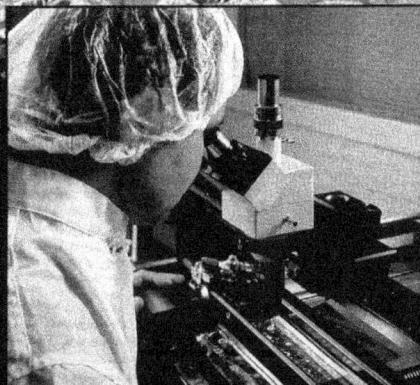

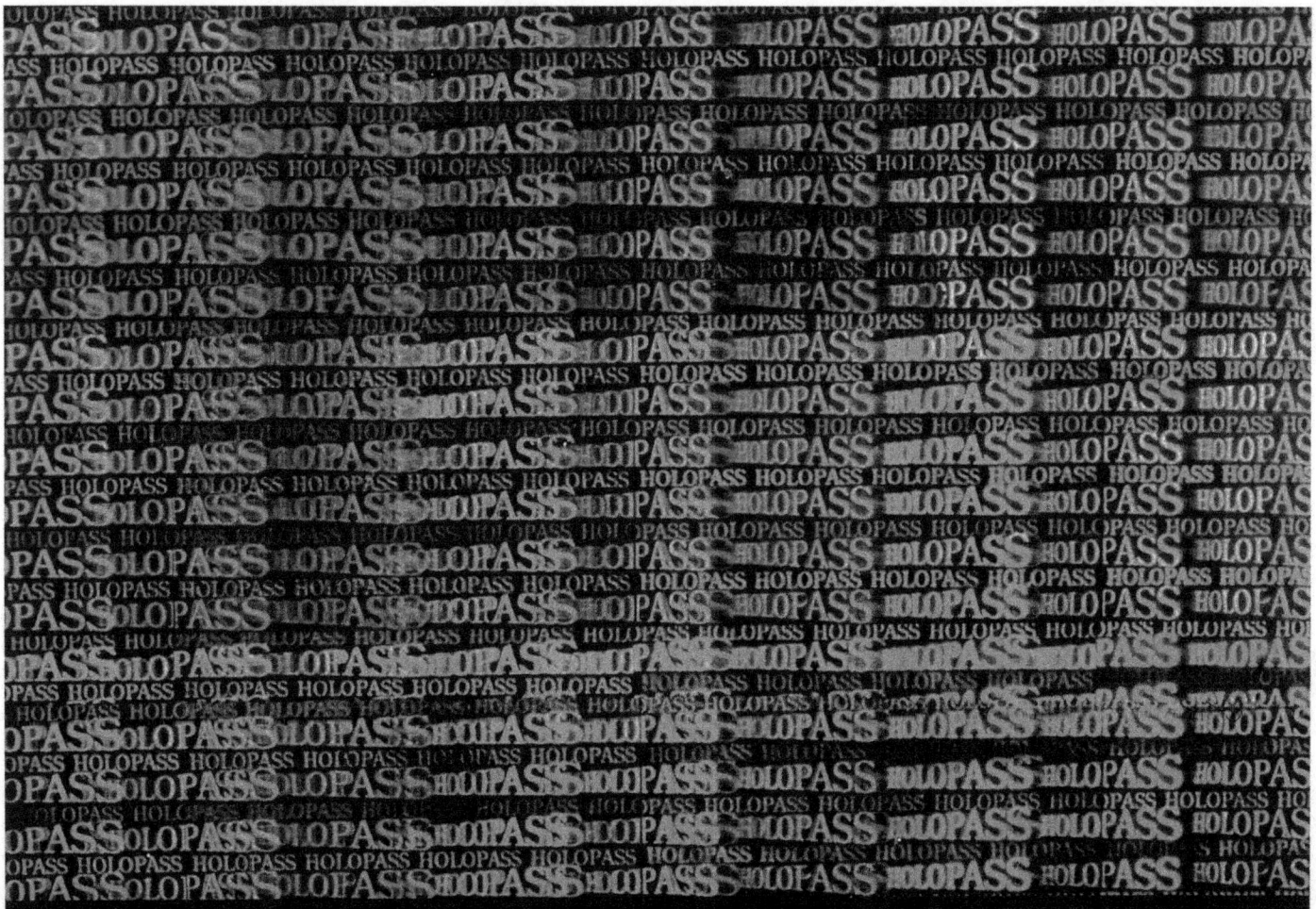

50 YEARS
OF DEPENDABILITY

PUBLICATIONS

PACKAGING & DISPLAYS

MERCHANDISING

SECURITY

PROMOTIONS

A holographic source with a thorough understanding of the process from origination to the finished product is what you will find when you choose Astor Universal. We take the mystery out of working with holograms. Simply tell us what you want to achieve and we'll make it happen.

With 50 years of hot stamping foil manufacturing experience and over 15 years of holographic experience, it's difficult to find a more qualified source than Astor Universal. We can help you develop ideas, provide samples, and work out application logistics.

Call today and let us help you develop a holographic product catered specifically to your application. You simply can't find a more dependable source than Astor Universal.

ASTOR UNIVERSAL
A *MARKEM* Company

Strategic Locations Throughout The World

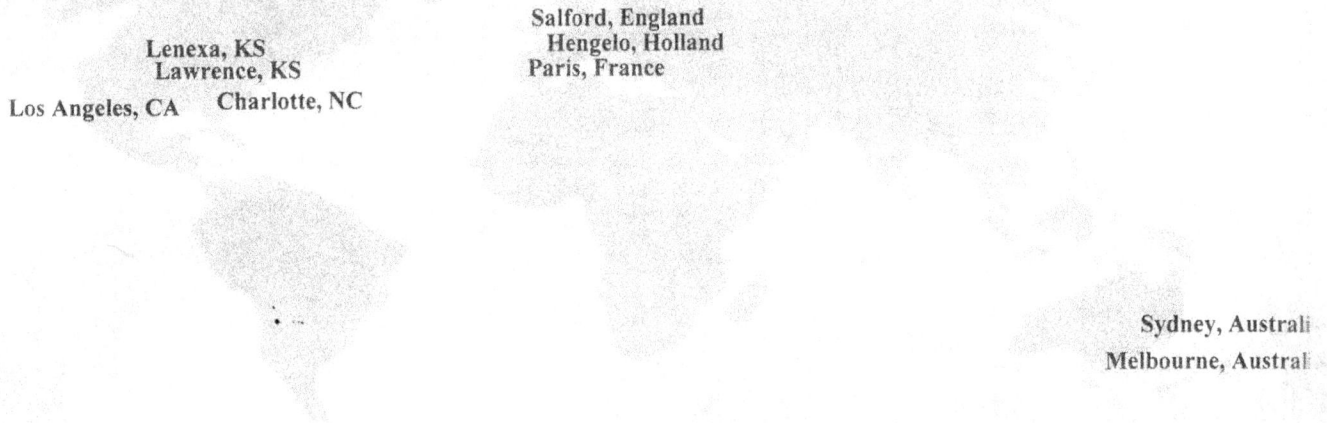

Lenexa, KS
Lawrence, KS
Los Angeles, CA   Charlotte, NC

Salford, England
Hengelo, Holland
Paris, France

Sydney, Australi
Melbourne, Australi

- ☐ Precision recombination
- ☐ Embossing from in-house origination or independent sources
- ☐ ISO 9001/BS 5750 Certified

- ☐ Secure facilities
- ☐ Over 15 years of holographic experience
- ☐ Both wide and narrow web production

All of our holographic products are available in hot stamping foils or thick films for lamination. We offer a complete line of holograms, including custom images (2D/3D, 3D, computer-generated, and Dot matrix), stock images, or our large selection of holographic patterns.

Choose a company that understands holograms from the first to the final step - choose Astor Universal!

ASTOR UNIVERSAL
A *MARKEM* Company

Lawrence, Kansas
Toll Free: 800-255-4605
Tel: 913-842-7674
Fax: 913-842-9748

Salford, England
Tel: +44 161 789 8131
Fax: +44 161 787 8348

Villejuif Cedex, France
Tel: +33 1 46 77 89 89
Fax: +33 1 47 26 48 76

Hengelo, Holland
Tel: +31 74 2 555 625
Fax: +31 74 2 505 309

Sydney, Australia
Tel: +61 2 663 0628
Fax: +61 2 662 1020

Melbourne, Australia
Tel: +61 39 562 1200
Fax: +61 39 562 1405

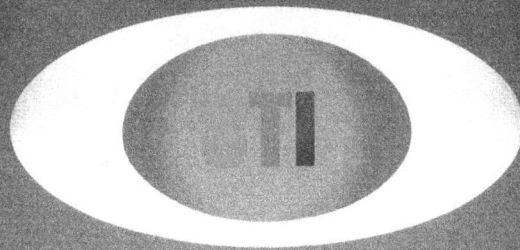

IS SHEDDING
NEW LIGHT ON THE
HOLOGRAPHIC INDUSTRY

**Forget what you know about conventional holography.
Today, computer technology has merged with lasers to
create three-dimensional images which
have true color capabilities.**

**Computer Graphics can be output into holographic originations.
Computer animation, morphing and modeling are now
being used extensively by STI to create some of the
most exciting images you will see.**

The Holography Marketplace cover hologram was originated by the Simian Company and produced by STI.
For details about the components of this image take a look at the copyright page.

SECURITY HOLOGRAPHY

Any true security hologram is a combination of different and unique holographic features. If you are considering the development of a security hologram, STI can provide you with unique technology that is not available through any other commercial source. In addition to supplying products, STI provides consulting to companies and governments to create or maximize their security programs. The STI Research and Development Team has brought many innovative and original products to the marketplace.

COMMERCIAL HOLOGRAPHY

Innovative holographic images are the future of commercial holography. Holograms that are designed to operate in a number of different retail lighting conditions are critical to the success of a commercial hologram. We can work with you to design a cost effective holographic solution for all your packaging, printing, and product enhancement needs.

EUROPE

STI-EUROPE • HUTTONS YARD, MAPLEDURWELL,
BASINGSTOKE, RG25 2LP U.K.
PHONE: +44 (0) 1256 346 208
FAX: +44 (0) 1256 292 38
ATTN: RENÉ CHARON

HOME PAGE at http://www.stiovd.com

Linda Law Holographics

425 New York Avenue
Suite 202, Huntington,
NY11743

Phone: (516)673-3138
Fax: (516)673-9127
e-mail: llholo@i-2000.com

Computer Graphics for Holography
2D/3D Graphics, 3D Animation
Graphics for the Holography Marketplace cover created by Linda Law

Holographic transfers applied with *heat or pressure* in stock or custom *shapes and sizes* for permanent adhesion to all surfaces. Sealed edges prevent delamination in all weather...washable...dry cleanable.

Patented product *and process*
Distributors sought

Holography Presses On

Jan Bussard
Box 193 Spring Lake, MI 49456
Phone 616/842-5626 Fax 616/842-5653

A figurine is removed from its mold. Photo provided by J Farina

DM: Although most of my model making experiences are based on the film industry (motion pictures, television, and television commercials), I have found the process of making models for holography quite similar. To me the only difference between sculpting a forced perspective landscape measuring 16 feet by 32 feet and 'a 2 inch by 3 inch [holographic] origination is the design, scale and detail required.

AM/LS: The primary difference is that holograms are relief models with a forced or false perspective. This means that we are creating a full-round 3D illusion with a depth, for most of our clients, of no more than 3/4 of an inch.... This type of modeling produces some unique and difficult challenges. An additional process unique to holography modelmaking is the painting of the models, which is done mostly by airbrush.

2. How do you usually interact with the other parties involved in the production process?

JF: The interaction of the people who are involved in a project can become complex. Often, we work for the company who originates the hologram. This origination facility will, in turn, typically work for a large company who commissions the hologram. The level of communication between all parties involved in the project will affect the quality of the final hologram.

DM: One must have very strong communication skills to effectively deal with directors, producers, art departments, model shops, machine shops, etc.

AM/LS: Whoever contracts us to do the model is the one we remain in direct communication. It is very important to establish a rapport with the client so information can flow freely.

3. What basic design elements should a potential client know about (size, detail, placement, color, etc.) before contracting your services?

JF: It should be the goal of the modelmaker and client to reach a reasonably complete mutual understanding concerning the various requirements of the final model. We usually work from graphic images supplied by the client via express courier or facsimile. The specifications of the model including size, depth, and surface finish are then verified by our facility. Details can be as sharp as the material will allow, but cost will be in some measure related to the amount of detail.

DM: All art directors have specific needs and requirements. A recent project we modeled for American Bank Note Holographics was very successful because their art department was organized and quite specific about what they wanted, and we provided high quality models (of a variety of D.C. Comics characters) to fit their needs.

AM/LS: They have to know exactly what they want. Everything has to be decided upon before the model maker starts the job. They need to provide us with artwork, or in the case of us designing the model, approval of our drawings. We need to know the size of the model (this usually differs from the overall size of the hologram), the depth of the model, different textures and painting requirements (if needed). The more time a client can give to the model maker, the better the final project. We also produced high quality models of D.C. Comics characters for American Bank Note Holographics and agree that the project was successfully completed due to the high level of communication we had with them.

4. What size models can you create?

JF: Although we can produce models in any size and depth specified, our workshop concentrates on smaller sculptures, typically for holograms measuring 4" x 5" or less. Since the majority of commercially produced holograms fall in this category, we have organized our production methods to produce cost effective models in this size range. Our recent model of the Anheuser-Busch logo is a highly detailed sculpture measuring less than I inch by 1 inch by 118 inch, while our miniature topographical map of the United States, which was produced for Holographic Design Systems, measures about 3 inches x 4 inches x 114 inch.

DM : We can create any size model.

AM/LS: We can create any size model with any depth specified by our clients. One of the smallest models we were asked to produce was 112 inch in diameter with a 11 16 inch depth.

5. How detailed can your models be?

JF: From a technical point of view, the level of detail possible is related to the type of modeling material used. Materials softer than plaster may not be suitable for fine detail work.

DM: We can create a model with as much detail and realism as the client wants.

AM/LS: We strive for the finest, most detailed models

possible. One example is a cigar model we did for Crown Holografx where we had to realistically recreate the texture of a tobacco leaf and the ash particular to this brand of cigar. To effectively copy the texture of the tobacco, we used the real tobacco leaf from the cigar wrapped around the modeled form. Using fast setting rubber for the mold, we were able to recreate the exact texture of the cigar. The creation of the ash presented its own unique challenges. We decided that carving the ash directly in plaster would work best. The painting of the ash was done by hand to create the subtle shades of a burned ash. Our model was so successful it was virtually indistinguishable from an actual cigar.

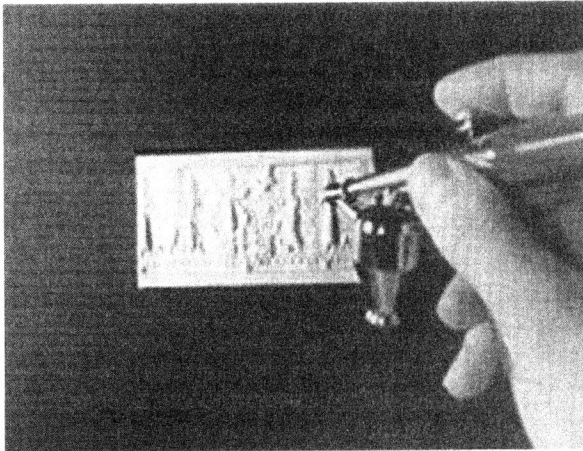

Finishing a highly detailed relief Photo provided by J Farina

6. Do you provide all the components in the scene or just the primary subject? Can you combine 2D graphics with your 3D models?

JF: Two dimensional graphics are often components of~ holograms either as backgrounds or overlays. It is our experience that the holographer usually prefers to provide these. Usually, we are only responsible for the sculpted component which serves as the primary subject of the hologram. However, we can also provide computer graphics or photo-etched designs (a procedure used to make highly detailed designs in thin metal sheets, such as copper or nickel).

DM: We create all the components in the scene.

AM/JS: We can create all the 3D components in the scene. Usually the client will assemble the elements when they shoot the hologram.

7. What are the pros and cons regarding using physical models and miniatures rather than creating stereograms from video or cinematography?

JF: Fully three dimensional holograms offer a visual experience entirely unlike any other medium. The same cannot be said of stereograms, which share much with cinematography. Expense may be another factor to consider. The cost of commissioning a professional film shoot can be two or more times the cost of hiring a modelmaker.

DM: The computer is a great tool for designing models and producing graphic work, but when the client requires texture and depth it is more cost effective to make a [physical] model than it is to use a computer.

8. What are the typical time frames involved on your end?

JF: Hologram projects show considerable differences in terms of complexity, which affects the amount of time needed to create the model. Typically, our models are completed in 2 - 4 weeks. Shorter or longer turnaround times are possible.

AM/LS: Anywhere from 1 week to 2 months. The more time a client can provide, the better the model. We are usually asked to turn a model around in 1-2 weeks, sometimes less. It is very rare in other industries to have such a quick turn around time.

9. What are the typical costs involved?

JF: Project cClmplexity will also affect cost. Highly detailed models or models that require painting in shades of gray will involve additional work. Multi-channel holograms often require more than one model, and the price will increase proportionately. Although lower or higher costs are possible, most of our models are priced between $2,000 and $4,000. Less costly work may be obtainable, but the client should consider ifthis is an area which should be compromised.

DM: If a client has clear specifications and requirements, it will cost less than if we are approached with a concept have to go through the design and approval process. A simple cut logo can range from $600 to $800 or more depending upon the complexity, whereas a sculptural model can go anywhere from a thousand dollars to thousands of dollars.

AM/LS: Model prices vary widely depending on the complexity and the size. An average model ranges from $2,000 to $4,000. When a company is putting together a bid for a project it would be best to consult the modelmaker beforehand about the price of the model to include in the bid. Making models for holograms is intensive, specialized and time consuming. We want to emphasize that if you want the best product, you.have to be willing to fund its creation.

10. What materials do you usually use? What tools?

JF: The preferred material with which we supply our customers is the hard plaster marketed as Permastone, which has a proven track record of success in holography. It can hold whatever level of detail the model maker can giveit. However, the requirements of a model vary greatly, and we are prepared to supply our customers with any material which can be pulled from a mold, such as polyester resin. We use Super Sculpey for some applications. Also, while working for the British Museum, we developed a coating system which produces better-quality holograms regardless of laser wavelength. It produces brighter highlights and darker shadows, while polarization is maintained. The types of tools we use are considered proprietary, although they range from X-Acto blades to customized miniature knives.

DM: This depends upon the client's needs and the end result. We use clays for sculptural objects and acrylics and metals for structural objects. [To combat vibration] we utilize materials that resonate at different frequencies. We utilize hand sculpting, computer generated laser cutting, and EDM machining.

AM/LS: Depending on the model, we decide what material would work best. Typically, the model is started by using oil based clays (non-hardening). We then use rubber molds to cast plaster, hydrocal, or polyester resin. The choice of the material depends upon the size, detail and fragility of the model. Usually we use dental and jewelry-making tools and miniature files. We also create tools designed for specific projects.

11. What "tricks" do you use to create (or exaggerate) image components?

JF: Forced perspective is a common technique used by modelmakers, but it has its limitations. Since many holograms have a very shallow depth, models are often made with foreground areas enlarged, and background areas reduced. This will give the impression of greater depth.

The model maker has greater freedom to use forced perspective when parallax is reduced, as in an embossed rainbow hologram. Since the viewer has a limited range of viewing angles, the artificial change in scale between foreground and background will not be as noticeable.

AMlLS: A model for a hologram has a front and sides, but remains flat on the back. Since this type of perspective does not exist in nature, we have to visualize which elements (i.e. a hand, head, etc.) are closest to us and which elements are far away. By increasing the size of the elements closest to us and decreasing the elements furthest away, we can create the illusion of 3 dimensional form in space... By painting light and dark shadows onto the models, we can heighten the illusion.

12. What final advice would you give to our readers?

JF: I would like to ask everyone to make an extra effort regarding what is really the most important (and often least appreciated) part of the process: the creation of the sculpture. Holograms will become far better when we are able to stop thinking in terms of making holograms of objects. Instead, we should make objects for holograms.

AM/LS: Since the quality of the hologram depends on the quality of the model, using the best model will produce the finest hologram.

3

Laser Fundamentals

This chapter could be titled "from photons to etalons". It begins with a review of the fundamentals of electromagnetic radiation and then proceeds to a detailed explanation of how the lasers that are most commonly used in holography (CW lasers) work.

Laser Fundamentals

Electromagnetic Radiation

To sufficiently understand the operation of lasers, their many advantages and their necessity in the production of holograms, one must first comprehend certain properties of our physical world.

The entire universe consists of only two things: matter and energy. Matter is all things that have physical substance; energy is the mover, or potential mover, of physical substance. Matter is the stuff we see, smell and feel. It has mass and occupies space.

Energy, on the other hand, is more abstract. It is most often invisible, though sometimes not. Yet, it is everywhere. It lurks in the crevices of every molecule and sweeps the skies with its magnificence. A master of transformation, energy facilely converts itself from one of its many forms s to another, all without sacrifice.

Energy is the driving force behind all forms of motion: the motion of our car, the motion of planets, the motion of atoms. Nothing moves without it. Matter, without energy, is reduced to a dark, frozen lump of nothingness. In a dynamic universe, matter both possesses energy and is affected by it.

Energy not only changes form, it is easily passed from one object to another. Interestingly enough, no matter how many times it transforms or transfers, the amount of energy involved in any given transaction never changes. The law of conservation of energy, one of the most important laws in the universe, dictates that energy is never created or destroyed; it can only be transferred to another object or converted into a different form of energy.

Because the amount of energy in the universe remains fixed, phrases such as "energy shortage" and "depleted energy" are misnomers. You can not lose energy, nor can you be in short supply. The amount of energy in our environment is so great that it is beyond our comprehension. The discomforts in past decades from "energy shortages" were created only by our inability to either convert energy to a usable form or distribute usable energy to where it was needed.

Energy is measured in Joules, in honor of the British scientist James Joule. One joule is roughly the amount of energy required to lift an apple from your kneecap over your head.

A glass of apple cider has 502,092 joules (equal to 120 Calories-the Calorie is another unit of energy often used when referring to the content of food) of food energy. A. gallon of gasoline has over 200 million joules of energy.

The process of applying energy to matter is called work (also measured in joules). Work is the mechanism that transfers energy through a system. It is produced by applying a force on an object such that motion occurs over a distance. For example, when you pick up a book. you have performed work on the book. The heavier the book. the greater the force that is necessary to raise it-therefore. the more work done. The farther you raise the book, the greater the distance in which the force must be applied. Again, more work is accomplished.

Energy is formally defined as the ability to do work. It can be classified in two categories: stored energy and motion energy. Stored energy is more commonly called **potential energy.** When raising a book in the air, work must be performed on it. While suspended in air, however, the book has the ability to perform work in the opposite direction-courtesy of the earth's gravitational field. This "stored" energy may be released simply by dropping the book .

The potential energy an object possesses due to its position in a gravitational field is called, predictably enough, **gravitational potential energy** (or GPE).Any object raised above the earth's surface has gravitation potential energy. Water behind a dam has a significant amount of GPE that may be converted (as mandated by the law of conservation of energy) into electrical energy.

Other forms of potential energy are also commonplace. A coiled spring is the good example of **mechanical potential energy.** By performing work on the head of a Jack-in-the-box, one can push it into the box. With the lid secured, the box has stored energy (mechanical potential energy) hence, the ability to do work. By unlatching the lid, the stored energy is released and work is performed in the reverse direction on our friend Jack.

The axiom that "opposites attract" is especially true for electric charges. Electric charges that are positive attract those that are negative, and vice-versa. Equally, two electric charges of the same type (both positive or both negative) repel each other. The attractions and repulsion o f electric charges are caused by invisible **electric fields** produced by each charge. An electric field permeates the territory around each charge, affecting all other charges that occupy its space. The larger the charge, the more influence its electric field exerts on its occupants. We encounter electric fields to varying degrees throughout a typical day. We witness them when we use our dryer, for it is electric fields that cause static cling.

To separate two opposite charges-or unite two like charges-requires work. Like the Jack-in-the box, when work is applied to bring like charges together (or to separate opposite charges), **electric potential energy** is created. Remove whatever constraint that holds the stored energy (in the Jack-in-the-box the constraint was the lid, in electric charges it is usually a non-conductive material like air or plastic) and work is performed in the opposite direction.

A familiar device utilizing electric potential energy is the battery. The stored energy in a battery can be released by placing a conductive path between the positive and negative terminals. The performance of battery is stated in terms of voltage (also called electric potential, abbreviated with a V and measured in Volts). Voltage, an important element in laser operation, is the ratio of electric potential energy (EPE, not to be confused with electric potential) to the amount of charge (abbreviated in equations with a q,

measured in **coulombs**). In equation form:

$$V=(EPE)/q$$

Another common source of voltage is the generator. Generators are machines that convert various types of energy into electric potential energy. Generators in dams convert gravitational potential energy from elevated water. The energy is transported to homes and businesses, readily available for those who wish to do a little work. Since generators produce higher voltages than batteries, they are used to supply power to all gas lasers and most others.

Matter is the greatest repository of energy. Atoms arranged together have binding electrical forces (called **bonds**) that act much like infinitesimal coiled springs. When bonds are broken, stored energy is released. This stored energy in molecules is called **chemical potential energy**. The gas we pump into our cars and the food nourishing our bodies are two common forms of chemical potential energy being utilized in our lives. Forces holding the nucleus of an atom together store an astounding amount of **nuclear potential energy**, as witnessed on July 16, 1945, when the Manhattan Project unveiled the atomic bomb.

Matter in motion possesses **kinetic energy**. An object will gain kinetic energy when work is done to it. An object will lose kinetic energy when work is done against it. The amount of kinetic energy an object gains or loses is exactly the same as the amount of work done on or against it.

For example, the engine of a train converts chemical potential energy into kinetic energy and performs work on the train. The train will move; it now has the amount of kinetic energy equal to the net gain of work done on it. If you turn off the engine, the train eventually stops, even if it is riding on a perfectly level set of tracks. This is because friction (between the wheels and the track; between the wheels and their axles; and between the air molecules and the front of the train) is performing work against the train.

When a train in motion hits a stationary object in its path, work is performed on the object. The object will move. Some of the train's kinetic energy is transferred to the stationary object. If one removes all sources working on and against a moving train on perfectly flat tracks engine, friction and objects in its path-the kinetic energy of the train will never change. The train will continue to move forever at a constant velocity.

Other kinds of motion energy include heat, sound and electromagnetic radiation.

Heat occurs from the motion of molecules. The faster the molecules move, the more heat generated. A common source of heat is friction. In our previous example, friction performed work against the train. The kinetic energy of the train transformed itself to frictional heat in its wheels, ax- les and tracks (to a lesser degree, the air molecules). The train eventually stopped because its kinetic energy was entirely transformed into frictional heat. In most energy exchanges in nature, heat is part of the transaction.

Sound is another form of motion energy that occurs when a disturbance in a medium (commonly air) produces molecules to vibrate back and forth creating "sound waves". Each molecule receives the wave, vibrates back and forth and returns to its original position, but not before imposing a similar disturbance on its neighbor. The neighboring molecule repeats the same maneuver, as does each successor, thus creating a chain of disturbances that allows sound energy to propagate through the medium. Eventually, the sound waves hit an eardrum causing it to vibrate. The vibrating eardrum creates signals to the brain that enable us to "hear" the sound energy.

One of the most important forms of motion energy is **electromagnetic radiation.** It exists everywhere through- out the universe and comes in many forms. Radio and television waves can be transmitted hundreds of miles through the air enabling music, images and conversation to magically appear in our homes. Microwaves, used in radar and modern cooking devices, ensure safe travel and a fast meal. Infrared radiation warms our skin and other vital regions of the universe. Visible light, the only form of electromagnetic radiation that we can see, enables our world to have definition and beauty. Ultraviolet radiation burns our skin and cures our plastics. X-rays help doctors diagnose problems in our bodies while gamma rays are found in many forms of radioactive decay.

Although we perceive and apply them differently, all forms of electromagnetic radiation are essentially the same phenomenon. Only the amount of energy per fundamental unit distinguishes a microwave from a beam of light.

The fundamental unit of electromagnetic radiation is called the **photon,** an infinitesimally small "packet" of energy. Radio waves have relatively low energy per photon. Microwaves have more energy per photon than radio waves but not as much as infrared radiation. A photon of visible light has more energy than a photon of infrared radiation but less than ultraviolet radiation. X-rays and gamma rays carry the most energy of all.

Electromagnetic radiation is created by accelerating or decelerating an electric charge. The greater the acceleration (or deceleration) of an electric charge, the more energy it will produce. It would take a much greater deceleration of an electric charge to create an x-ray photon than a microwave photon. Electric charges that are stationary, or those moving at a constant velocity, do not create electromagnetic radiation.

An electron, the most fundamental unit of negative charge, is the most common vehicle for creating electromagnetic radiation. For example, a radio station produces radio waves by accelerating electrons up and down a transmission antenna in a process called **oscillation.** An antenna is limited in its ability to rapidly accelerate and decelerate electrons, however. This is why antennas do not create visible light. Electron activity in atoms is the most prolific manufacturer of visible light. As explained later, this activity will be the basis from which lasers are created.

Properties of Electromagnetic Waves

Water waves make a good model for the study of electromagnetic waves because they are commonplace, exhibit comparable properties, and move slowly enough to carefully observe. There are a few profound differences between the two (for example, water waves must propagate in a medium where electromagnetic waves need not), but not enough to impugn our comparison.

A wave is created by a disturbance in a medium (for electromagnetic waves, a disturbance may be created in empty space). In a swimming pool, a swimmer resting in the shallow end of the pool slaps his hand on the water. The disturbance creates a wave that moves from the shallow end to the deep end (for this example and all fictitious pools in this section, allow the edges to absorb all waves that hit it, thus eliminating the effect of reflected waves). The wave has a "crest" (high point) and a "trough" (low point).

A closer look at the wave would reveal that the water molecules do not travel with the wave. In fact, if you measured the net movement of all the water molecules due to the wave, it would total zero. One may ask, "If the water molecules aren't moving in the direction of the wave, what is?" The answer is energy.

If the swimmer slaps his hand many times in regular intervals, a wave with a series of crests and troughs is created. Each pair of one crest and one trough is called a cycle due to their tendency to repeat. If the intervals are fast. the crests and troughs (or cycles) will appear to be closer together. If a sunbather sitting halfway between the deep and shallow end of the pool had a watch, she could count the number of cycles that pass by her each second. This value. the number of cycles per second, is called the frequency (for space economy, we use the letter f in equations) of a wave. The unit of one cycle per second is more commonly called a Hertz (abbreviated Hz) in commemoration of Ger- map physicist Heinrich Hertz. Because the frequencies of electromagnetic waves are quite high, larger units such as megahertz (one million hertz, abbreviated MHz) and gigahertz (one billion hertz, abbreviated GHz) are commonly used.

The velocity (v-measured in meters per second) of a wave is directly related to the medium in which the wave is travelling. If the pool was drained and filled with molasses, the velocity of the waves would be less than those moving in water. For the remainder of this section. it will be assumed that all waves generated in our fictitious pool are travelling at identical velocities.

The distance between two successive crests (or two troughs) on a wave is called the wavelength (A) which is measured in meters or subunits of meters. The most common subunit for wavelength measurement of electromagnetic waves is the nanometer (abbreviated nm) which is one-billionth (1×100^9) of a meter.

When the swimmer slaps the water with slow intervals, the distances between the crests and troughs of the waves are large, hence the wavelength is long. The sunbather times very long cycles per second on such waves. The frequency is small. However, when the swimmer rapidly slaps the water, the crest and troughs seem to bunch together. The wavelengths are short. The sunbather counts many cycles per second. Waves with higher frequencies, therefore, have shorter wavelengths and waves with lower frequencies have longer wavelengths. The relationship between the two can be summarized in the equation:

$$f = v/\lambda \text{ or } \lambda = v/f$$

It is important to remember that, as long as you adjust their numerical values per the above equation, frequency and wavelength are interchangeable. In the study of light it is common to use either term.

The swimmer may also notice that when he slaps the water with more force, the crests become taller and the troughs become deeper. The height of the crests (or in many cases, the depth of the trough) is called the **amplitude** of the wave. If the swimmer continues producing waves, one long, unbroken string of crests and troughs will span the entire length of the pool. This is called **continuous wave** transmission (often called just **CW** transmission).

But, the swimmer could decide to produce one wave, rest (thus saving his energy), and then produce another wave-followed by another rest period. By saving his energy between waves, the swimmer could slap the water harder and produce waves of greater amplitude. This is called **pulse** transmission. Lasers transmit in a similar manner; they are either continuous wave or pulse.

In a V-shaped swimming pool with two shallow ends converging to one deep end, two swimmers at rest (swimmer A and swimmer B, equal distance from the deep end) start slapping the water at exactly the same time and with exactly the same intervals. Not only would both waves (wave A and wave B) have the same frequency but, as they passed the sunbather, all the crests of wave A would pass exactly at the same time as all the crests of wave B. Similarly, all the troughs of wave A would pass at exactly the same time as the troughs of wave B. The two waves are said to be **in phase.**

Two waves being **in-phase** or **out of phase** refers to a comparison of the two waves at a given point. In the example above, the given point is the exact spot where the two waves pass the sunbather. Two waves can be out of phase at a given point for a variety of reasons. The disturbances could have started at different times. The frequencies of the two waves could be different, or the distance travelled to the given point (called the path length) could be more for one wave than the other. If the two waves were in different mediums, they could have different velocities.

Two waves having identical frequencies and velocities that are out of phase, will be so to the same degree at all points. For example, two waves at 60 Hz that are 80° out of phase at the sunbather will be 80° out of phase at the deep end of the pool and all points in between.

Two waves with equal frequencies and velocities starting at the same time can be out of phase if their path lengths are different, e.g., if swimmer A was slightly further from the sunbather than swimmer B. Holographers use this fact to create their holograms.

When wave A and wave B meet at the deep end, they will join together and this phenomenon is called interference. How they interfere depends on what degree the two waves are in or out of phase (called their phase relation-

ship). If wave A and wave B are in phase, the crests of the two waves will meet and combine to form one large crest for each cycle whose amplitude is the sum of the two individual crests. The troughs of wave A and wave B would also combine to form one large trough in each cycle whose amplitude is the sum of the two individual troughs. This is called constructive interference (see diagram A).

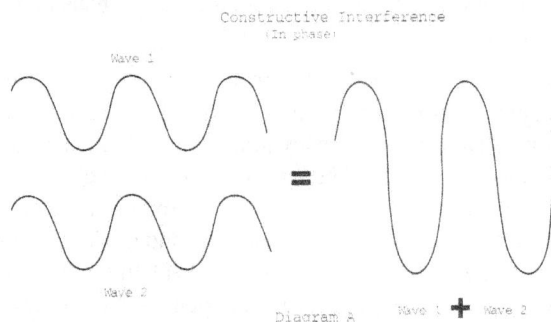

Diagram A

If the two waves are 180° out of phase, the crests of wave A would merge with the troughs of wave B (and vice versa), cancelling each others' amplitudes. The result is a combined wave with little or no amplitude. This is called destructive interference (see diagram B).

Because the phase relationship of two waves can change from one point to another, the two waves can be in phase when they pass the sunbather, but out-of-phase when they hit the deep end of the pool. The ability of the two waves to stay in phase while they travel the length of the pool is called coherence. Waves that stay in phase for a long time are said to be very coherent.

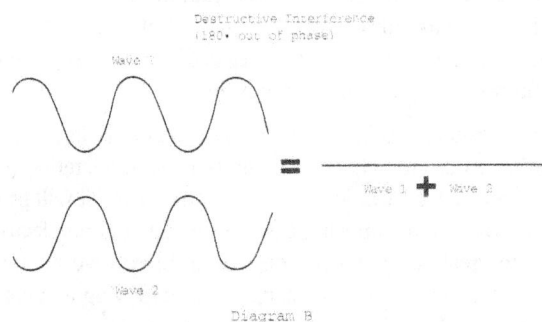

Diagram B

Suppose the two swimmers agreed to create continual constructive interference at the deep end of the pool. They would start slapping the pool at the same time while trying to maintain exactly the same frequency. If both swimmers have extremely good timing, they can keep the two waves coherent for a long period. But the swimmers, like other producers of waves (lasers, for example) aren't perfect. Their frequency may be slightly off.

The difference in frequency of any source (source is a common term for a device or system that produces waves) is called its bandwidth (abbreviated f, and is measured in Heitz). The coherence of a source can be determined by measuring how long the waves stay in phase. Be it swimmers or a laser, the distance that the source can guarantee the waves will stay in phase is called its coherence length (abbreviated L; it is measured in meters or subunits of meters). The coherence length of a source is of great importance to holographers. It is directly related to bandwidth by the equation:

$$L = v/\Delta f$$

Lasers are used to produce holograms primarily because no other source offers enough coherence.

Although electromagnetic waves exhibit exactly the same wave properties as the water waves described above, there are some notable differences between the two. Water waves propagate in two dimensions on a plane. Electromagnetic waves tend to propagate in three dimensions. As mentioned earlier, electromagnetic waves can propagate with or without a medium. Electromagnetic waves move extremely fast; water waves move relatively slowly. The interference of a water wave is determined by its amplitude-with electromagnetic waves, it is a function of its intensity.

In empty space, electromagnetic waves move at a velocity of 300 million meters per second (3×10^8 m/s, also known as the speed of light- it is abbreviated in equations with the letter c) . In his 1905 paper on special relativity, Albert Einstein correctly defined the speed of light as the absolute fastest velocity possible- a cosmic speed limit, so to speak. In air, the velocity of electromagnetic waves is just slightly less than "c". In most applications where electromagnetic waves are travelling through air, it is acceptable to use "c" as the velocity of the wave. Therefore, for electromagnetic waves travelling in either free space or air:

$$f = c/\lambda \text{ or } \lambda = c/f$$

and

$$L = c/\Delta f$$

Electromagnetic waves are a union of electric and magnetic fields that are at right angles (90°) to both each other'" and the direction of their movement (see diagram C). When electromagnetic waves propagate, there are infinite amount of directions they can travel. A laser is designed to channel waves such that their propagation is substantially in one direction .

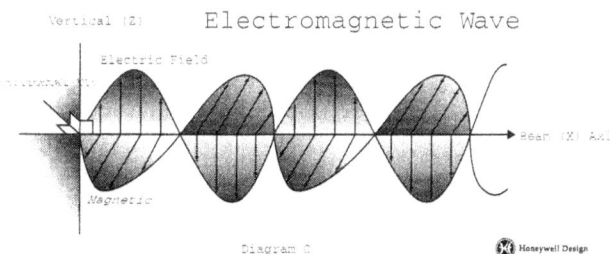

Diagram C — Electromagnetic Wave

This unidirectional propagation of electromagnetic radiation is generally referred to as a laser beam.

Even while moving in a common direction, each wave can have its electric field and magnetic field oriented differently. The electric field can point straight down on the

first wave, sideways on the second. There are an infinite amount of directions the electric field (or magnetic field, which stays exactly at a right angle to the electric field) can be pointing on the beam "axis" for each wave moving in the same direction.

Many properties of waves are more consistent if their electric and magnetic fields are properly aligned. The ability of electromagnetic waves to be aligned in the same orientation on the beam axis is called polarization. The human eye is not sensitive to polarization and can not distinguish between polarized or unpolarized light waves. Some insects, like bees, are more sensitive and use polarization to determine direction. In holography, where consistency of the source's waves is critical, polarization is essential.

Because a reference point is needed in defining polarization, the electrical field is used to identify the position. If the electric field is travelling directly on the xz plane, the wave is defined as vertically polarized. How close the beam is to being polarized in the vertical position can be described by its polarization ratio. A laser beam with a 100:1 polarization ratio is very close to being polarized in the vertical plane- a 500:1 ratio is closer still.

P.olarization can be achieved by several means, including reflection, transmission, scattering and birefringence. Birefringence is a phenomenon that occurs in certain crystals-such as calcite- and other materials. Such materials limit the absorption of waves to those with specific electric field orientations. Sunglasses use this effect reducing the amount of glare received by the eyes. Crystals with maximum). birefringence allow only one orientation to be absorbed and transmitted. The beam exiting the crystal is polarized.

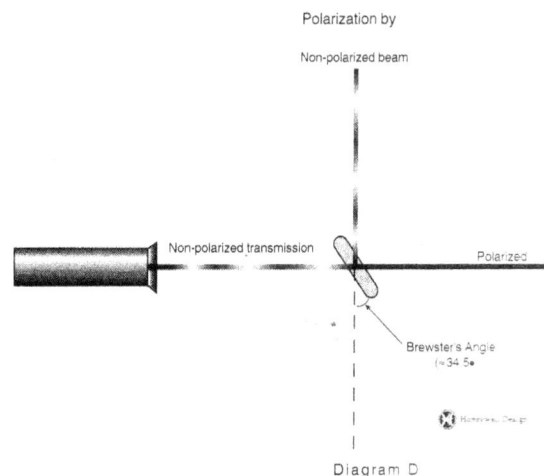

Diagram D

Scattering occurs when an atom deflects a photon a\\'ay from it. Scattering of electromagnetic radiation in the earth's atmosphere produces partial polarization. Bees use this type of polarization for navigation. Polarization by scattering is not an effective source for most photonic applications. For lasers, polarization by transmission is more relevant.

If an unpolarized beam strikes a non-conductive, transparent target at a specific angle (called **Brewster's** angle, see diagram D), a polarized beam will pass through the target. This process is called polarization by transmission. Waves in the beam that do not have the selected electric field orientation are reflected from the target. Brewster's angle varies for different wavelengths and materials, but for most lasers it is in the neighborhood of 34.5 degrees.

Particle Properties of Light

On October 19, 1900, Max Planck introduced a concept that was to revolutionize science. In an effort to resolve conflicts between scientific theory and experimental evidence, Planck suggested that **energy** (abbreviated in equations with the letter **E**) is not continuous, but instead comes in discrete little "packets" called **quanta**. Further, an "energy packet" of light was directly related to its frequency by the equation:

$$E = h \cdot f$$

where h is defined as **Planck's Constant** and is equal to 6.63 x 10.34 Joule Seconds.

Although the mathematics seemed to work, and it did resolve current conflicts between theory and experimental data, the implications in Planck's hypothesis were rather hard to accept. Clearly, light exhibited wave properties such as **diffraction** (bending of a wave around a corner), interference and polarization. Waves are inherently continuous and not discrete. Frequency, for example, used in Planck's equation is a wave phenomenon. Yet the energy in the same equation describes light as discrete packages of $h \cdot f$. How could light possibly consist of particles and demonstrate properties of waves?

The numerous experiments and the profound mathematics that followed are extremely significant and detailed. The final result of two decades of scientific fervor was a new definition of the laws of physics now known as **quanturn mechanics**.

At the core of quantum theory is the concept of **duality** which states that light, electromagnetic radiation, energy and even matter is both a wave and a particle. Electromagnetic radiation itself is composed of minute "wave packets" called photons that demonstrate properties of both continuous waves and discrete particles.

In terms of wavelengths, the energy of one photon is expressed as:

$$E = h.c/" \text{ or} " = h.c/E$$

As stated earlier, all electromagnetic radiation is the same phenomenon. Only the energy per photon is different. The fundamental unit of electromagnetic radiation is the photon. One can classify all forms of electromagnetic radiation by the wavelength (or frequency) of the photon. Because the wavelength involved in the common forms of electromagnetic radiation is small, it is usually measured in nanometers (1 x 10-9 meters, abbreviated nm). In visible light, the wavelength of a photon determines its color. Red had the longest wavelength (740-622 nm), followed by or-

ange, yellow, green (577-490 nm) blue (489-430 nm) and violet (429-390 nm). White light is a mixture of all colors. Photons with wavelengths greater than 740 nm produce infrared (below red) radiation. Photons with wavelengths less than 390 nm create ultraviolet (beyond violet) radiation .

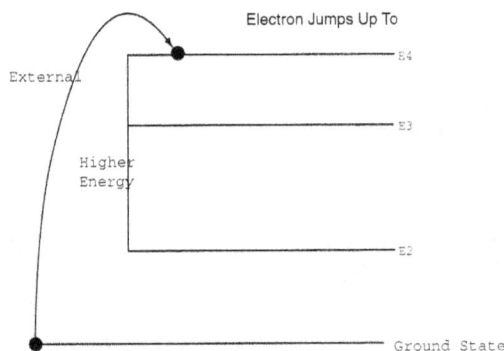

Diagram E

Photons can be created by transferring energy to an atom. In an atom, electrons reside in various positions and energy states. If the atom is stable, the electrons are defined as being in their **ground** (lowest level of energy) state. When external energy is transferred to the atom, the electrons get "excited" and respond by jumping up to higher, or excited **energy states** (see diagram E).

Quantum mechanics dictate that an electron cannot reside between two energy states. It has to jump up all the way to a higher state or not jump at all. The electron will stay in the excited state for a very short period and then spontaneously drop back down to a lower energy state.

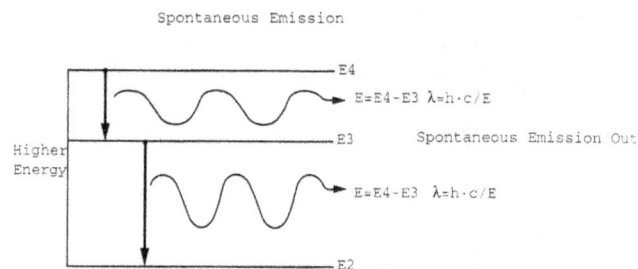

Diagram F

When the electron drops one or more energy steps (also called **transitions**), it releases the amount of energy difference between the two states (see diagram F) in the form of a photon. The wavelength of the photon is determined by the amount of energy released- the energy difference of the two energy states-through the formula" = h.c/E. If the energy released is small, the wavelength of the photon will be large and vice versa. This process is called s**pontaneous emission.**

There are two principal mechanisms that enable energy to be transferred to an atom: absorption and collision. Absorption occurs when a photon bumps into an atom. If the photon's energy (determined by the wavelength of the photon using E=h.c/A) matches the energy difference between a lower energy state where an electron resides and a higher energy state (called an **energy band gap**), the atom will absorb the photon. The photon energy is transferred to the atom, kicking the electron up to the higher energy state. If the photon's energy does not match any of the energy differences in two excited states of the atom, scattering occurs, redirecting the photon without otherwise altering it.

Collision occurs when a moving particle (an electron, ion, atom or molecule) smashes into the atom with the proper amount of momentum. Some or all of the particle's kinetic energy is transferred to the atom, again raising the electrons of the atom to a higher energy state.

Laser Theory

Albert Einstein was the first to recognize the significance of Planck's concept of quantized energy. He used Planck's E=h.f equation to derive his explanation of the photoelectric effect in 1905, for which he later received the Nobel Prize in Physics. In 1916, Einstein predicted another phenomenon now known as **stimulated emission**-the basis for all laser technology. The same year, Einstein also released his most prized work, the general theory of relativity. The theory of stimulated emission- went unnoticed until the late 1940's.

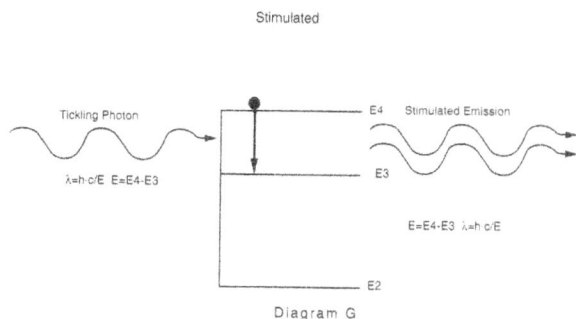

Diagram G

An excited electron in a higher energy state will spontaneously drop to a lower energy state and emit a photon. The wavelength of that photon is determined by the difference of the energy between the two energy states (A = h.c/ E). If, before the electron drops to the lower energy state, a photon with a wavelength identical to that which is about be produced by spontaneous emission passes by the exited atom, it will stimulate the electron to drop. This stimulation (also called **"tickling"**-see diagram G) will force an emission of a photon that is identical to one passing. This process is stimulated emission. Both photons have the same wavelength (and therefore the same frequency), are in phase, coherent, and travelling in the same direction. It is important to note that the "tickling" photon is not absorbed by the atom; it must only pass closely.

In an environment with many identical excited atoms, photons can multiply rapidly through stimulated emission. Two photons quickly become four which quickly become eight. Since atoms and photons are extremely small, eight photons can become billions in a reasonably short distance. This process of multiplying photons is called **amplification**; it is the essence of a laser. The term LASER itself is an acronym for **L**ight **A**mplification from **S**timulated **E**mission of (electromagnetic) **R**adiation. Billions of photons, with the same wavelength, in phase, coherent and travelling in the same direction can be a very useful tool.

For lasing to occur, it is essential that there are more atoms with electrons in a higher energy state than those at the lower energy state-a condition called a **population inversion**. An atom with its electrons in lower energy states will absorb a passing photon instead of duplicating it. If there is more absorption than stimulated emission, ampli- fication will not occur. How can a baker increase his inventory if he has five hungry children and only four cookies in the oven? A population inversion assures continuous multiplication of photons.

To create a laser (see diagram H), many atoms of one type must be contained in a given space. The atoms used for lasing are called the active medium and are housed in a container. Since the energy states of the active medium create most or all of the stimulated emissions, it is the **active medium** that determines the possible wavelengths produced by the laser (λ = h • c/E). **Solid state** lasers have active mediums made from matter that is solid at room temperature. Neodymium and chromium ions (an ion is just an atom with either an excess or deficit of electrons) are common active mediums used in solid state lasers. The active medium in **liquid lasers** are, of course, liquid and those in **gas lasers** are gas. Common active mediums used in gas lasers are neon, argon and ionized cadmium.

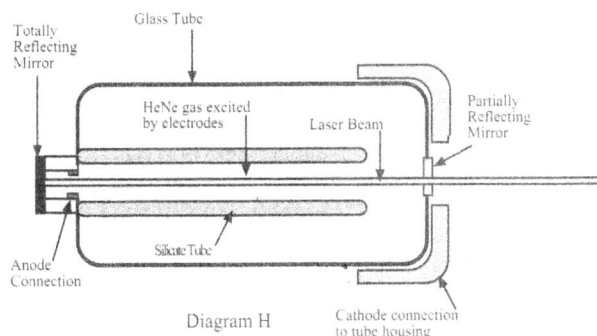

Diagram H

A **pump** transfers energy to the active medium, raising their electrons to an excited (higher energy) state. Brisk and continuous pumping will create and maintain a population inversion. As stated earlier, energy can be transferred by either absorption or collisions. In solid state lasers, an optical pump produces a flood of photons with energies that are easily absorbed by the active medium. Common optical pumps are flash lamps, laser diode arrays and other gas, liquid and solid state lasers.

The vast majority of gas lasers use electron collisions to

pump energy to the active medium. Normally, electrons will not flow through a neutral gas. If a large voltage (from 1,000 to 20,000 Volts, depending on the gas) is applied to the gas, the gas will "break down" and allow a **discharge** of electrons to rush through it. This initial voltage is called a **spike**, and is applied to the gas through two electrodes (an **anode** and a **cathode**) on opposite ends of the container. Once electron current is flowing through the tube, the voltage is automatically reduced (by 90 to 4,000 Volts, depending on the gas). The discharge of electrons and other charged particles collide with the atoms in the active medium, enabling energy to be transferred. Both the spike voltage and the operating voltage necessary to operate the laser are furnished by a power supply, usually an external box that transforms either. 117V AC or 220V AC to the required voltage.

Many active mediums are not efficient at receiving energy from the pump. In such cases, the active medium must be combined with a transfer medium-a substance that compensates by efficiently collecting energy from the pump then passing it on to the active medium. Solid state transfer mediums should be good absorbers; gas transfer mediums require the ability to efficiently receive energy from collisions.

The transfer medium must be compatible with the active medium in two ways. First, it must have some common higher energy states that provide a channel to efficiently pass energy to the active medium. Second, it must be chemically compatible, allowing peaceful coexistence with the active medium as well as the other components inside the tube.

There are typically 5-20 transfer medium atoms for every active medium atom. Inert helium makes a good transfer medium and is used in helium-neon (**HeNe**, pronounced Hee-Nee) and helium-cadmium (**HeCd**, pronounced Hee-Cad) gas lasers. Common transfer mediums in solid stare lasers are yittrium atrium garnet (called "**YAG**"), yittrium lithium fluoride (**YLF**, pronounced "Yelf"), sapphire and ruby.

The container that holds the active medium (and transfer medium, if applicable) must protect it from elements that may interfere with the lasing process. In solid state lasers, the crystalline transfer medium encompasses the active medium atoms, forming a strong durable solid structure that serves as the container. Because the transfer media in solid state lasers house and transfer energy to the active medium, they are called **hosts**.

For gas lasers, the container is almost always a long cylindrical **tube**. Tubes are made of various materials; ceramic and glass are the most common. Inside the tube is an equally-long, yet very narrow (3 millimeters) bore that allows lasing to occur in a straight, usually horizontal path.

During lasing, a wide variety of spontaneous and stimulated emissions occur throughout the tube with photons propagating in every conceivable direction. All emissions except those travelling down the bore exit from all sides of the tube with limited or no amplification. Those photons travelling down the length of the bore continue

to multiply through stimulated emission. By making the bore long enough, one could continue the lasing process until adequate amplification was achieved. At this point, a billion or so photons exit the bore in the form of a laser beam.

Unfortunately, in order to achieve sufficient amplification, the bore would have to be forty feet long. A forty foot laser would be extremely awkward to both transport and operate. A shorter bore is needed to make the device more practical.

By passing photons back and forth (called **optical feedback**) several times through a shorter bore, adequate amplification can be attained. This is accomplished simply by placing mirrors on both sides of the bore. If both mirrors are 100% reflective, extremely high amplification is achieved. However, no photons exit the laser (photon exiting the laser is called **transmission**). This, of course, has no value whatsoever.

However, if one mirror could reflect some of the photons back into the bore while allowing the remainder to exit, both amplification and transmission could occur. Such a mirror found on lasers is called the **output coupler**, or **OC**. The second mirror is known as the **high-reflector**, or **HR**. A perfect high-reflector will provide 100% reflection. In actual lasers, there is a small amount of light transmitted from the HR, referred to as **leakage**.

Power is the amount of energy a source produces each second. It is measured in units of Joules per second, more commonly called **Watts**-in honor of the British scientist James Watt. The power in a laser reflects the amount of photons per second exiting the laser.

Laser designers strive to maximize the power produced by the laser. However, if the OC allows too many photons to exit, the number of photons returning to the bore may not be adequate to provide significant amplification. This, in turn, limits the amount of transmission. Therefore, the amount of transmission and reflection provided by the OC must be properly balanced to compliment both the amplification process in the laser **cavity** (area between the two mirrors) and the transmission from the cavity. In most lasers, output couplers allow 1-3 % o f the photons to be transmitted.

It is common to have multiple wavelengths lasing inside of the cavity. Because there are a variety of paths of energy states for which an excited electron can travel back to the ground, there are a variety of energies (and therefore, wavelengths) that can be emitted throughout its descent. The electron can also bounce down two steps at a time-maybe three- each time emitting photons of higher energy and lower wavelengths.

Certain transitions are more dominant, however. The more spontaneous emissions produced at a given wavelength, the greater probability of stimulated emission. The more stimulated emissions, the more amplification. The most dominant wavelength in a given laser is called the **primary wavelength**. The second most dominant wavelength is called the **secondary wavelength**.

Photons with undesirable wavelengths can be eliminated

from the cavity by putting special thin film coatings on the mirrors. Such coatings only allow a specific range of wavelengths (for example 430-460 nm) to reflect back into the cavity. Thus photons with undesirable wavelengths will not multiply.

The lasing process described above is continuous wave transmission. Lasers that produce this kind of transmission are called, expectably, **continuous wave lasers** (or just **CW** lasers). Lasers that provide pulse transmission are called **pulse lasers**.

A continuous wave laser can be converted to a high energy pulse laser by installing a **Q-switch** in the laser cavity. Q-switches are devices that enable the active or transfer medium in the cavity to collect maximum pump energy before beginning the process of stimulated emissions. The Q in Q-switches is an abbreviation for "quality factor" to represent the quality of a feedback system. In the "low Q mode", a Q-switch limits optical feedback in the cavity for a very short period. If optical feedback is blocked, continuous stimulated emission will not occur.

During this period, the active medium continues to collect energy from either the pump or the transfer medium. A high percentage of electrons are elevated to higher energy states creating a large population inversion. When the Q-switch opens (high Q mode), a rapid and powerful episode of stimulated emission occurs in the cavity until the Q-switch is again closed. The active medium begins receiving energy, and again its electrons begin to elevate in preparation for the next high Q mode.

The repetitive bursts of lasing in the cavity result in a string of powerful energy pulses departing from the output coupler. The average length of time of a pulse is referred to as its **pulse length** (measured in seconds and subunits of seconds). The number of pulses per second is called its **repetition rate** (or more commonly, **rep rate**, measured in Hertz).

There are four types of Q-switches: chemical, electro-optical, acousto-optical and mechanical. The first three are common and found in a variety of applications. Mechanical Q-switches are seldom used because they tend to be slow, noisy and produce unwanted vibrations.

Although the overwhelming majority of holographic applications use CW lasers, a new branch of holography called **pulse holography** is emerging. This kind of holography uses high-energy pulse lasers to expose the recording medium.

The high energy pulses produced by a pulse laser significantly reduces the exposure time of the recording medium (usually silver halide) to less than one hundred nanoseconds Reduced exposure times yield some significant benefits.

Perhaps the most important benefit of pulse holography is that vibration is no longer a factor. In CW holography, one or two minute exposure times require the interference patterns on the medium to stay constant until the medium is exposed. Vibrations in the optical train or table can cause the interference fringes to overlap, which distort the image of the hologram. Vibrations are a primary enemy to a CW holographer. Expensive vibration-dampening optical benches are needed to keep vibration to a minimum.

In pulse holography, vibration is of little concern. Not much can happen to their interference patterns in less than one hundred nanoseconds. Because of this, motion can be captured in a hologram. This enables the holographer to produce holographic images of real people, pets and other animated subjects.

Pulse lasers capable of sufficient energy are extremely expensive, in excess of $50,000. The most common pulse laser used today is a Q-switched, frequency doubled Nd: Yag (neodymium Y AG) which   produce high energy pulses (0.5- 1 joules per pulse, with a pulse width of 15 nano seconds) at 532-nm.

Pulse holography has developed a cult following in the 1990's. It is used mainly in artistic holography. High cost and other production limitations has prevented pulse holography from being adopted in commercial applications

The laser tube and mirrors are held by a mechanical support structure almost unanimously referred to as a resonator. The title, nevertheless, is wrong. A **resonator** is a system consisting of a laser cavity and mirrors that enables rapid bi-directional optical feedback. It oscillates. Most Physicists will readily admit that the term is incorrect; how- ever, there is no other term available other than "mechanical support structure". Conforming to the majority, I will use the term "resonator" in this paper with the knowledge it is incorrect.

The resonator has the task of keeping the bore straight and aligned with the mirrors. This is not an easy assignment. The lasing process produces an ample amount of heat: The heat creates thermal expansion, which tends to shift the mechanisms that hold the mirrors (mirror mounts) and bore. The bore itself, being long and quite narrow, is extremely susceptible to thermal distortions.

When the laser cools down, the components of the laser tend to contract. Even the best resonators will sometime fail to keep the bore and mirrors in line. Because it is easier to align the mirrors than the bore, alignment devices are placed on the mirror mounts.

Such devices, called **tilt plates** or **xy plates**, enable the operator to change the positioning of the mirrors either sideways (horizontal, or "x" position) or up/down (vertical, or "y" position) without-changing the z position (frontwards and backwards-this would change the cavity length which can only hurt the laser's performance-see section V). Proper adjustment of the tilt plates enables maximum amplification inside the cavity, producing maximum laser power. The resonator must also provide mechanical protection from the routine bumps and bruises that may occur.

The tilt plates and mirror mounts are secured in the **resonator** by three or four resonator rods, which span the length of the laser. Resonator rods are the backbone of the resonator. They give mechanical support to the entire laser head and the hardware that holds the tube.

On large lasers, the resonator rods are made from carbon

graphite and are generally one to two inches in diameter. Carbon graphite has a very low **coefficient of expansion,** which means it will have minimal movement (expansions and contractions) when temperatures fluctuate. In smaller lasers, invar is generally used. Invar has a larger coefficient of expansion than carbon graphite, but provides equal strength at one-fourth the thickness.

In polarized lasers, **Brewster windows** are attached to the ends of the bore. The windows, non-conductive and transparent panels placed at Brewster's angle, seal the bore and polarize the photons that pass through it. One of three methods may be used for sealing the windows to the bore: epoxy, frit and optical contact. Sealing the bore and securing the Brewster window by epoxy was one of the first methods used on gas lasers. A space-grade epoxy glue is evenly distributed on a quartz Brewster stub and meticulously fastened on the bore. Prit sealing involves heating glass between the **Brewster stub** and the bore.

Perhaps the most effective method of sealing the windows, yet hardest to do properly, is optical contact. Optical contact requires a precise mechanical fit between the Brewster stub and the bore. The bore is heated, microscopically melting the Brewster stub directly to the bore.

In low-powered lasers, such as air-cooled ion lasers, HeNe and HeCd lasers, the Brewster windows are almost always made from fused silica. Fused silica is preferred due to its low absorption of electromagnetic radiation, which enables the highest possible transmission. In higher powered lasers, such as large frame ion lasers, fused silica is susceptible to solarization. **Solarization** occurs when an excessive amount of the transmitted photon energy is absorbed by a Brewster window, changing its optical properties. Two of the more common effects of solarization are thermal lensing and color centering.

Thermal lensing is caused when photon energy absorbed by the Brewster window is converted into heat. Heat circulating in the window changes its optical properties. It also warms the air surrounding the Brewster window, distorting the optical properties of the air. Both the window and the surrounding air act as a randomly shifting lens that causes a slight variation the direction of the beam. When the shifted beam hits the mirror, it may not reflect precisely down the center of the bore. Part of the beam may "clip" upon entering the bore, causing a significant reduction in power.

The effects of thermal lensing are very similar to those of a misaligned mirror. It is common for an unaware operator to try to correct the malfunction by adjusting the tilt plates. Often, it will work-temporarily.

Unfortunately, heat energy is not stationary. The distortions in the lens and its surrounding air can change, causing the beam to shift again. Or, if the operator turns the laser off, thermal lensing may not occur again until hours after restarting it. All previous adjustment of the tilt plates are no longer valid. The laser is now legitimately misaligned.

Thermal lensing is extremely frustrating if not detected.

An operator who finds his large frame ion laser constantly out of alignment may find it necessary to inspect his Brewster windows.

Color centering is the result of extreme solarization. In color centering, the Brewster window absorbs enough energy from the laser beam to change its molecular structure. When this occurs, the Brewster window will lose its transparentness.

Because most of the energy of a beam is in its core, the molecular restructuring is generally restricted to the center o f the Brewster window. Photons will no longer pass through the damaged region, producing a beam that has no light in its center. This "donut" shaped beam is unusable in most applications. To help reduce the effects of solarization, thermal lensing and color centering, Brewster windows on large frame ion lasers are generally made from crystal quartz.

Additional Properties of Lasers

Any light source that delivers exactly one wavelength is said to be **monochromatic**.

Ideally, stimulated emissions from a group of identical fuel atoms should produce very distinct wavelengths (or frequencies) lasing within the cavity; for example, a HeCd laser would produce amplification at 325 .0 and 441.6 nm. By using proper coatings on the mirrors, the 325.0 nm wavelength can be removed from the lasing process, thus producing a monochromatic beam at 441.6 nm. In actuality, however, this is not the case.

In the same manner the two swimmers discussed earlier produced slightly different wave frequencies, the lasing inside of the cavity produces minute variations of photon frequencies (or wavelengths) in the transmitted beam defined as the laser's **bandwidth** (Δf).

Several factors contribute to variation of photon frequencies inside the cavity, including the motion of an atom at the time it emits a photon.

Variation of photon wavelengths in a laser, called its **linewidth** ($\Delta\lambda$, measured in nanometers) is exactly the same phenomenon as a lasers bandwidth (Δf). Quantitatively, however, the two will not have the same numerical values. Linewidth can be numerically converted to bandwidth (and vice versa) by the following equations:

$$\Delta\lambda/\lambda = \Delta f/f \text{ or } \Delta\lambda = \Delta f \cdot \lambda/f \text{ or } \Delta f = \Delta\lambda \cdot f/\lambda$$

Often, it is helpful to eliminate frequency entirely from our equations. This may be done by inserting the expression f=c/'A. (see section II) into the above equations and applying some basic algebra. The results are:

$$\Delta\lambda = \Delta f \cdot \lambda^2/c \text{ and } \Delta f = \Delta\lambda \cdot c/\lambda$$

Coherence length (L) can also be expressed in terms of wavelength and Iinewidth:

$$L = c/\Delta f = \lambda^2/\Delta\lambda$$

Waves moving back and forth in a confined region create constructive and destructive interference similar to that in Section 2. If the frequency of the waves matches the resonant frequency of the region, constructive interference will occur throughout the length of the region. A set of non-moving waves, complete with crests and troughs, form in the region. These "standing" waves now dominate.

By changing the length of the region (or the frequency of the waves), the two frequencies no longer match. Destructive interference is introduced, and the standing waves disappear. If you continue to change the length, you will find other discrete distances (called **harmonics**) that will enable the frequency of the region to match those of the waves. Destructive interference is again replaced by constructive interference, and the standing waves reappear.

By increasing the length of the region, you enable more standing waves to exist within its boundaries. By decreasing the length of the region, less standing waves exist.

A laser has waves (photons are waves) travelling back and forth in a confined region (the laser cavity). Because many frequencies exist within the lasers bandwidth (M), some of them will match the resonant frequency of the structure. Standing waves will form within the cavity. Only those frequencies creating standing waves will continue to lase. These frequencies are called **longitudinal modes**.

Within the bandwidth are a set of distinct longitudinal modes spaced equally apart. There no frequencies lasing in between them. The distance between each mode is called the **longitudinal mode spacing** (abbreviated m in terms of frequency, measured in Hertz-see Diagram I).

Longitudinal

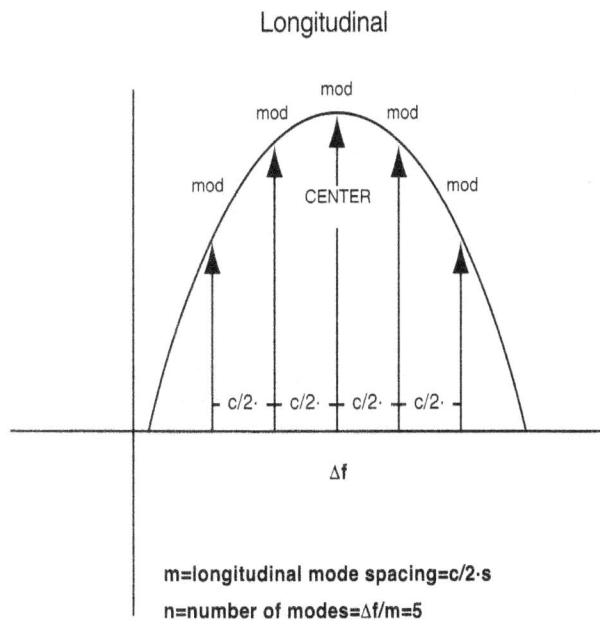

m=longitudinal mode spacing=c/2·s
n=number of modes=Δf/m=5

Diagram I

the **cavity length,** it is measured in meters) and can be calculated with the following equation:.

$$m = c/2 \cdot S$$

In terms of wavelengths, longitudinal mode spacing (**M**, in nanometers) can be calculated:

$$M = m \cdot \lambda^2/c = \lambda^2/2 \cdot S$$

The number of longitudinal modes (n) in the bandwidth can be determined by the equation:

$$n = \Delta f/m$$

In terms of bandwidth or linewidth, the number of longitudinal modes should be identical. The equation in terms of linewidth is:

$$n = (2 \cdot S \cdot \Delta\lambda)/\lambda^2$$

Because the mode spacings are very close together (generally less than 1/1000 of a nm), very small changes in the cavity length can cause the modes to move. In argon ion and other lasers that generate a significant amount of heat. the cavity can expand or contract enough to cause the longitudinal modes to literally jump over each other. This phenomenon is tailed mode hopping. In many laser applications, such as holography, **mode hopping** can cause undesirable effects.

Another mode that manifests itself inside the cavity is the **transverse electric and magnetic mode**, more commonly called the **TEM mode**. Light will propagate with distinct and defined geometrical paths. The most fundamental path will produce a clear, uninterrupted spot when projected on a target. Other paths, or "modes" will have dark irregularities (called **nodal lines**) separating the spot. The nodal lines can be either vertical or horizontal (see Diagram J) .

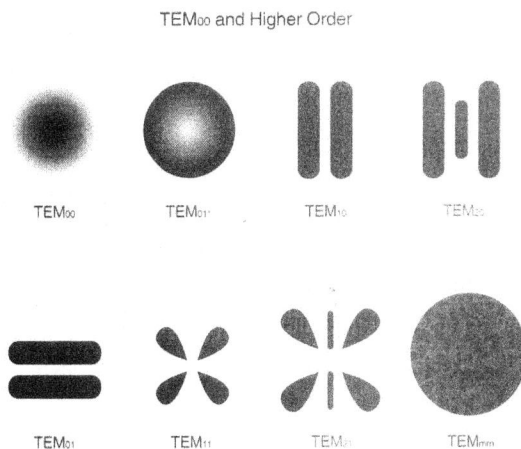

TEM₀₀ and Higher Order

Diagram J

These patterns are classified with subscript using the form TEMvh (where "v" designates the number of nodal lines in the vertical direction and "h" designates the num-

The longitudinal mode spacing is determined by the separation of the cavity mirrors (abbreviated S, also called

ber of nodal lines in the horizontal direction)--for example TEM_{10}, TEM_{20}, TEM_{11}. One of the more interesting patterns is the TEM_{01*}, a mode that produces a large circular node in the middle of the beam. Because of its distinct pattern, TEM_{01*} mode is often referred to as the "**donut mode.**"

Lasers can be built to produce only the fundamental mode TEM_{00} by reducing the ratio of bore diameter to the diameter of the TEM_{00} beam, usually to less than 3:1. This can be achieved by properly selecting mirror combinations that encourage TEM_{00} transmission.

Generally, long thin bores produce TEM_{00} mode more readily than fatter ones. In holographic applications, TEM_{00} transmission is essential.

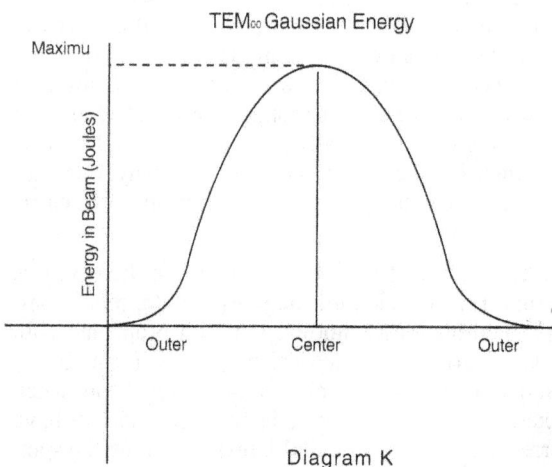

TEM$_{00}$ Gaussian Energy

Diagram K

TEM_{00} beams can be focused down to the smallest possible size spot known as the **diffraction limit**. In theory, the diffraction limit of a beam is its wavelength. In real applications, the smallest possible size of a focused spot is slightly higher. In holography, the ability to focus down to a "tight" spot is not an advantage since the beam is expanded.

The beam of a **multimode laser** has a combination of many modes resulting in an uneven energy distribution in its cross section. Multimode beams (commonly written **TEM$_{mm}$**) deliver more energy than, but have none of the advantages of those modes listed above. A TEM_{mm} beam can also be focused down, but not nearly as tight as TEM_{00}, Although multimode lasers are used in a variety of applications where laser power is the primary concern, they can not be used for holography.

Mechanical instabilities in the optics and tube can cause the beam to wander. A laser's **beam-pointing stability** (also called **angular drift**, measured in microradians) measures how much the laser beam drifts from the beam axis. Beam pointing stabilities of 60 microradians or less are considered good.

Types of CW Lasers Used in Holography

The three primary types of CW lasers used in holography are argon ion, HeCd, and HeNe lasers. Each has

distinct advantages that are related to the holographer's needs. Typically, the type of recording material, size of the hologram, and operator budget determines which laser is best suited.

In embossed holography, photoresist is the primarily medium used for recording images. The use of photoresist enables mass-produced holograms. Photoresist chemically etches the holographic image onto a glass plate. The optically-engraved glass plate (called a master) is made conductive, and then electroplated-which produces a shim. The shim is placed on an embossing machine for mass stamping of embossed holograms

Because photoresist is extremely sensitive to wavelengths between 420-450 run, HeCd lasers (which lase at 441.6 nm) are most often chosen. HeCd lasers are much more cost-effective to operate than comparable lasers, enabling originators of holograms to keep their operating costs down.

Artistic holography isn't constrained by the necessity to mass produce. This gives the holographer freedom to choose from a variety of emulsions to produce their holograms. In such cases, most holographers prefer to use emulsions sensitive to the primary wavelength (514.5 run) of an argon laser. Argon lasers provide an attractive combination of high power and long coherence length, enabling holograms that are both large and visually striking.

In most forms of holography, the coherence length of the laser determines the size of the hologram. Generally, a 10 cm coherence length laser produces holograms that are 10 cm x 10 cm (4" x 4",1 inch = 2.54 cm). Seasoned hologra- phers routinely extend the size of their holograms without increasing their coherence length, some being able to shoot a 6" x 6" while using a source with 10 cm coherence length.

Novice holographers typically can't justify the hefty expense of an argon laser. A HeCd laser, though much more attractively priced, still costs over ten thousand dollars. For "weekend shooters" on a budget, the HeNe laser provides a cost-effective alternative to the higher-priced argon and HeCd lasers.

Certain laser parameters are essential to all forms of continuous wave holography. The beam must be TEM_{00} mode. A polarized beam (at least 100: 1) is necessary in all forms of holography except dot matrix; the low exposure times of dot matrix holography, typically 5 milliseconds per dot, reduce the need for polarization. The laser can not generate excessive heat or vibration, two mortal enemies of the holographer.

With the exception of dot-matrix holography, high power is very desirable. Less photon power requires longer exposure times. Not only do longer exposure times raise the cost of labor (professional holographers do not work for free), but they increase the odds of something going wrong. As every holographer will gladly tell you, things do go wrong.

Argon Ion Lasers

Before the advancement of the helium-cadmium laser, argon ion lasers were the preferred choice for almost all forms CW holography. The argon laser provides a generous amount of power, polarization and coherence-three of the more prized parameters in holography.

Argon ion lasers produce lasing at many wavelengths between 454.5-528.7 nm-usually eight to ten-and can be equipped with a **prism wavelength selector** in the cavity to allow the operator to select a specific wavelength. Of primary interest in holography are the 514.5, 488.0 and 457.9 nrn wavelengths. The 514-nm line is the most powerful, followed closely by the 488.0 nrn line. Large-frame argon lasers can provide nearly 10 Watts of power at 514.5 nm and over 6W at 488.0 nm. Most argon lasers have a polarized beam with a polarization ratio of 100:1 or greater. The combination of high power and high coherence length make argon ion lasers the best laser for the serious artist who can afford its substantial price.

In embossed holography, the powerful 514.5 and 488.0 nm wavelengths of an argon laser have little effect on the photoresist used to record the holograms. Because photoresist is extremely sensitive to wavelengths between 420 and 450-nm, only the 457.9 nm line can effectively expose it. Unfortunately, the 457.9 nm wavelength is relatively weak in comparison to the more powerful 514.5 and 488.0 nm lines- less than 20% of the relative power. Large frame ion lasers produce approximately 1W of power at 457.9 nm.

Argon ion lasers can be equipped with another intracavity device called an **etalon**. An etalon is a wedge-shaped piece of high quality optical glass which, by means of constructive and destructive interference, can eliminate longitudinal modes from the laser cavity. The etalon acts as a separate laser cavity inside of the main laser cavity. When the beam enters the etalon, it is reflected inside the wedge to form its own optical feedback before exiting. By adjusting the angle where the beam enters and the temperature (which changes its index of refraction and cavity length) of the glass, the etalon can choose which longitudinal modes will survive (through constructive interference) and which ones will not (through destructive interference). The beam " exits the etalon, plus or minus a few longitudinal modes, to resume amplification in the main cavity.

By reducing the number of longitudinal modes, the etalon reduces the bandwidth. Lowering the bandwidth raises the coherence length. In essence, the etaIon allows the holographer to control how much coherence length the laser beam will have. While most lasers refer to their coherence length in centimeters, an argon laser with an etalon can attain coherence lengths of many meters. The trade-off, however, is power. An etalon will produce losses of 30% or more, depending on degree of bandwidth reduction.

The active medium in an argon laser is a pure argon gas. Because argon is a good energy absorber, it needs no transfer medium. Argon atoms are excited by passing a high-current density discharge through a ceramic tube. An initial spike of a few thousand volts breaks down the low pressure gas (approximately I torr), then the voltage drops to 90-400 volts while the current jumps to 10-70 amps (dc). The discharge current is concentrated in a small-diameter bore in which stimulated emission takes place and must be high enough to ionize the argon gas (hence, the

title argon ion laser). An external magnet placed immediately outside the tube produces a magnetic field parallel to the bore axis that helps confine the discharge to the bore.

The conditions inside an ion-laser plasma tube are extremely harsh. Highly charged electrons and ions violently collide with the tube and bore eroding the surfaces, and contaminating the gas. A sizeable amount of UV radiation is generated in the cavity which, over time, tends to damage the Brewster windows and other optical surfaces.

Many times, microscopic particles from the walls of the bore can peel off and fall into the beam path. These pesky flakes can cause minute losses in power called drop offs. More advanced materials used in the inner cavity of argon ion laser tubes have reduced the amount of lasers experiencing drop off problems.

The average lifetime of an argon ion laser tube is directly related to the how high the operator sets his tube CWTent. Those trying to achieve maximum power will find an average tube lifetime of around 2400 hours. More prudent operators can expect approximately 3000 hours per tube.

Although argon lasers are capable of producing a significant amount of optical power, the energy efficiency of the argon laser is actually quite poor. Because stimulated emission takes place in energy transitions far above the ground state, much energy is required to pump the electrons up to an excited state.

The inefficient conversion of energy creates a substantial amount of heat. To remove this heat, metal disks are brazen inside the tube to allow it to transfer from the bore to the outside of the tube. A metal duct outside of the tube circulates water which transports excessive heat away from the tube. To keep the temperature of the tube low enough to operate, large-frame ion lasers require a water flow rate greater than three gallons per minute.

Once the water exits the laser head, it must either be recirculated or dumped down the drain. To recirculate water, a heat exchanger is placed in the water flow loop. The exchanger works much like a radiator, extracting heat from water, exiting the head and transferring it to a different medium (usually air or other water). Cooler water exits the heat exchanger and is sent back to the tube.

Although heat exchangers can be relatively effective, their high vibration makes them unappealing in sensitive applications like holography. Most holographers prefer an open-cycle system in which the water flows from the tap to the laser, and then down the dta)n.

The power supply must also be water cooled, although water circulating in the middle of high-voltage circuitry is not a favorable combination. Condensation and leakage jeopardize the performance and safety of the unit.

Due to the high-current requirements needed to produce stimulated emissions, argon lasers operate from 208VAC, three-phase line voltage. The extreme conditions of the laser tube tend to induce nontrivial expansions and contractions in the laser cavity, making argon lasers susceptible to mode hopping and poor beam pointing stability.

Though not perfect, argon ion lasers offer the best combination of power and coherence length for artistic holography. The price of a large-frame ion laser generally starts at $30,000. Re-tubing costs range from $13,000 to $15,000.

Helium Cadmium (HeCd) Lasers

The 441.6 nm line of a HeCd laser exposes photoresist used in embossed holography ten times more effectively than the 457.9 nm line of an argon laser. Unfortunately, for many years HeCd laser were only capable of delivering a maximum of 40 mW TEM_{00}, The marginal power resulted in unusually long exposure times for Originators of 2D/3D (also called multiple plane), 3D and composite holograms. A IW argon laser at 457.9 nm, comparable to a 100 mW HeCd laser in its effective exposure rate, could expose photoresist 2 1/2 times'faster than the most powerful HeCd laser. Further, a HeCd laser could only provide 10 cm of coherence length, adequate enough for a standard 4" x 4" master. As noted earlier, only the more resourceful hologra- phers could use a HeCd for the more illustrious 6" x 6" image. In comparison, Argon lasers can provide an unlimited amount of coherence length- a lbeit at the expense of power- by having an etalon installed.

As for originators of dot matrix holograms, in which the exposure time is trivial, HeCd lasers were readily adopted. A HeCd laser costs less to buy and operate, is easier to use, has lower maintenance, and lasts longer than a comparable argon laser. HeCd lasers do not suffer from common argon laser malfunctions such as thermal lensing, color centering, power supply leakage and mode hopping. Further, HeCd lasers operate off of standard II 7VAC, need no water to cool the tube and, therefore, no special plumbing is required in the facility.

In the last four years, HeCd lasers have quadrupled their power output. A typical HeCd laser can deliver more than ISO mW TEM_{00} and provide 30 cm coherence length- more than enough for a 6" x 6" hologram. The effective exposure on photoresist for HeCd lasers now meet 0; exceed large-frame argon lasers, while saving the average holographer in excess of $800 per month per laser on their electricity and water bills. This improvement has created a profound change in embossed holography marketplace. Today, seven out of every eight lasers bought for commercial embossed holography is a HeCd.

The active medium used in a HeCd laser is cadmium. Standard HeCd lasers use naturally reoccurring cadmium that consists of a blend of three isotopes: Cd_{112}, Cd_{114} and Cd_{116}. It is abundant in nature, and therefore, costs pennies per gram. The stimulated emissions of cadmium provide a primary lasing wavelength of 441.6 nm, with a secondary wavelength of 325 .0 nm. One manufacturer has patented a 353.6 nm HeCd laser. With naturally reoccurring cadmium, the spectral bandwidth is 3.0 Gigahertz, which translates into 10 cm coherence length. Standard HeCd lasers can produce up to 120 mW TEM_{00} at 441.6 nm.

By processing naturally reoccurring cadmium in a manner similar to processing nuclear grade uranium, anyone of the three isotopes can be isolated, producing **isotopically enriched cadmium**- more commonly called single **isotope cadmium**. Because the technology and equipment required to produce single isotope cadmium is restricted, the processing of it is extremely expensive. This significantly raises its price. Currently, single isotope cadmium sells for over $1,600 per gram.

An average HeCd laser consumes from five to eight grams of cadmium. Using single isotope cadmium can increase the price of the laser many thousands of dollars. The highest powered HeCd laser with naturally occurring cadmium is priced around $18,000. An equivalent laser with isotopically enriched cadmium sells at $25,000.

Having only one isotope lasing in the cavity produces a third of the spectral bandwidth (1.0 GHz). One third of the bandwidth nets three times the coherence length. In essence, the use of single isotope cadmium raises the coherence length from 10 to 30 cm. As an added benefit, single isotope cadmium also produces more power-about 30% more at 441.6 nm. Single isotope HeCd lasers deliver up to 170 mW, TEMoo at 441.6 nm. This translates into 1.7Wof equivalent power when compared to the 457.9 nm line of a large frame argon ion laser in effectively exposing photoresist.

Similar to argon ion lasers, the pumping process in HeCd lasers (see diagram M) is accomplished by a high density discharge that breaks down helium gas (the transfer medium). The discharge is produced by an anode and a cathode and confined to a long, narrow glass bore. The spike voltage is typically 13,000 volts and after breaking down the gas (ionization), is reduced to approximately 2,000 volts.

After breakdown of the helium gas, cadmium placed near the anode in a **cadmium reservoir** is heated to about 250 C by a heater wrapped around the reservoir. In approxi- mately five minutes, the cadmium starts to vaporize. Through a natural process called **catephoresis**, the cadmium vapor migrates uniformly from the anode, through the bore, and towards to the cathode. Because catephoresis exists, a unifotm distribution through the bore is possible. Otherwise, the lasing process would be impossible to control.

It is critical that the cadmium is properly "trapped" before reaching the cathode. Cadmium ions being deposited on the cathode would drastically alter the electrical properties o f the laser, affecting the laser 's performance . A cold trap is placed centimeters from the cathode to stop the advancing cadmium. Another cold trap is placed in front of the Brewster window, to eliminate the deposition of cadmium on the window.

The strict regulation of both the helium and cadmium vapor pressures are vital to the performance of the tube. The amount of cadmium in the bore can be easily detected by measuring the tube's voltage. A feedback circuit is placed in the head that adjusts the cadmium heater should the tube voltage read too high or too low.

Helium must also be kept at a proper pressure. Helium, being a very small atom, can diffuse out of the tube; the amount is seldom of consequence. During operation, though, the cold trapping process tends to trap the smaller helium atoms under the larger and more massive cadmium atoms. The helium atoms get buried, and the tube pressure eventually lowers. To replenish the lost helium, a refill bottle is placed in the tube with a supply of high pressure helium. When the tube senses low helium pressure, a heater wrapped around the refill bottle is switched on. The elevated temperature raises the kinetic energy in the bottle, creating even higher helium pressure in the refill bottle. The higher temperature also widens the atomic spacing of the container. The combined effect creates an increased diffusion rate of helium from the refill bottle to the tube.

Helium-cadmium lasers that are stored risk subtle migration of helium atoms from refill bottle into the tube. This creates an excess of pressure in the tube, which can make the laser difficult to start. HeCd lasers, when stored, should be started at least once a month and operated from 2-4 hours to stabilize the system.

A typical HeCd laser tube typically lasts 4000 hours. Single isotope cadmium laser tubes, because manufacturers tend to limit the amount of cadmium in the reservoir, last about 3500 hours.

HeCd lasers have quickly reached a position of dominance in the embossed holography marketplace. The combination of high power, effective exposure rate, low cost of operation, and ease of use make it the laser of choice for embossed holography.

Helium Neon (HeNe) Lasers

The HeNe laser was the first gas laser to be commercially available, brought to market in 1961 . Over 30 years later, the HeNe laser is still the most commonly used laser today. Supermarkets use HeNe lasers to scan the bar codes on packages for quick and efficient customer check out. High schools and universities find their low price, ease-of use, good beam pointing stability and long tube lifetimes extremely attractive. HeNe lasers operate from a 117VAC source and are air-cooled.

An average HeNe laser costs a few hundred dollars,

making it an affordable tool for those who normally could not afford the expense of an argon ion or HeCd laser. HeNe lasers are low powered, typically delivering between 0.5 and 1 mW, TEM_{00} at 632.8 nm. More expensive models are available, delivering up to 35 mW, TEM_{00} at 632.8 nm. The beam is generally polarized with a coherence length between 20 and 30 cm. An intracavity etalon may be installed for greater coherence, but the corresponding loss of power tends to create extremely long exposure times. The average lifetime of a HeNe laser tube is about 15,000 hours.

The first HeNe laser ever demonstrated emitted an 1153- nm wavelength, but almost all HeNe lasers are utilized at the 632.8 nm line. HeNe lasers emitting green, yellow and orange wavelengths are also available, but their low power makes them ineffective in most commercial applications.

The active medium in a HeNe laser is neon. As in HeCd lasers, helium is the transfer medium. The laser tube (see Diagram N) consists of a glass envelope (bulb containing the cathode) with a narrow bore through its center. The bore can be anywhere from 10 to 100 cm in length, depending on how powerful the laser is.

A 10,000 volt,(dc) spike breaks down the two gasses in a narrow capillary tube. The voltage drops to between 1,000 and 2,000V with a current of a few milliamperes. Electrons in the discharge pump both helium and neon atoms to excited states. The more abundant helium atoms collect most of the energy, then transfer it to lower energy or ground-state neon atoms through a series of inelastic collisions.

This transfer of energy is very efficient for two reasons. First, both the helium and the neon gasses have two higher energy states with comparable energy values. Second, both pa'irs of matching higher energy states are characterized by prolonged delays before allowing their electrons to drop. Energy states that hold their electrons longer (up to many milliseconds) are called **metastable states.**

HELIUM-CADMIUM LASER

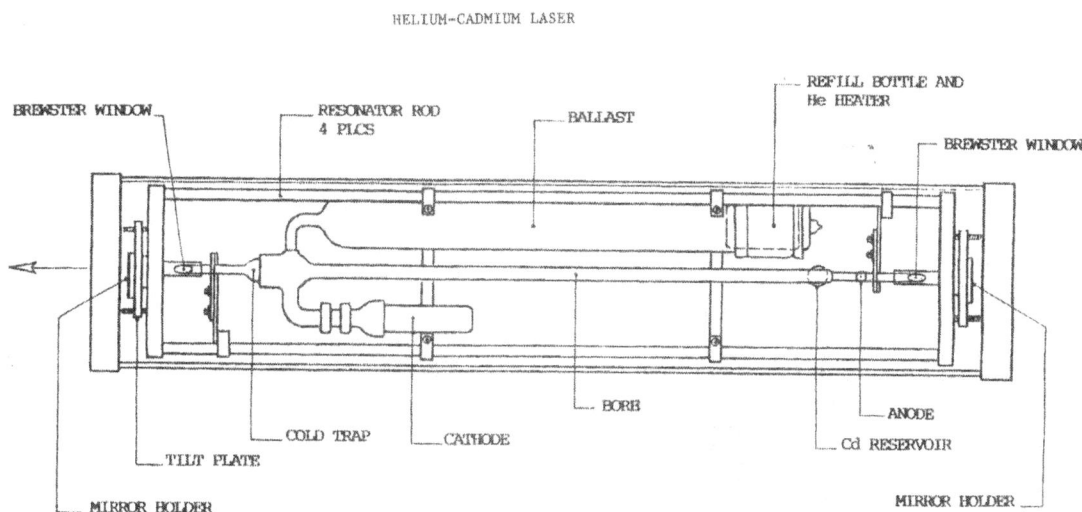

Diagram M

When the abundant helium atoms in the tube are excited by the discharge current, more of them will have electrons in one of the two metastable states than in other higher energy states. A leaky bucket in a rain storm that holds water longer than a comparably sized peach basket, will more likely have rainwater in it the next day.

A sizeable population of helium atoms with electrons in the metastable state is built up. The excited helium atoms wander in the tube, collide with non-excited (ground state) neon atoms and transfer their energy to the them. Since the higher energy levels of the two gases match, the amount of energy transferred to the neon atom is just enough to raise its electron to a metastable state. Soon, the majority of neon atoms have their electrons in metastable state. A population inversion soon crevelops and lasing begins.

HeNe lasers are being challenged by low-powered laser diodes that emit at the 650 and 670-nm wavelengths. Laser diodes are extremely small, light, use very little current, need extremely low voltages, and are less than one fourth the price of a HeNe laser. Laser diodes have inherent properties that are detrimental to holography, however. Low coherence lengths (less than 2 cm), and wavelength instability make the use of laser diodes in holography unfeasible.

The low cost HeNe will always be the perfect laser for novice holographers. The combination of affordability, high-reliability and ease-of-use make this laser perfect for the production of budget holograms.

Editor's Acknowledgment - Special thanks to Michael Fisk for his contribution to this chapter:

4

Lasers - Current Trends

This chapter features a discussion of the newest CW and pulsed laser systems available to today's holographers. Experts in both fields contribute valuable information. Of special note is an article about the laser systems used to produce large format holograms.

For shoppers with a limited budget, it also includes information about acquiring surplus, used and repaired laser equipment.

New CW Lasers Increase Productivity for Holographers

by Paul Ginouves

Until recently, successful use of holography in R&D or production applications usually required the use of an ion laser. The ion laser, in turn, required hands-on operator attention to maintain optimum alignment and achieve single-frequency operation. Fortunately, recent developments in laser technology have eliminated both these requirements. Specifically, ion lasers have undergone significant maturation in terms of ease of use and implementation, resulting in true "hands-free" operation, even for applications as demanding as holography. In addition, the diode-pumped solid-state (DPSS) laser has now reached a performance level necessary for holographic use.

This article examines the "holographer-friendly" features that define the latest generation of ion lasers. It introduces DPSS laser technology, including its advantages and its current limitations. In addition, it will provide a look at a commercial holography application where the benefits of both laser types are critical enabling factors.

Ion Laser Maturity

The first argon ion laser was made in 1964. Ion lasers opened the door to practical holography and have dominated this application because of their low-noise, high-power output characteristics. Indeed, photopolymers and other materials have been developed with sensitivities that specifically match ion laser wavelengths. As listed in Table 1, argon lasers can produce several watts of narrow-line output at "several wavelengths in the UV through green. Krypton lasers have slightly lower output power, but can produce over 2 watts of single-frequency output in the red (647.1 nm) as well as lower powers at other visible wavelengths. Mixed gas (krypton/argon) lasers produce a combination of these output wavelengths. (See top Coherent photo on the last page of the color picture section in this book.)

Output Line	Small Frame	Large Frame
Argon Ion (all lines, multimode 5W 20W)		
457.9	0.2W	0.8 W
488.0	0.9W	4.2W
514.5	1.2W	5.4W
Krypton Ion (all lines, mutimode 1.0W 4.6W)		
413.1	0.15W	1.1W
647.1	0.5W	2.1W

Table 1: Typical single freqency outputs for the higher power visible wavelegths of small frame and large frame ion lasers.

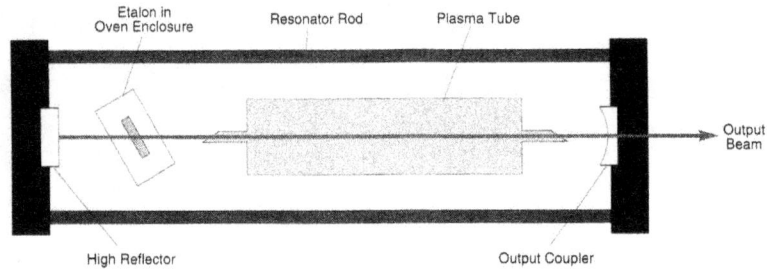

Figure 2 Schematic of key elements of a single-frequency ion laser head.

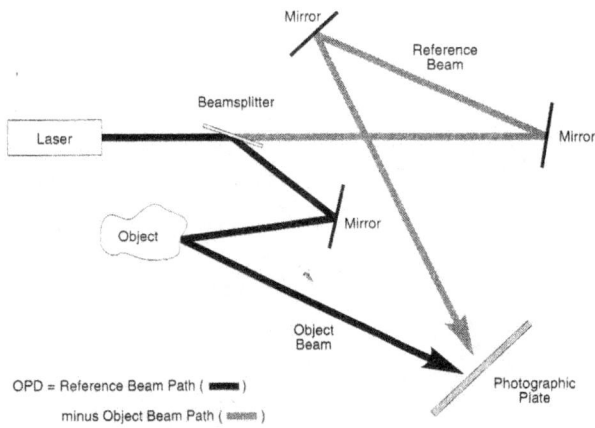

OPD = Reference Beam Path (▬▬)
      minus Object Beam Path (▭▭)

Recording a Transmission Hologram

Figure 3 Laser coherence length must be several times greater than the OPD to produce high-quality holographic images

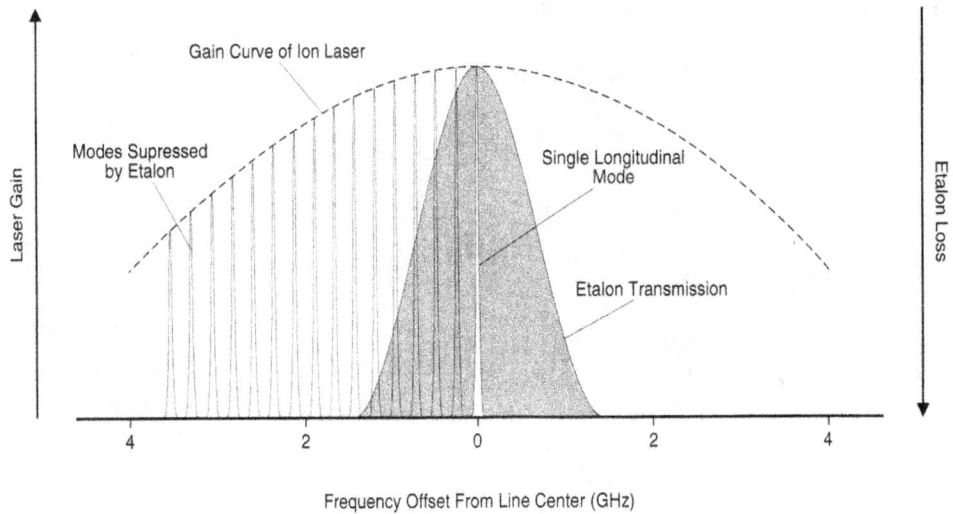

Figure 4

The basic elements of a single-frequency ion laser head for holography are shown in Figure2. A prism is used as a wavelength filter to select the wavelength line of choice (e.g. 488 nm). Each laser output line has a typical width of 0.004 - 0.01 nm. This means that the coherence length of the laser is less than 50 mm. Unfortunately, successful holography requires a coherence length many times longer than the optical path difference (OPD), as defined in Figure 3. For most holographic applications, the linewidth of the laser output must be narrowed by using an intracavity etalon - a simple optic that serves as a narrow-bandpass filter.

Figure 4 illustrates how the etalon achieves narrow-line operation. Each laser line actually consists of many discrete wavelengths, or longitudinal modes. These modes satisfy the equation, $nl = 2L$ where n is a very large integer, l is the wavelength, i.e., the distance between the rear mirror and the output coupler. As shown in Figure 4, the loss due to the etalon restricts laser oscillation to a single longitudinal mode.

Automated Stabilization

Ion lasers continue to evolve. In the past five years, manufacturers have made tremendous progress in the area of active stabilization. Early ion lasers had serious stability problems, particularly when operating narrow-line, i.e., with an etalon. Waste energy from the lasing process generates many kilowatts of heat. Consequently, minor changes in cooling-water flow or tube current can cause measurable changes in the overall head temperature . For this reason, in high-end products, the cavity mirrors are supported on a low-expansion alloy (Super-invar) structure. But the distance between the cavity mirrors is typically greater than 1 meter. Thus, even with a low-expansion resonator, small changes in temperature have the potential to change the alignment of, and distance between, the cavity mirrors. Angular misalignment results in a reduction in output power, poor mode structure, and higher noise. For these==== reasons, the laser required constant attention and adjustment in order to maintain optimum performance.

In 1989, one major laser manufacturer, Coherent, successfully eliminated this alignment problem using an approach known as PowerTrackTM. With this system, internal photodiode sensors detect changes in the beam power. Signals from these sensors are interpreted by a microprocessor that directs actuators to make compensating angular adjustments on the mirror mounts. Alignment is thus continuously maintained with no operator intervention - assuring maximum power at a given current, highest transverse mode quality, and lowest optical noise.

With single-frequency operation, however, there was still the problem of mode-hopping: the sudden shift from one longitudinal mode to another. This is accompanied by a drop in output power and fluctuations in coherence length. This phenomenon occurs because temperature affects both the etalon thickness and the laser cavity length, thus changing their wavelength characteristics. If the head temperature drifts, the cavity mode may drift off the center of the etalon transmission peak (see Figure 4). This causes a drop in laser power. As the drift becomes larger, the next cavity mode will eventually come close to the etalon center and the laser output will suddenly "jump" to this mode. Because of the associated power and wavelength fluctuation, mode-hops during hologram mastering, or HI --> H2 transfer, have the potential to completely ruin the hologram. This has long been cited as the major drawback of ion lasers in holography.

To deal with this problem, a number of stabilization schemes have been employed by laser manufacturers, with varying degrees of success. For instance, in 1995. Coherent introduced the v-TrackTM system on its SabreR ion lasers. The principle is remarkably simple. The etalon is enclosed in a stabilized oven to maintain constant temperature. The cavity mirrors are supported on mounts that permit both angular and translational adjustments. To maintain single-mode operation, the laser's microprocessor senses any temperature-induced mode drift and directly compensates for cavity expansion or contraction by lengthening or shortening the distance between the mirrors. This stabilizes the length of the cavity.

This automated system virtually eliminates the effects of temperature drift in the laser head, ensuring hours and even days of mode-hop-free operation. Also, warm-up time is reduced to a few minutes. The benefits are greatly increased throughput for any holography application, as well as recording longer exposure holograms, such as rainbow holograms, with consistently high quality.

DPSS Lasers

Like many other areas of photonics, laser technology is inexorably moving towards all-solid-state solutions. For hQlographers , the area o f most current interest is DPSS (diode-pumped solid-state) technology. These lasers have been under development since the mid-1980's and are now delivering suitable performance for a number of holographic applications, most notably for non-destructive testing (NOT) and H ---> H2 transfers of masters.

A DPSS laser uses a gain medium of Nd:YAG (neodymium yttrium aluminum garnet) or Nd: YV04 (neodymium yttrium aluminum vanadate). These are efficient lasing materials that produce near-infrared (1 064 nm) output. This near-IR light can be frequency doubled to the green (532 nm) by a non-linear. crystal such as KDP (potassium dihydrogen phosphate). Many lasers incorporating Nd:Y AG or Nd:YY04 use high-energy flashlamps to optically pump these materials. Such lasers are used in holography for generating pulsed masters of moving objects.

In DPSS lasers, the pump energy is supplied by laser diodes. These are much more powerful versions of the same types of laser found in CD players. They efficiently convert electricity to light and do not require water cooling. Also, their output can be tuned to match the absorption profile of the crystal gain medium so that very little of their pump light is wasted. In multiwatt lasers, these laser diodes can be located in a compact power supply and the pump light supplied to the head via fiber optics. This results in an extremely compact, high-power laser head (see

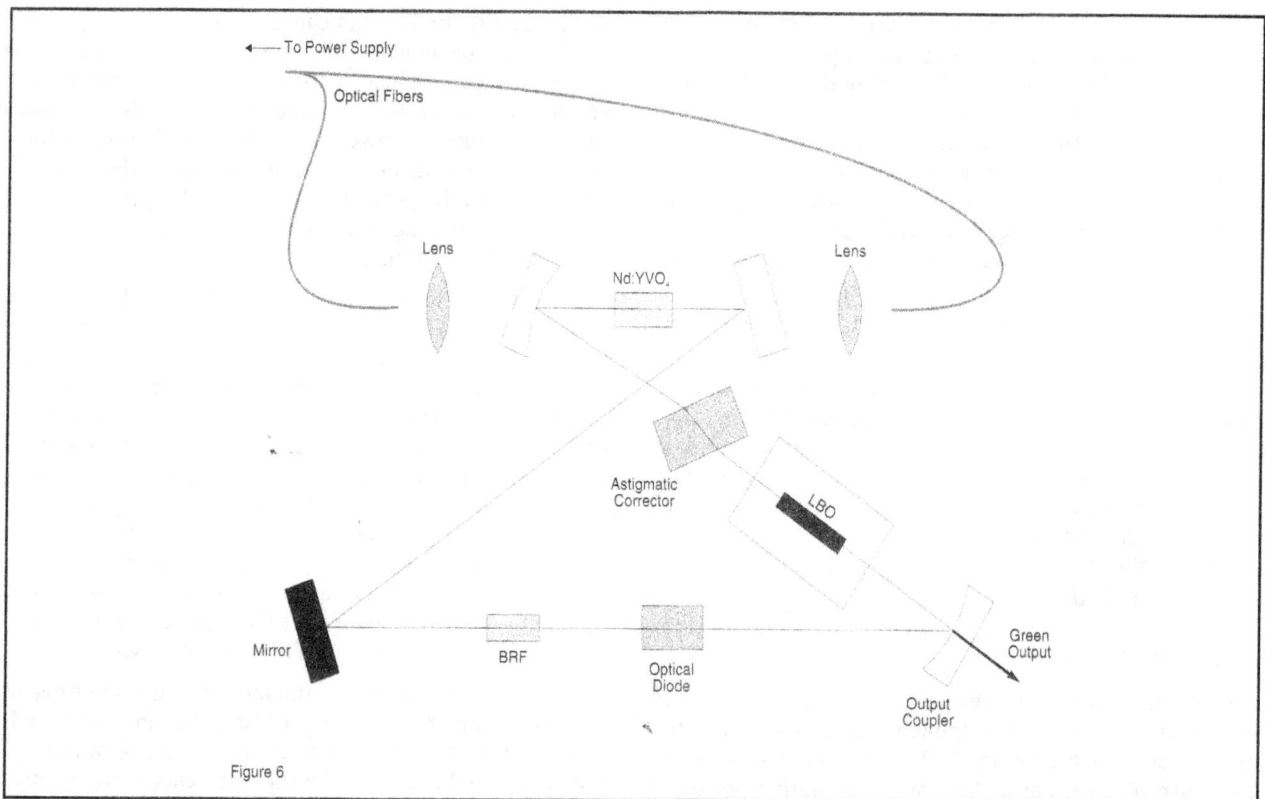

To Power Supply

Optical Fibers

Lens

Nd:YVO₄

Lens

Astigmatic Corrector

LBO

Mirror

BRF

Optical Diode

Green Output

Output Coupler

Figure 6

photo) and minimal heat loading, requiring no external cooling .

Until very recently, no DPSS laser offered the appropriate characteristics for the holography marketplace. Some companies concentrated on developing high powers (up to I watt by 1995) but with multi-mode output and poor coherence. Other lasers were designed to capitalize on the potentially long coherence length of Nd:YAG and Nd:YV04. Originally producing 10s of milliwatts of green output, these single-mode lasers had reached the 100s of milliwatts by 1995. In 1996, however, there was a significant increase in performance with the development of a laser (Coherent's VerdiTM) which produces 5 watts of cw, single-mode output. (See bottom Coherent photo on the last page of the color picture section in this book.)

As shown in Figure 6, this laser uses a ring configuration for efficient single-mode operation and overall compactness. The compact cavity results in favorable mode-spacing, making it easier to select a single mode using the inter- nal etalon. The probability of temperature-induced mode hopping also is greatly reduced. Nonetheless, to ensure mode-hop-free operation, active stabilization is employed. This closed electronic system is completely transparent to the user. In fact, the only control on this laser is the power switch.

This type oflaser offers several important benefits for holographers because of its all-solid-state construction. It is rugged, reliable, compact, requires no cooling-water, operates from a standard 110V supply and is highly portable. Furthermore, the spectral linewidth of the output is much narrower than an ion laser - making for a much longer coherence length.

Lasers like this are particularly well-suited to NDT applications, as high power and long coherence enable faster testing of larger parts. In addition, because of the laser's portability, there is the potential to test these larger parts in the field instead of hauling them into a laboratory. The laser is also well-suited for transferring (copying) H 1 holograms made with high-power pulsed Nd:YAG lasers. Transferring a hologram with the same wavelength used in the original recording greatly simplifies the transfer process, and can increase production yields.

What about limitations of these new lasers? At this time, the only significant limitation of DPSS lasers is that they are restricted to a single visible output wavelength: 532 nm. Ion lasers will therefore continue to be the preferred lasers in applications requiring other visible wavelengths.

Three-Color Holography -An Enabled Application

The practical benefits of both DPSS and stabilized ion lasers are well-illustrated in an emerging commercial application - high-quality, full-color display holograms.

The holograms typically used for advertising or artistic display purposes are reflection holograms. The hologram is mounted on black paper or cardstock and is front-illuminated. Each ofthese holograms was created with a single laser wavelength. When viewed with a white-light source, such as a light bulb or sunlight, most of the light passes through the hologram and is absorbed by the backing material. But light that is at the same wavelength as the original laser satisfies the so-called Bragg condition, and is diffracted back towards the viewer to form a virtual

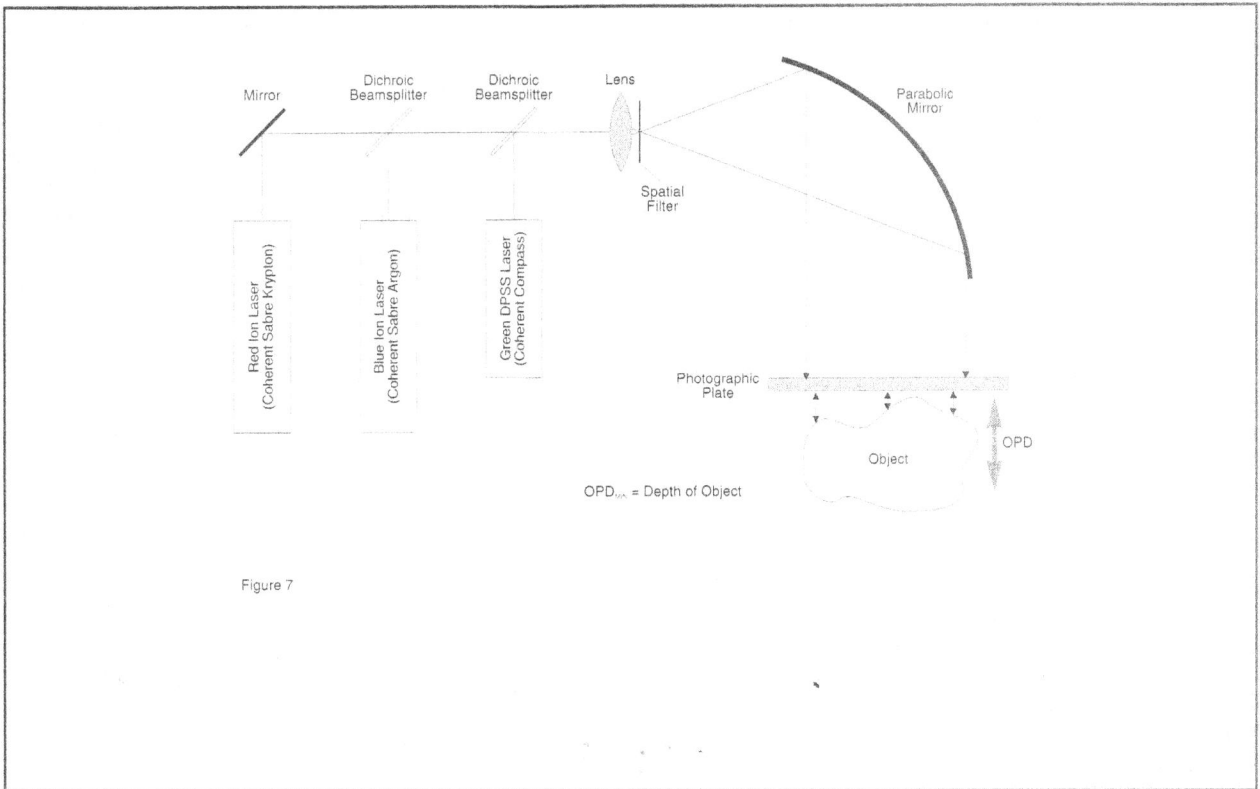

Figure 7

OPD$_{///}$ = Depth of Object

image. (Tilting the viewing angle shifts this wavelength and changes the color of the image - analogous to the color change seen when varying the incident angle of a diffraction grating.)

The principles offull-color holography have been known for many years. A silver halide emulsion plate is simultaneously exposed using three laser wavelengths (red, green and blue). The resultant zone-plate diffraction pattern is actually the sum of three separate patterns created by the different laser wavelengths. When this type of hologram is ... viewed using a white-light source, three wavelengths now meet the Bragg condition, corresponding to the original laser wavelengths. True color can be faithfully reproduced by balancing the relative intensity of the laser beams, the relative sensitivity of the photo emulsion, and the color temperature of the viewing light source.

HaLOS Corporation (Fitzwilliam, NH) is pioneering the commercial market for this type of holography. According to Chief Scientist, Dr. Qiang Huang, "Until recently, there was neither the laser performance nor the ideal photo emulsions necessary to produce the consistent quality needed for commercial viability." High-resolution color holograms that show sharp detail require silver halide emulsions, with exceptionally fine grain structure. Explains Huang, "At HaLOS, the key is a new emulsion developed by Slavich of Russia. This combines ultrafine grain size with photosensitizers optimized for the three laser wavelengths."

To record a master, HaLOS combines the red beam from a krypton-ion laser at 647.1 nm with a blue argon-ion beam at 476.5 nm and a green DPSS laser beam at 532nm as shown in Figure 7. Dichroic filters are used to efficiently combine the beams. The resultant "white" laser beam is then used in a Denisyuk setup in which the recording plate acts as the beamsplitter. In this configuration, the effective OPD is the distance from the plate to the furthest observable part of the object.

Huang explains that actively stabilized lasers are critical. "Trying to economically produce this type of hologram at a consistent quality would not be realistic if we had to constantly monitor and tweak three separate lasers." While citing the availability of hands-free ion lasers as a critical economical advantage, Huang echoes the sentiments of many cutting-edge holographers in stating that, "We really like the DPSS laser. This simple tool represents the future in lasers for holography. Our ideal situation would be to have red, green and blue versions of these lasers."

Conclusion

To summarize, it is unlikely that ion lasers for holography will ever be as simple and'reliable to operate as a light bulb. However, recent progress in automated operation and active cavity stabilization systems has certainly moved ion lasers a long way in the direction of that ideal goal. At the same time, a new laser technology that does have the potential for light-bulb simplicity and reliability has become available to holographers. With these tools, expect continued expansion in the applications of holography.

Paul Ginouves is Senior Product Managel; Coherent, Inc., Laser Group, 5100 Patrick Henry Drive, Santa Clara, CA 95054, e-mail, tech_sales@cohr.com

Chapter 4 - Lasers - Current Trends - Page   57

This article will outline some of the options available to holographers who wish employ pulsed lasers in their work.

Pulsed Lasers for Holography

by Ron Olsen

Laser hardware for holography in the nineties has evolved, improved, matured and become more reliable - all marketing cliches for "not changed a whole lot". The basics of a Q-switched high energy laser remain reasonably true to the designs of our "forephysicists" - only the pricing defies nostalgia. I will describe the perfotmance criteria for a pulsed holographic system and conclude with a description of our own custom Nd:YAG/Glass system at my company, Laser Reflections (will also describe briefly some of the options and features you may wish to consider depending on your budgets and your inclinations toward gadgetry and tweaking.

When asked to contribute in holographic system designs - I advise potential customers to begin with a serious assessment of what they hope to image routinely and how they envision their final targeted product. The subject matter, format size and presentation technique will dictate most of the critical laser parameters. I also ask for a realistic assessment of the budget. A lot ofpeople are shocked to learn that Neodymium based lasers can have a six figure price tag.

For those not yet familiar with the differences between long and short exposure holography, I offer a hasty summary as a prelude. During the holographic recording, allowable motion on the part of all subjects/objects involved (subject matter plus recording hardware) is - 1/10 of a wavelength of the light being used. The recording time (exposure in seconds) is dictated by many factors including: the emulsion's sensitivity (film speed), the wavelength of laser light (different emulsions are matched with dIfferent lasers each with their own sensitivity curves); the size of the holographic image being recorded tit takes more light to expose bigger plates); the distance from the subject to the film plane; and the reflectivity of the subject matter. There are two ways you can address the need to remain within the motion constraint while getting enough light to expose the holograph: 1) you can keep things stable (morgue-like) and deal with long exposures (tens of seconds) or 2) you can use a Q-switched laser which supplies you with ample energy in doses which fast enough to be oblivious to normal motion. As you might guess, I preach the latter approach and am an evangelist for Q-switched Neodmium YAG and Glass lasers.

Our system routinely images human subjects on 14"x30" silver halide plates at distances to 36" . Our end-products are reflection "holographs" up to 14"x24" for playback with low-voltage halogen lighting. I offer this case study as a starting point because it covers most of what someone designing your laser recording system (like me) will need to know at the outset. Keep in mind that you really can'tjust go out and buy a pulsed laser holographic recording system "off the shelf" - the holographic camera

and laser flash attachment will vary greatly as a function of your own imaging parameters and budget constraints.

Preferred Subject Matter

The realities of today's pulsed studio recording technology dictate subjects which can be carried through a normal doorway but are otherwise not limited by reasonable (sub-ballistic) motion. We image tanks of fish, pouring liquids, and scantily clothed women.

If the subject matter is not thusly transportable, then the camera must be designed for portability. We have designed a portable recording system which could be easily set-up in hotel suites, the catacombs of coliseums or on animal game parks (for night imaging under safelights) but the problems of imaging outside the relatively benign studio environment are not easily overstated.

I advise people to image people. So much of commercial imaging involves human subjects and they are actually the easiest, and I would argue the most gratifying, of holographic subjects. With few exceptions, they listen and respond to directions - very unlike horses and foxes - all of which we have imaged in our studio. We use two-side light- ing, and working with real-life models allows us to optimize the available laser light. We light our subjects at a very low energy level with green light at 1Hz to check for shadows, highlights, etc. For imaging human subjects, the frequency doubled output of a Neodymium based laser is a must. I do not even consider ruby lasers as I believe their wavelength is inappropriate for this application.

Materials including Ti:Sapphire and Cr:LiSAF are potentially candidates for pulsed holography but it is somewhat premature to consider them for any except academic and true R&D environments. This becomes a recutTing theme for me, "How much time do you want to spend with the laser and how much time do you want to devote to actually producing holographs?" Unfortunately, this same question has been the primary impetus for holographers adapting the ruby laser and there I draw the line. If the end- product is unacceptable, the relative ease of producing such a product is inconsequential.

Preferred Holographic Recording Emulsions

We use silver halide for all our mastering work for the simple reason that it works better than anything else. It requires only standard photographic processing; it supports extremely high resolution; it is commercially available from several sources, and it is three orders of magnitude more sensitive than other commercial materials. I would not care to be involved in holography or to design a holographic recording system were it not for the commercial availability of silver halide emulsions. We master almost exclusively on glass plates - the cost saving associated with working on film is not really an issue for image capturing and the advantages of recording on plates include fewer optical surfaces and simplified processing.

Format and Size

We do lensless imaging of subjects which are between 18" (minimum) and 36" (maximum). Because we are doing 1: 1 imaging within this very limited range, the size of our format dictates the amount of subject matter we can include in an image. We use Agfa Millimask plates which are commercially available as 14" x 24" and 14" x 30" plates. These formats are reasonably consistent with our ability to light subjects - we could accommodate wider than 14" but not much taller than 30". The use of longer plates for mastering (in the portrait format) allows us to produce images which are viewable by people of all heights. When we use these same plates in the landscape format, we sacrifice vertical parallax for extreme horizontal parallax - not typically prudent for public display. Recently we have begun working with Russian FPR emulsions, but as of this writing their availability in format sizes larger than 11" x 16" is unknown. For those who wish to put more into less- the option for reduction lenses does exist but is tied to the exponentially restrictive pricing of quality large optics. We do some minor reductions and enlargements in the process of producing copies - at present we are limited to ~1.5 x enlargement and reduction. This degree of reduction allows us to put the majority of a smallish folded adult in a 14" x 24" reflection transfer.

Mastering and Copying Geometries

Our old friend, the inverse squared law dictates that as you double the distance between the subject and the emulsion - you reduce by four times the amount of light incident on the plate. While with enough laser energy you may be able to record transmission images displaying up to several meters of scene depth (for backlight viewing via laser) - you have to consider this distance as being tied to the eventual product playback light source (as will be described later) and the set-up used for producing copies. Studios which can produce copies are oftentimes limited" by the size of their optical tables as to these maximum allowable distances. Our own copy table limits us to 36".

Final Applications

We do not differentiate masters produced for different end-products but our system is certainly excessive for holographers who intend primarily to image 2" square images of chess pieces. Reprographics - which is to say copies produced from the masters - can be produced on silver halide (low volume highest quality); photopolymer (medium volume, excellent quality); and rainbow embossed (high volume, good quality and least expensive). As a potential customer for an expensive pulsed holographic recording system, you should know what markets you are addressing and what your eventual customers are willing to pay for advantages like increased viewing and larger format.

Recommended Lighting

If you intend to produce reflection images which are viewable with normal halogen lighting, you should remember that your masters may be extremely impressive (up to several meters scene depth) but your eventual product, reconstructed by halogen, will be limited to no more than 24" of scene depth (optically imaged on silver halide or photopolymer) - no more than 3-4" when embossed. It is nice to discuss the next generation of holographic illuminators including semiconductor lasers - but the cost and complexity of such sources eliminates their serious consideration. LED's also offer great promise for display but here again it is just a bit premature.

Laser Hardware

Because no established laser manufacturer offered a pulsed Neodymium based recording system as herein de- scribed, I designed and built our own from commercially available laser component hardware. All of the described imaging requirements have a corresponding laser perfor- mance parameter which can be specified as follows:

•	wavelength: ~530nm (this relates to imaging any- thing alive as pertaining to water absorption)

•	energy: ~1J (relates to fonnat, distance between subject and recording emulsion, reflectivity of the subject) - here I am using 14" wide x 30" long plates and African American subjects at 36" as the supreme test) I model around silver halide emulsions from Agfa with a measured sensitivity of ~20 microjoules per cm2

•	pulsewidth : 10-20ns to eliminate any possible motion considerations and to promote optimum frequency doubling (the laser fundamental output is at Imicron and a non-linear, peak-power dependant crystal is used to generate the output at 0.5micron)

•	linewidth: ~0.005cm-l (single longitudinal mode for a ISns pulse at S32nm) The narrower the linewidth, the larger the potential for scene depth (somewhat like camera lens f-stops). There exists an inverse relationship between coherence length and pulsewidth - assume for practical pictorial imaging purposes that less than 10 nanosecond exposures are counterproductive and that longer than 50ns is unwarranted.

•	repetition rate: single shot for full energy output, -1Hz for low power stages. How fast can you load a plate and light a subject? We are pretty good at it and consider I shot every five minutes to be pushing the limit. As I described, we use the low energy output of the Nd:YAG stages (E-1mJ) at ~1Hz to light the subjects prior to unpacking and exposing the plate. If this low energy, higher repeti- tion rate mode is not chosen or unavailable, you can substitute a co-linear visible beam Gust make sure that it is indeed co-linear).

•	beam spatial profile: this is a figure of merit which applies to the physical properties of the beam. It is usually skirted deftly by manufacturers who use terms like "near-Gaussian" or "flat-top". Because the system I am describing is not available commercially within the U.S. this specification can be put into hardware terms as follows: for a 1J output at -S30nm, specify a final rod diameter of not less than 1/2". The diameter of the final amplifier goes a long ways towards determining the spatial profile of the beam -

as does the amplifying medium. Nd:Glass is the bees knees when it comes to uniform beam output - YAG, YLF and ruby can be described in progressively grimmer anatomical terms. If you attempt to pull too much energy out of a small diameter rod you will introduce structure in the beam which will result in a non-uniform holographic exposure.

Things to Know About When Shopping

Injection-seeding is the ultimate in narrowband, long coherence length Q-switched laser operation but you need a dedicated oscilloscope (-$20k) and a means of shuttering the system so that it can continue to operate at a minimum of 5Hz for proper seeding. We do not recommend injection seeding for pictorial holography.

Standard commercial Nd:YAG lasers are best utilized as front-ends for ultimate amplification in glass. I think it perhaps worth noting that the output of an Nd:YAG operating at 1.32 microns is probably superior to ruby in producing high quality red light for holography.

A physically long beam path can compensate for a lot of minor problems in beam quality - don't get too cute with studio compactness.

If possible, purchase a commercially standard laser which can be later upgraded with an additional amplifier. Systems from Positive Light, Continuum and Quantel are modular in design allowing straight-fOlward upgrades. By statting with a less than the ultimate custom system, you may learn that what you were initially convinced were expensive necessities, may indeed be overkill.

Standard Laser System Suppliers

Continuum, Santa Clara, California - Model Powerlight with electro-optic Q-switching injection seeding (I know what I just said but passively Q-switched lasers are not an option among mainstream providers). Install it on a 2' x 8' or larger honeycomb breadboard for possible upgrade and to facilitate beam delay lines, etc. as will be required. Energy specification 500m] at 532nm. Model 61C E at 532nm 750m]

Custom Laser System Suppliers

Positive Light, Los Gatos, California. - Producer of custom solid-state laser systems to include some of the world's most elegant laser systems in Nd:YAG, Nd:YLF and Nd:Glass.

Continuum, Santa Clara, California - Producer of standard and custom pulsed lasers with an emphasis on high energy Neodymium YAG and glass systems. Specify a two-stage Nd:YAG laser built on a breadboard for later energy upgrade. Standard commercial front-end Surelight II may be customized

Quantel, Orsay, France - Continuum with a French accent.

The nsLooks system at Laser Reflections - Using commercially available hardware from any of the above, you can copy the basics of our laser. We use a passively Q- switched Nd:Y AG oscillator operating in the TEM_{00} mode. Two etalons and a passive Q-switch (BDN dye) define a single longitudinal mode - we have a 500MHz scope and a fast photo-diode to look at the temporal profile but have not used it as a diagnostic for more than a year.

A preamplifier of Nd:YAG provides -100m} at 1Hz as an input to the Nd:Silicate Glass amplifier which we use in a double-pass configuration to supply us with just over 2} at 1064nm. A 40mm long KD*P doubling crystal is -50% efficient and we work routinely with 1] at 532nm. The frequency doubling without the final amplifier stage in operation is extremely inefficient - allowing us to work at a manageable level (lm}) for image composition - we do not have an alignment laser within the system. We take great care to underfill our final amplifier, assuring a beam spatial profile which is "very nearly Gaussian".

Much of the historic rap against Nd:YAG lasers concerned beam spatial profiles - but the dark days ofNd:Y AG crystal growing are long past - to the point that except for fanatics like myself - commercial Q-switched Nd:Y AG lasers as they leave the production floor would be adequate for most pictoral holography.

We split -10% of the energy for the reference beam (vari- able via a half-wave plate and a dielectric polarizer) which is delivered to the plate by a large float glass mirror at Brewster's angle. The remaining 90% of the energy goes to the two ground-glass diffusers which assure that the illumination beams are eye safe.

We work under red safelights and to maximize the studio session, we often put aside halfofthe transmission masters for processing the day following a shoot - at this point we have that much confidence in our technique.

In an eight hour studio recording session we can shoot and process as many as eight 14" x 30" masters each of which can be subsequently processed and presented for customer viewing. Using the same laser, directed to a second table, we produce image-plane reflection copies on glass or film. We can have product up to 14" x 24" ready on the next day - but normally we request three days for delivery of the first whitelight viewable transfers.

We have been recently asked recently to offer a laser system which will accommodate full color portraiture - my answer is that this puts the cart before the horse: we first need to know the characteristics of the panchromatic film and its suitability to exposure by Q-switched laser sources. A three-color narrow linewidth pulsed laser system while technically feasible (I mentioned my inclination for an all Nd:YAG system) is never the less extremely intimidating as a day-to-day production tool.

Of more immediate interest to us at Laser Reflections is the development of a black & white portrait system - this is a project we hope to find funding for within calendar ·97.

During the past year Laser Reflections has produced more than J00 masters and two hundred in-house copies. Recent sittings in- clude aCiOr James "Scolly" Doohan, football star and FOX-TV personality Ronnie Lott, San Francisco Mayor Willie Brown and Playmate of the Year, Julie Cialini. For filrther information see their ad and listing in the Business Directory.

Alternative Sources for Lasers
Suitable for Holography

For those that cannot afford to purchase a new laser from the original equipment manufacturer (or an authorized distributor), there are alternative sources for lasers suitable for holography. Specifically, there are companies that specialize in selling never-been-used "surplus" equipment and others that resell used and/or refurbished hardware, at prices far below retail. Besides offering low prices, these companies often have access to "obsolete" lasers and related equipment which holographers still find desirable.

It is also possible, but often less practical, for knowledgeable buyers to find decent deals on lasers from various other sources - such as optical laboratories, universities, manufacturing plants and corporate auctions. It is worthwhile to hunt for bargains at these places if you have the time and the expertise. Unfortunately, these sources do not provide the technical support, selection and service needed by most professional holographers and hobbyists.

For these reasons, we recommend that you contact those companies whose primary business is the resale and/or repair of electro-optical equipment. (See the Business Directory in this book for a listing of such companies) These. companies have developed a close working relationship with the laser industry and consequently can provide you with a high level of customer service.

Since many of these companies are staffed by experienced laser aficionados, they are often able and willing to answer technical questions before and after the sale. Sales-people are typically encouraged to spend some extra time with novice shoppers in order to ascertain their specific needs and steer them to the right equipment. (As in most industries, the factory's sales force is typically geared to service the larger corporate accounts and don't have the resources to answer a lot of basic questions from small businesses and hobbyists interested in purchasing single units.) In addition, most of the companies we know about in this business have one or two people on the staff who are especially familiar with the lasers required by holographers.

Another good reason to shop with resale businesses is product availability and selection. Most of these companies carry a wide selection of lasers from different manufacturers in various price and power ranges, rather than only a single product line. Prices usually range from a few hundred dollars for low powered units to thousands of dollars for more industrial gear. Although most holographers are hunting for HeNe and Argon lasers, it is also possible to find an assortment of more esoteric hardware, such as Heed, Nd:Yag and ruby lasers.

Most notably, these companies often stock models which the manufacturer has discontinued, even though they have proven extremely useful to holographers in the past. These companies also sell components (laser heads, power supplies, optics, etc.) for customers interested in assembling their own units and spare parts that might otherwise be unobtainable. Several US companies mentioned that they routinely customize equipment to make it suitable for countries with different electrical systems.

Finally, it is important to consider warranty protection. Gas lasers do have a finite life expectancy and are rather delicate pieces of equipment. Reputable companies should provide some sort of guarantee, especially on used equipment. (Some buyers might prefer buying refurbished lasers, as the life expectancy can be more easily ascertained). Every company we surveyed offered a warranty, ranging from 30 days to a full factory equivalent. Obviously, it benefits the customer to have one, especially if the laser is being shipped.

Where does this surplus and used laser equipment come from? One company we interviewed specializes in purchasing large quantities of surplus components directly from the OEM and resells both pre-assembled packages or individual parts. This "factory fresh" equipment is typically obtained from excess inventories of discontinued models, spare parts and production overruns. Another company we asked acquires large lots of used equipment from corporate users who are upgrading their equipment or switching technologies. (For example, many inexpensive HeNe gas lasers became available when supermarket chains purchased new bar-code scanners built around solid state diode lasers.) Other businesses purchase older units from a variety of sources and refurbish them to factory specs. All these companies are potential sources of good equipment at discounted prices.

Since the primary reason for buying surplus, used, or repaired lasers is price, we surveyed several businesses to find out how much a customer could reasonably expect to save. Steve Garret, owner of Midwest Laser Supplies, replies, "The amount of money saved depends on the cost of the laser, however, one can often save (an average of) 50% off the price of new equipment." At the time of this survey, his company was selling a seven mWHeNe laser and power supply for 1/3 the cost of a similar new unit, and a new Argon for 30% below the factory price.

Martin Hasa, ofMWK Industries, concurs, "The clear advantage ofpurchasing a surplus or used laser is cost. A new HeNe laser purchased directly from a manufacturer or one of their distributors will usually cost about three to six times more than a similar unit purchased from us. For example, in our 1996 catalog we list a surplus ten mw HeNe and power supply for 1/3 the cost of a similar system offered by a major distributor." Dennis Meredith, owner of Meredith Instruments, agrees, "Surplus equipment often costs between 114 to Y2 of what you would pay for a comparable new model, assuming the manufacturer would deal with you. For instance, we sell a new 10 mw HeNe unit that would probably cost double or more, elsewhere."

Substantial savings are available for refurbished equipment, too. For example, at the time of our survey, Don Gillespie, owner of El Don Engineering, was offering a used Argon laser for Y4 the cost of a new unit and used HeNe for 60% below the original price. Both units were fully warrantied. Martin Hasa concludes, " Though the price of new laser equipment can be far beyond the means of the average individual, used (and surplus) laser equip- ment makes holography an affordable pursuit for people, companies and institutions on a tight budget".

Lasers Used to Produce Large Format Holograms

by Simon Edhouse

The telephone rings. It's a farmer calling from Western Queensland, in outback Australia, wanting to know if I can help him by creating a hologram of a large hawk to hover over his field to scare away unwanted birds. "It must move across the field , and be able to change its color and appearance totally." Unfortunately, I have to tell him that I don't think we can be of assistance with his project. Then there was the private investigator who found his way to our remote studio location, knocked on the door, and asked if I could make a hologram of him to appear like he was sitting in his car when he was not actually there. It is bizarre moments like these, when the general public's perception of holography collides so spectacularly with the reality of the medium, bringing a kind of comic relief, that makes the frustrations and tediousness of holography somehow seem worth while.

It seems that a hologram has become a generic concept simply meaning anything that is sufficiently futuristic and impressive. Unfortunately, somewhere between the hype associated with the technology and the true nature of holography, something has become lost- a general understanding of the wonderful and real potential of the medium.

Creating a Large Format Studio

Dr. David Ratcliffe founded Australian Holographics (AH) with the very deliberate intention to develop large format holograms as its main priority. David appreciated the artistic possibilities that large format holography seemed to promise but could not find a studio that had the kind of technology that he wanted to develop. The market for large holograms was then almost non-existent, and the financial risks were great. These risks were compounded. by the fact that David used many unconventional and untested techniques in his method of producing large holograms; an aspect that would, however, later bear fruit.

The AH project necessitated building a large climate controlled studio incorporating a 6 x 5 meter optical table weighing some 25 tons. The system had to allow for both the creation of large-depth scenes for mastering, in addition to affording the space required for the effective production of ultra-large format rainbow transmission and reflection hologram copies. The resulting studio now produces transmission holograms up to 1.1 x 2.2m and reflection holograms up to 1 x Im. A heavy sand-filled cavity steel construction was used for the table. The suspension system was constructed around nine Firestone air bags connected to a standard pneumatic set-up with needle-valves, ballast tank and compressor. Overhead towers were designed to carry large transfer mirrors at heights of over three meters above the table. These towers were constructed from hollow steel tubes filled with sand. Over the years, lifting systems for the large glass film holders evolved from hand operated, to mechanical and finally to pneumatic.

Figure 1 - Australian Holographic's Large Optical Table

The firstAH laser, which we still use faithfully today, was a Coherent small frame 6W (all-lines) Argon laser which produces at 514.5nm SF at around 2W. For reflection work this was supplemented with a large frame Russian-Kryp- ton laser from PLASMA. This gives around 1W SF at 647nm fairly reliably, if you don't mind changing the seals on the . water circulator from time to time.

Different recording materials and chemistries have been used over the years. We started with Ilford materials but changed to Agfa when Ilford stopped production. TheAgfa now seems to be superior to the older Ilford emulsions. Since our work is almost exclusively large format we tend to always work with film. For Rainbows, we use 8E56 green material and for reflections and transmissions, either the 8E56 or the 8E75 depending on what color laser light we are using. We also have cause from time to time to use large glass plates made in various places in the former Soviet Union.

Chemistry varies depending on the application. For Rainbows we have consistently used an Ascorbic Acid developer and a simple Potassium Dichromate bleach. Tests by Adelaide University have shown excellent diffraction efficiencies with these chemicals (reaching 70% with optimized drying, and in some cases higher) and for the standard Rainbow hologram we haven't found better. For reflections, it is difficult to cite a single formula. For Argon transfer holograms a Pyrogallol developer is often used as this allows a uniform image in the yellow with appropriate choice of bleach and presoaking in water. For Krypton or dye transfers, usually we'll use another developer, sometimes with shrinking bleach, sometimes without. Ofcourse for uniform color control we frequently use pretreatment with TEA.

Creating a new Pulse Laser

In 1992 David relocated to Europe to establish a base in Lithuania and to develop a network of contacts with laser and optics manufacturers in the old Soviet Union. Having worked with many Russian scientists in his time as a mathematical physicist he understood very well the strong po-

HIGH IMPACT HOLOGRAPHY

Founded in 1982, Light Impressions was the world's first embossed holography company and continues to enjoy premier status for innovation and quality.

From a world class manufacturing facility in the UK with sales support in every part of the globe, we are ideally positioned to bring you the benefit of our many years of experience whatever your holographic needs might be.

Over the years, we have produced many of the industry's most significant orders and we are dedicated to finding better ways of serving our customers across all market sectors.

Light Impressions has recently been formed a new division, *Light Impressions Systems*, to supply the increasing numbers of customers who wish to become producers in their own right and wish to be certain of production of the highest quality.

For product enquiries in North America please contact:

Light Impressions International Ltd

American Sales Office

430 West Diversey Parkway,
Suite 501,
Chicago,
Illinois 60614,
USA

Tel: 1 312 665 1579
Fax: 1 312 665 1679

For further information on Light Impressions and Light Impressions Systems please contact:

Light Impressions International Ltd

Worldwide Sales Office

5. Mole Business Park 3,
Leatherhead,
Surrey.
KT22 7BA
England

Tel: 44 1372 386677
Fax: 44 1372 386548

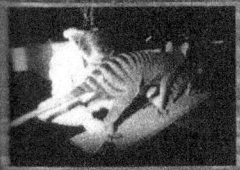

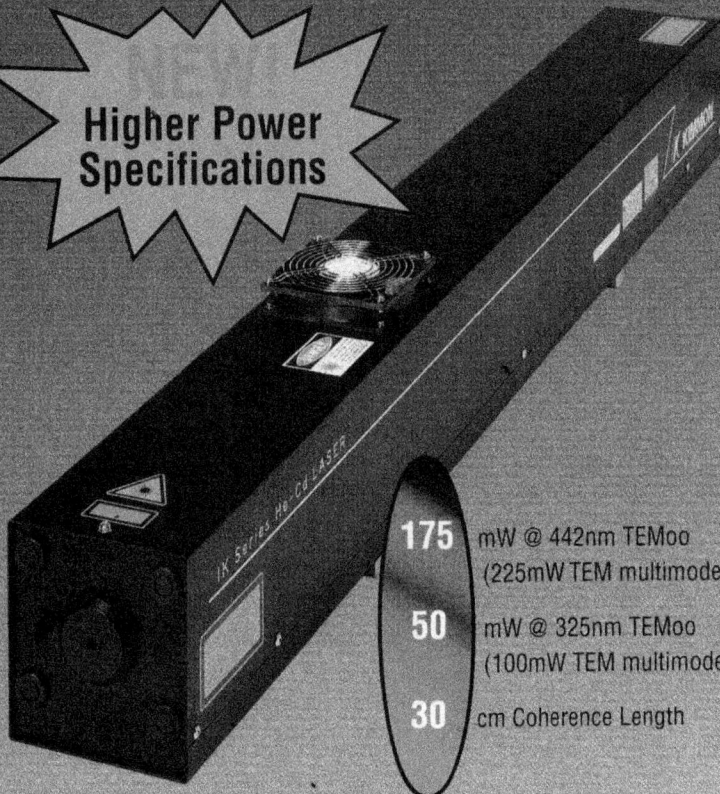

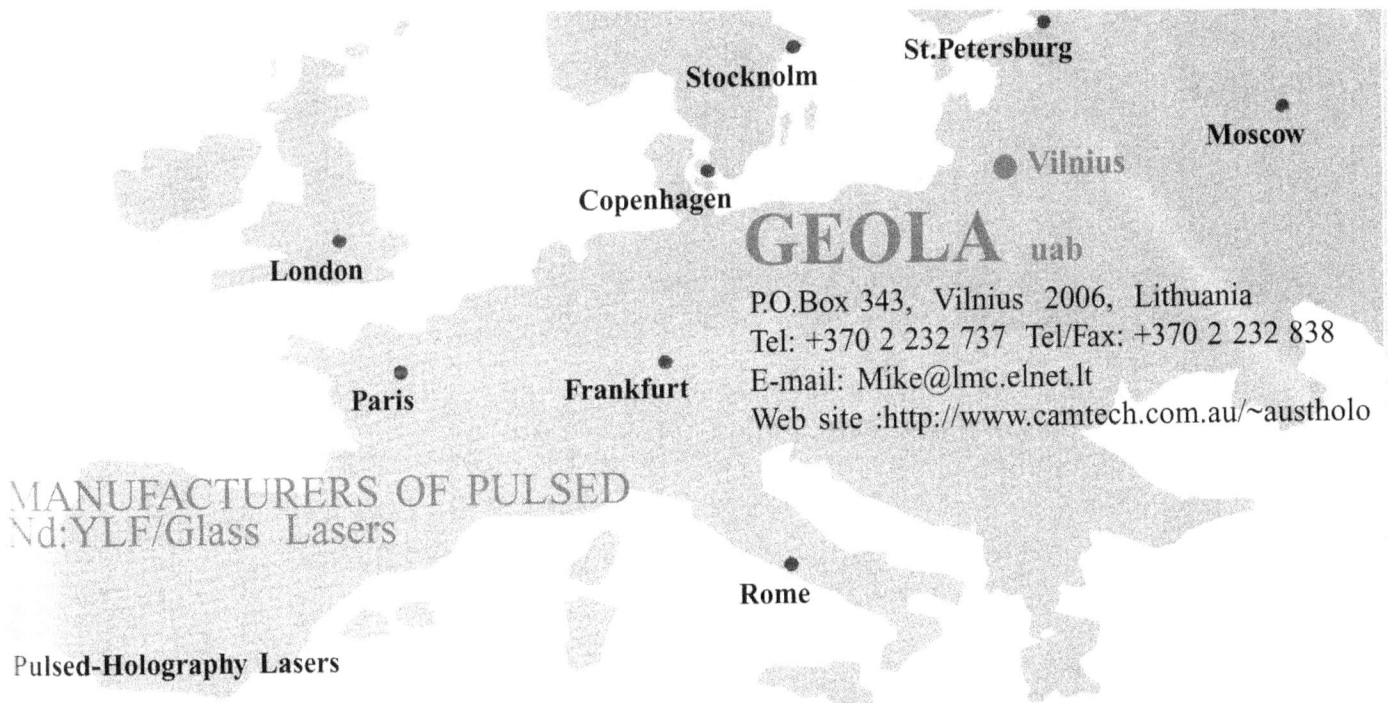

GEOLA uab

P.O.Box 343, Vilnius 2006, Lithuania
Tel: +370 2 232 737 Tel/Fax: +370 2 232 838
E-mail: Mike@lmc.elnet.lt
Web site :http://www.camtech.com.au/~austholo

MANUFACTURERS OF PULSED
Nd:YLF/Glass Lasers

Pulsed-Holography Lasers

GEOLA have just released a series of new Holography lasers. The G5J is a Nd:YLF/phosphate glass system giving over 5 Joules at 526.5 nm. This laser incorporates the latest advances in the technology including SBS phase conjugation, suspended INVAR base, fully interlock-protected electronics and a digital user interface. One and two Joule models are also available. Prices start at $35,000 for a 1J system*.

Pulsed Holography Systems for Portaiture and Large-Format

GEOLA makes *standard* (2+1 or 4+1 beam) and *custom* pulsed Holography camera systems. A restricted range of such systems is available for rental. Full systems with technical support can be dispatched to any location in the world for Holography projects that cannot be taken to Vilnius.

Optics Sales

GEOLA supplies a comprehensive range of high quality low-cost optics for both intra- and extra-cavity use. A custom and standard product range is offered. Absolutely all your holographic optics...

Custom Work

GEOLA accepts contracts for the design and manufacture of custom lasers in Nd.

PULSED STUDIO RENTAL

GEOLA operates a modern Large-Format Pulsed Holography Studio in Vilnius town centre which is available for rental to Holographers. Our prices are extremely competitive and allow an artist with a small or moderate budget to be exeptionally productive. We have access to extensive costume houses and props used for television, film and theatre and we maintain a database of actors and actresses. We cater for all types of holography shots from the simple portrait to large laser transmissions of purpose-built sets with image volumes exceeding 100 cubic metres.
Access to Vilnius by major Airlines is usually no more than 2 hours from most large European cities and accomodation in the historic town centre is cheap and very pleasant.

*We reserve the right to change our prices without notice.

The Foreign

Specialist in manufacturing a whole line of holographic products:

- -Watches
- -Key Chains
- -Calculators
- -Glasses

As a custom designer, we provide a full range of holographic services for:

- -Advertising
- -Promotions
- -Displays
- -Packaging
- -Premium and incentive
- --Gifts

We custom-make any holographic product according to your specifications. We are a French-managed company specializing in holograms with high quality standards, and offer Hong Kong very competitive prices.

Please contact us for details.

Dimension

H OLOGRAPHIC PRODUCTS

Head Office:
1901 Manley Commercial Bldg.,
367-375 Queen's Road, Central
Hong Kong
TEL: (852) 2542-0282
FAX: (852) 2541-6011

Show Room:
The Peak Galleria
Level 2, Shops 29 & 42
The Peak, Hong Kong
TEL: 2849-6361

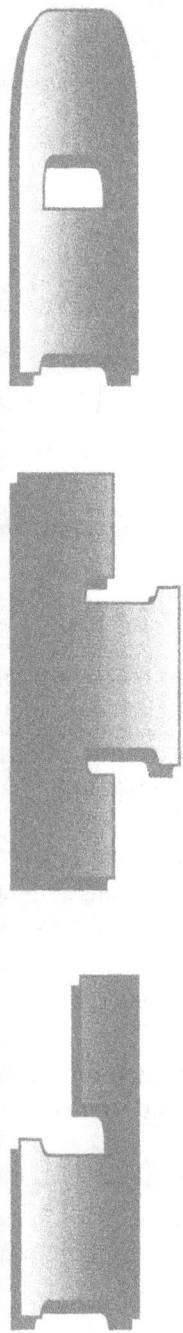

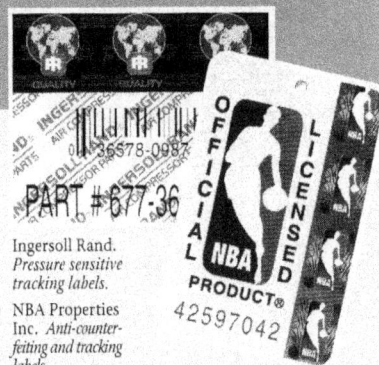

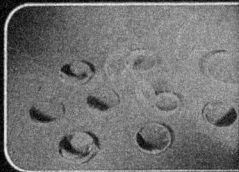

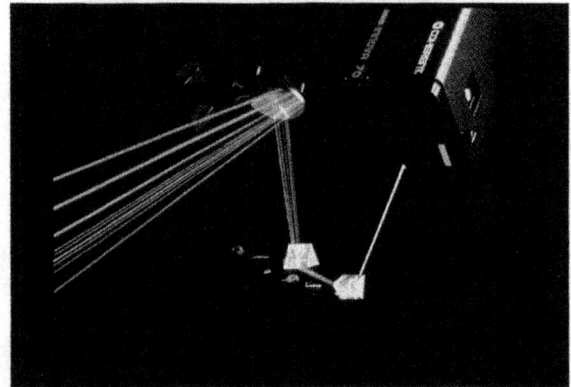

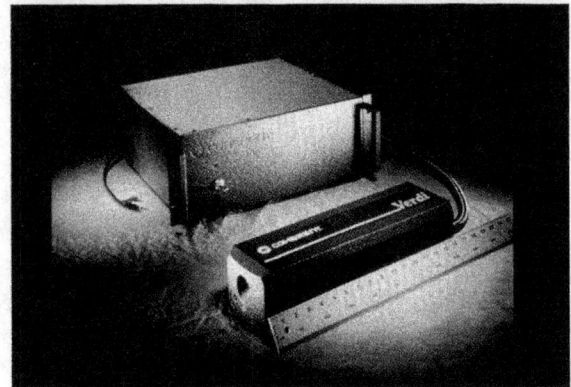

Above: Three successful applications of embossed holograms by Crown Roll Leaf, Inc.
Right Top: A full color reflection hologram by Holos Corporation.
Right Middle:Two CW lasers from Coherent, Inc.
Right Bottom:A large format hologram installation by Austrailian Holographics.
Below: A large format hologram by Australian Holographics produced for an educational exhibition.

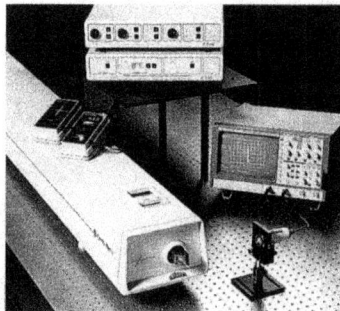

EMBOSSER II SERIES SPECIFICATIONS

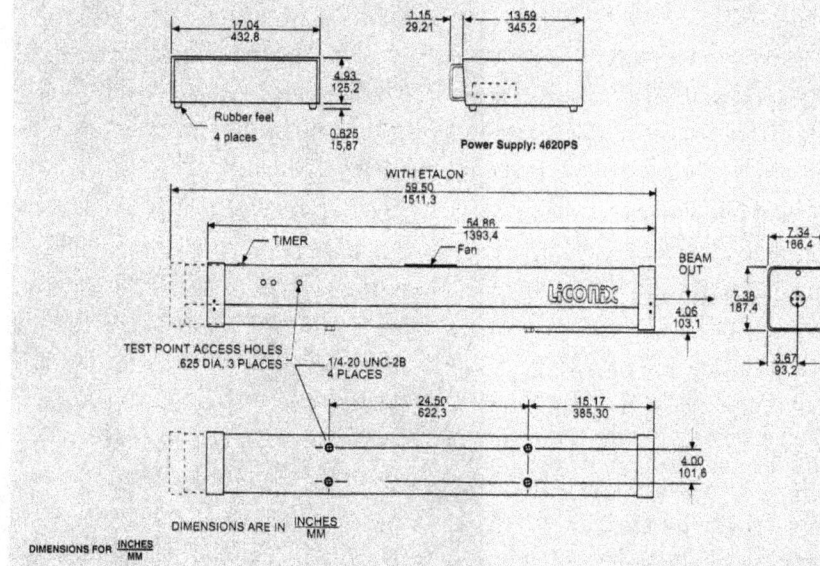

DIMENSIONS ARE IN INCHES/MM
DIMENSIONS FOR INCHES/MM

EMBOSSER II SERIES LARGE FRAME HeCd LASERS

EMB II Model	λ nm	POWER mW	MODE TEM	DIAMETER 1/e², nm	DIV. mrad.	COH. LGTH cm
3620N	325	20	oo	1.2	<0.5	10
3630NX	325	30	oo	1.2	<0.5	30
3630NX*	325	30	01*	1.5	<1.0	30
3640NX*	325	40	01*	1.5	<1.0	30
3650N	325	50	mm	1.5	<1.0	10
3660NX	325	60	mm	1.5	<1.0	30
3675NX	325	75	mm	1.5	<1.0	30
7620N	354	20	mm	1.5	<1.0	10
7630NX	354	30	mm	1.5	<1.0	30
46120N	442	120	oo	1.3	<0.5	10
46150NX	442	150	oo	1.3	<0.5	30
46170NX	442	170	oo	1.3	<0.5	30
46170N	442	170	mm	1.5	<0.8	10
46215NX	442	215	mm	1.5	<0.8	30
4650NE	442	50	oo	1.3	<0.5	50
4660NEX	442	60	oo	1.3	<0.5	50

SPECIFICATIONS

Power stability @ constant temperature	<3%/hr
Beam pointing stability @ constant temperature	10 µrad
Spectral bandwidth, FWHM, N style / NX style	3GHz / 1GHz
Polarization, plane vertical	>500:1
Optical noise, %p-p, DC - 10MHz	20±5
Power supply voltages, selectable	100, 117, 220VAC±10%
Weight, head N style / NE style	33lbs / 38lbs
Power supply	33lbs
Power consumption	<1k Watt

ENVIRONMENTAL SPECIFICATIONS

Operating air temperature	10C to 30C
Recommended operating air temperature	22C±3C
Storage temperature	-20C to 50C
Relative humidity	0 to 90%
Shock	20g

DANGER
LASER VISIBLE AND INVISIBLE RADIATION
AVOID DIRECT EXPOSURE TO BEAM
CLASS IIIb LASER PRODUCT

NX models employ isotopically enriched cadmium for enhance power in stereolithography and for enhanced coherence length, 30cm, in embossed holography applications. Horizontal mounting recommended. Specifications subject to change without notice.

**YOUR COMPLETE HOLOGRAPHIC
CONVERTING SOURCE
WITH OVER TWENTY YEARS EXPERIENCE**

~MASTERING~   ~EMBOSSING~

~SPECIALTY LAMINATING~   ~REGISTER DIE CUTTING~

~CUSTOM PACKAGING~   ~PREMIUM ITEMS~

~HOT STAMPING~   ~PRODUCT ASSEMBLY~

**WITCHCRAFT TAPE PRODUCTS, INC.
BOX   937   COLOMA,   MI.   49038
1-800-521-0731   FAX: 616-468-3391**

tential synergy that could, in principle, be realized by finding partners in Eastern Europe. After years of often frustrating work, experimenting with a variety of technologies and after collaborating with widely separated companies and institutions, he has finally established a strong presence in the region.

David soon realized that in order to maximize the potential of mastering for large format, it would be necessary to start working with Pulsed holography. This is because the nanosecond exposure time (typically 25ns) of a pulsed system allows one to shoot otherwise unstable images, and also does away with the need for a vibration isolation table. This frees the holographer from some of the limitations of a vibration isolation system and has the added benefit of theoretically allowing the laser/camera system to be taken to the subject, rather than always the other way around. This allows the possibility, for instance, of making holograms of scenes like the excavation of China's Entombed Warriors in situ, or perhaps a hologram of the inside of Tutankhamen's tomb. Because of the speed of the flash, it is possible for example, to generate a hologram that freezes the motion of a bullet as it travels through a glass of water, long before the water has even thought of leaving the glass. The implications of this innovation allow for a whole new world o f unique holographic images to be recorded: a bee- hive colony in flight, or a person in a shower showing every droplet frozen in time and floating in suspended three dimensionality.

The problem with available pulsed laser systems at the time was that they were not designed specifically, and hence not completely appropriate, for Artistic and Commercial holography, or that the systems were inadequate technologically. All systems, whatever their construction, were extremely expensive. David realized that by building a base in Lithuania he could bring together the superior Soviet optical technology with western design requirements and reliable electronics to produce the laser that we, and many others in the holography comrnuvity needed.

The design of a pulsed laser for holography is restricted by many issues. There are methods of obtaining just about any frequency in the visible spectrum, but there is clear evidence that solid state systems are the most appropriate. This leads to a choice between different laser crystals. Ruby was the first laser to be invented and although historically the favorite laser for pulse holography, it has many problems which actually make it rather inappropriate. Neodymium (Nd) as it turns out, is much more suitable. This crystal is a four-level laser and so is intrinsically far more efficient than ruby, which is a three-level. In practical terms, for this factor, this translates to much smaller power supplies.

There are many varieties of Neodymium. Nd:YAG is perhaps the best known. It has an emission at 1064nm in the infrared and can be frequency doubled with a non-linear crystal such as DKDP or KTP to produce an emission in the green at 532nm. For large energy applications an ND:YAG oscillator is traditionally paired with silicate glass amplifiers. Despite the fact that ND:Y AG/S ilicate Glass can produce a potentially good holography laser, there are significant reasons for choosing the lesser known material Nd:

YLF. Nd:YLF lases at 1053 and can be paired with Phosphate glass amplifiers for high energy. Its doubled emission is hence at 526.5nm. Phosphate glass has significantly more gain than Silicate glass and this is a great advantage. In addition YLF, with lower gain than YAG, allows more energy to be stored in the oscillator without risk of superluminesence, an effect that can be a problem when designing aYAG system. Also YLF has weaker thermal lensing than YAG, and hence, even though it has higher thermal conductivity, YLF can tolerate more optical pumping without propagation modification. This is extremely important when you want to design a laser that has a high frequency (i.e. with high thermal stress) low-energy alignment mode in addition to the normal high-energy, high-frequency mode.

After taking into account these and many other constraints, David proceeded to develop a laser along these lines. Initially the idea was to build a 51 system based on aND: YLF ring cavity oscillator with two Phosphate glass amplifiers. The oscillator design incorporated an LiF passive Q switch and etalon. Matching of the oscillator to the first amplifier was done by magnification of the far-field. This assured a perfect seed distribution. The first amplifier was two-pass and used SBS correction. This was vitally important in obtaining a constant beam divergence and propagation direction. The second amplifier was single pass and used focal plane translation incorporating vacuum spatial filtering between the two amplifiers. Frequency doubling was done by a DKDP crystal which gave approximately 60% conversion efficiency. Such a system, when matched with Western quality electronics, represented the best marriage of western design philosophy and Soviet optical know-how. This system, now marketed as the G51, can actually give up to 81 in the green. Its clear advantages are its almost perfect beam parameters, its long coherence, (typically 3-5 meters) and its almost perfect shot-to-shot reliability. Also important are, its capacity to produce a 1Hz low energy mode for alignn1ent with exactly the same beam parameters as in a normal high energy pulse, compact size, and near perfect output color at 526.5nm which in practical terms means great portraits without the need for skin make-up usually required when using a Ruby laser. More recently G51 has been supplemented by I and 2 louie systems. Currently a 11 system goes for around $US 35,000, which is comparatively cheap for a new holography laser.

Most frequently for pulsed holography we use the Agfa 8E56 material, and when using an Sm6 developer good results can be obtained. We have also used the green sensi- tive Russian plates, but while they are cheaper, we still prefer the higher resolution, higher sensitivity Agfa emul- sions. Currently we are testing plates from HRT Holo- graphic Recording Technologies, Gmbh in Germany which we hope will provide an alternative to Agfa. We are particu- larly interested in these plates as this company is offering sizes up to 50 x 60 cm. Specializing in large format, this is really the minimum master size for a meter square transfer reflection hologram, and although everything is possible using film, glass is easier.

GEOLA - The Laser and The Business

David's endeavors in Lithuania have resulted in the formation of GEOL A , short for "General Optics Laboratory", a business specializing in the manufacture of specialized NdlYLF/Phosphate Glass Pulse lasers for holography and also offering a state-of-the-art pulsed holographic mastering facility in Lithuania's historic capital, Vilnius. Geola has recently built new premises in Vilnius housing modem stylish offices, a studio environment and laser fabrication laboratories. The image room o f the pulsed laser studio has a ceiling height of six meters and allows for the creation of holograms with image volumes of up to 100 cubic meters and hologram formats of more than 2 meters square. This, coupled with the fact that virtually everything of importance in the lab is made in-house especially for this application, has given the lab the potential to make wonderful large depth holographic images. The studio is available for rent to holographers for artistic work.

Advances In Large Format Holography Over Time

Although pulse holography has added considerably to the capacity of Australian Holographics to generate new, kinds of holographic images, there are still many occasions where Continuous Wave (CW) mastering is desirable, if not essential. The requirement for stability in the CW mastering process, has a surprisingly beneficial aspect, in that it allows for the utilization of unstable curtained areas to effectively render invisible unwanted elements in the field ofvision. This trick is still unique to CW and is sorely missed during Pulse mastering, where the problem is that often too many things are visible and there are limited methods available to conceal them. Thus if a large object is required to apparently float unsupported in space, CW mastering, rather than Pulsed, provides the means to easily achieve this illusion. Many important elements involved in producing high quality large format holograms rest not so much with the traditional concerns of holography but rather with aesthetic concerns that relate to table layout, and lighting techniques that endeavour to feature the subject without visual distractions and to control glare and reflections that lead to non-linear noise. Over time, a vocabulary of devices are built up to deal with the changing demands of each project.

It is worth noting that it took fifty years from the appearance of the first photograph in 1826 to invent the roll-film camera, and another forty eight years to put perforations in that film to arrive at the 35mm camera in 1924. In the forty nine years since the invention ofholography, we have made great advances in the technology, but like photographs taken in 1924, where most photographers were still coming to terms with the medium, clever use of holographic technology has been limited to the small number of individuals who strive to push the envelope. Progress, however, is definitely being made. Consider, for example, that Denis Gabor's first hologram achieved a depth of only a couple of millimeters. Today in our studios, although limited by available film size, we make 2.2 x 1.1 meter holograms with depths of up to six meters. This allows images

from full size holograms of cars to a four meter model of a White Pointer Shark or a twenty square meter model of the Earth with the MIR space station hovering above.

Australian Holographics and Geola - Current Technical Work in Progress

Our recent exploratory work with simultaneous red and green flash pulse laser research (raman scattering) has given rise to some important results with a direct relevance to color holography (presented at the Cleo I Europe '96 conference). We have been able to achieve very significant non-linear raman conversion efficiencies, which approach the quantum limit whilst preserving the spatial and temporal coherence in the two-color beam. With experimental results showing such promise in this field, this project will hopefully lead us to large (>1m2) two-color image-planed pulsed reflection holograms. This is a technological solution which seriously addresses comments from the advertising industry that large format holography needs to be able to offer realistic product colors for advertisers. Incidentally, the new three-color PFG plates from Slavich similarly address this problem for small format CW work and have been producing excellent results in the Denisyuk format. T. H. lung, (Lake Forest College Centre for Photonic Studies) for example, showed some very nice work done with such plates at the recent Dennis Gabor Conference of Holography in Kecskemet this year.

We have also been collaborating with the optics department of Adelaide University to build a ring-dye laser for color transfer holography. This laser, which is being pumped by a 4W multi-mode multi-line Russian (PO-LARON) Argon laser, is a potentially great laser source and is of significant interest for color holography. To our knowledge such lasers, although used from time to time in holographic applications, have not been studied sufficiently. At present we are running the laser with DCM but plan soon to experiment with other dyes including R6G, which has a higher yield but the disadvantage of falling in a low sensitivity region of the Agfa 8E emulsions.

Other current research projects include experiments in\ underwater holography and high speed shutters for pulsed rainbow cameras for full daylight use. We are confident that making masters oflarge holograms underwater with a

frequency doubled Nd/YLF pulsed laser at 526.5nrn is quite feasible, and will lead to the development of a new camera system that could be used to make large holograms in oceanariums, and ultimately on the ocean floor. For daylight holography, high speed shutters will synchronize to the nanosecond flash of the pulse laser to all ow the film to be exposed effectively only by the laser flash and then closing before the film is affected by the sunlight. The resulting portable, if bulky holographic camera could be used to capture events like the split-second timing on the finish line in the men's 100 meters final at the 2000 Olympics. What an appropriate debut for a 21 st century technology that would be! So we have innovative current technology, and promising new technology on the drawing board, but it is all in the end just technology, that will hopefully one day seem commonplace and ordinary in the 21 st cen-

Figure 2 - Starting to stage a shot on the large optical table.

tury. The challenge, I believe, is to continue to use that technology as a creative tool box, and not see it as an end in itself.

The Marriage Between Art and Science

Over the last few years I have come to see our lab in Australia as something resembling a large photo-process-" ing machine, like the one down at the local shopping centre. The difference is we are not so automatic and we actually create pictures as well as develop and print them. The similarity lies in the fact that the lab is, after all, a fairly finite technological concern. Human creativity, on the other hand, the source of the ideas behind each 'hologram, is infinite. A hologram, like a photograph, is a product of technology, but in the end, is judged more on its creative content than its degree of chromatic balance or diffraction efficiency. In the final analysis, a hologram will be appreciated for the concept behind it, and people make concepts, not machines.

In recognition of this, we welcome collaboration with artists working in the field who can assist our growth by bringing unusual projects to our studios. Holographic artist Paula Dawson, for example, is a familiar face at our studio. We have a kind of symbiotic relationship where

Paula dreams up challenging projects, and together we try to work out how they can be achieved. The expansion of the potential of the holographic medium is in both our interests. We recently sponsored a complex project from Paula, that required a pulsed hologram of a heat plume rising through the shape of a cross to be overlaid on another holographic image recorded inside a four meter dome covered in detailed plaster relief and holding seven hundred and fifty evenly spaced franjipani flowers! (See photos on this page.) The result of all this work was a 30 x 40 cm hologram to be installed as a holographic shrine in a Catholic Church in Sydney. Through adventures like this we make discoveries about the nature and potential of holography, continuously adding to our box of tricks. This is something that would not be happening without people like Paula, who steadfastly refuse to compromise their often nearly impossible dreams.

The necessity for high quality 3D models led us to develop a business joint venture with the South Australian Museum. This arrangement gives us legal access to the Museum's vast collection of exhibits. As a natural progression from this activity we have become involved with all of the Australian Natural History Museums and various Science Museums and businesses in Asia and Europe to develop the concept for a thematic touring holographic exhibition called Prehistoric Lives. This exhibition will document key milestones in the development of evolution on earth. (See the last page of the color section in this book for a picture of one of these holograms.)

This will be done with thirty multi-channel rainbow transmission holograms and a series of five hexagonal laser transmission modules with a hologram on each face. The specific elements of the story, the choice of the actual prehistoric animals will be made by the international community of paleontologists and archaeologists via an interactive home page called Prehistoric Lives at MCM's website (http://www.mcm.com.au) being launched in November 1996. Projects like this will bring state-of-the-art holographic technology to a world audience, providing a new bench-mark exemplifying the power and new aestheticism of today's holographic technology.

For additional information contact the author. Simon Edhouse. in care of AUSTRALIA N HOLOGRAPHIC STUDIOS PO.Box 160. Kangarilla. S.A. 5157. AUSTRALIA TEL: +6188383-7255 FAX +61 8 8383-7244 URL: http://www.camtech.com. au/~austholo/

Figure 3 - Final work for a scene on the large optical table.

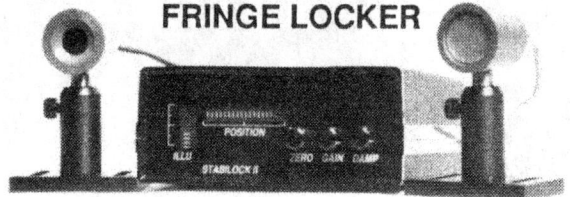
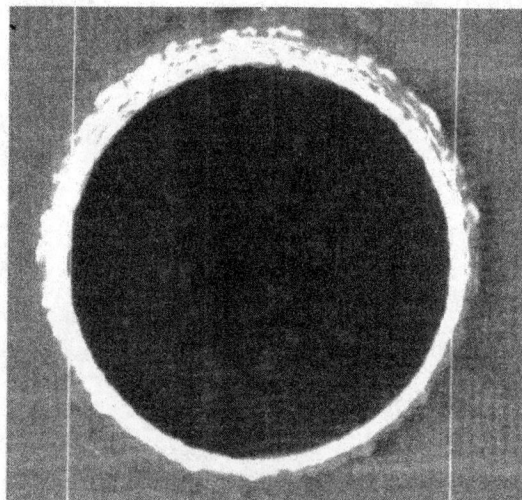

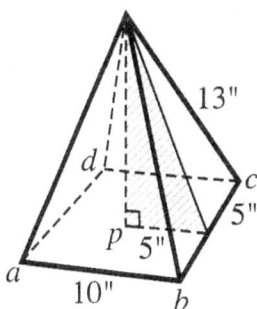

5

Optical Equipment

This chapter gives a short description of some optical equipment that may be of interest to holographers. First, information about aspheric lenses is presented. This is followed by a study that was done on the durability of various metals used to make pinholes. (Although it mainly deals with high powered lasers, the value of the study to holographers is that it presents the heat sink properties, reflectivity and melting temperatures of the metals used in pinhole manufacture.).

Aspheric Lenses for Holography

Aspheric lenses have one or more surfaces which are shaped to a surface of revolution about the lens axis, including conic sections but excluding a sphere. Commonly used aspheric surfaces include paraboloids, hyperboloids, and ellipsoids, they differ from spherical lenses in that the radius of curvature varies from the lens axis to the edge. This causes incident light rays to bend by different amounts depending on what portion of the lens it enters.

This property is used to minimize or eliminate spherical aberration, which results when light passing through different parts of a spherical lens focuses at different distances from the lens. The result is an image that is not as sharp as it would otherwise be. Spherical aberration is worst in fast optical systems, that is, where the diameter or aperture is large relative to the focal length. Hence, aspherics are effectively utilized in such systems.

Aspheric lenses are available in a wide range of diameters and focal lengths. Lenses as small as 1 to 2 mm diameter and focal length are used as laser diode collimators. At the other extreme, aspherics as a large as 16 inches in diameter are being used in holography. Materials can be either optical glass or plastic. The aspheric surfaces

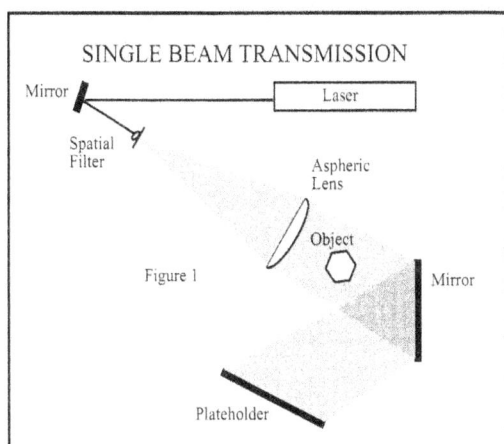

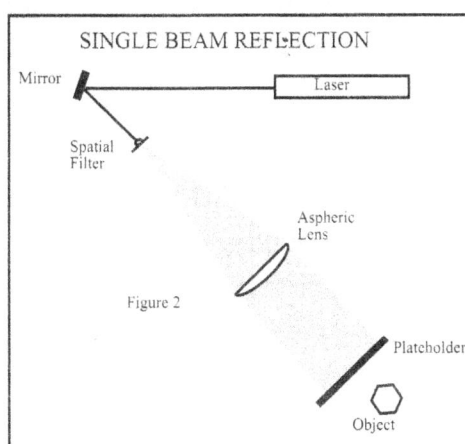

have traditionally been painstakingly "hand figured", whereby the proper amount of glass is selectively polished away from a spherical surface to obtain the desired shape. Increasingly, however, they are being molded from glass or plastics. This is much more efficient and economical for quantity production, and is suitable where high quality imaging is not a consideration.

Aspheric lenses have been used in holography to provide improved image quality. They may be used in single beam holographic setups shown in figures 1 and 2 to provide increased depth of field compared to setups not using them.

In split beam holography, they are used in place of aspheric mirrors to collimate the reference beam (figures 3 and 4). This significantly improves image quality by reducing the coma produced by the aspheric mirror as it receives and reflects the off axis beam.

The advantages of the aspheric lenses in holography, as already mentioned, are improved image quality and depth of field. The only disadvantage is higher cost compared to aspheric mirrors. The principal requirement in their installation is that they must be securely mounted to prevent any movement. This is true for any component used in holography; however, the large size and weight of these lenses makes this especially critical.

Further information regarding aspheric lenses is available from Control Optics Corp., Baldwin Park, California. See their advertisement and their listing in the Business Directory.

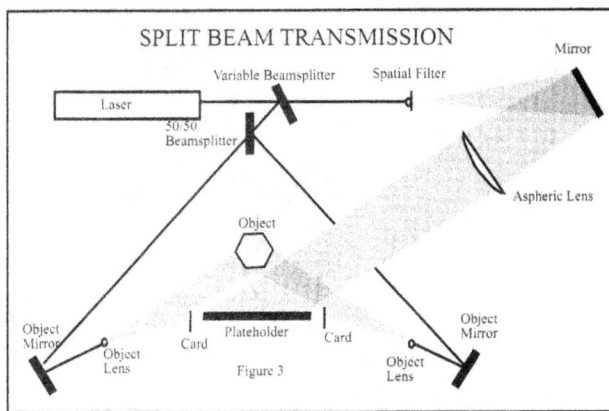

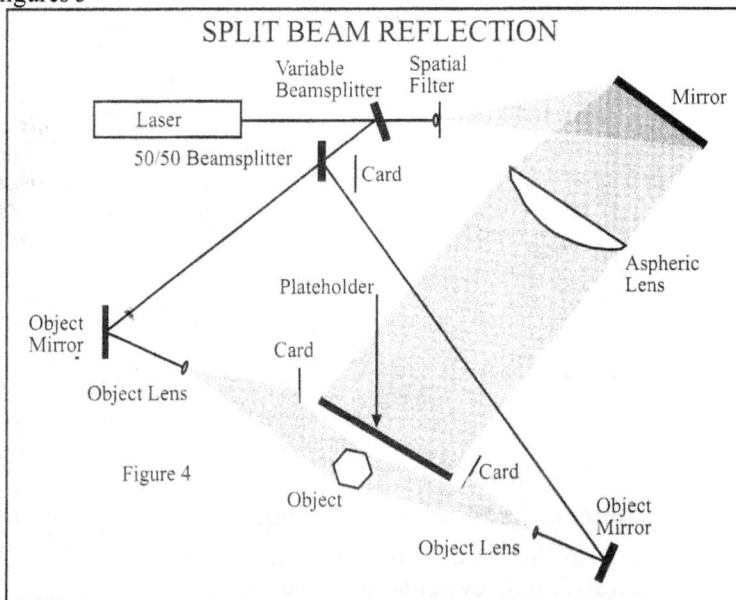

SPLIT BEAM TRANSMISSION

Laser — 50/50 Beamsplitter — Variable Beamsplitter — Spatial Filter — Mirror — Aspheric Lens — Object — Object Mirror — Object Lens — Card — Plateholder — Card — Object Lens — Object Mirror

Figure 3

SPLIT BEAM REFLECTION

Laser — 50/50 Beamsplitter — Variable Beamsplitter — Spatial Filter — Mirror — Aspheric Lens — Card — Object Mirror — Object Lens — Plateholder — Card — Object — Object Lens — Card — Object Mirror

Figure 4

High Temperature Materials For Laser Applications

by Gilbert R. Smith

The parameters relevant to selecting a material suitable for use as an aperture substrate in the optical transfer assembly of a laser, are melting temperature, thermal conductivity and surface reflectivity.

High energy and high power apertures are intended to be used in the optical transfer assembly of a system using a large laser as a source. The application of a high power laser with an optical aperture requires careful consideration of the aperture substrate material, with reference to the melting temperature. The laser power and the energy and the time duration of the beam pulse are also important. For applications where the beam pulse is continuously or intermittently generated in groups, the repetition rate and time between pulses are also important. A repetition rate should be selected that provides a packet separation time that is long enough to allow sufficient heat to be conducted away so as to avoid damage to the aperture disc.

The aperture substrate material should be one that is not deformed or partially vaporized by the laser beam. The material of a high energy or high power aperture disc should both conduct away sufficient heat and have a highly reflective surface facing the laser. In concurrence with this, it is advantageous to mount the aperture disc in a heat conducting holder. Table 1 outlines the melting tem-

Material	Melting Temperature
Alumina	1750°C
Copper	1083°C
Diamond	3550°C
Gold	1063°C
Graphite	3727°C
Elastaloy C-22	1356°C
Iridium	2443°C
Molydenum	2610°C
Platinum	1769°C
304 Stainless	1397°C
316 Stainless	1370°C
Tantalum	2980°C
Tungsten	3380°C
Table 1	

perature for a few of the more common pinhole aperture substrate materials.

Reflectivity

Table 2 outlines the normal incidence reflectivity for selected materials in the Nd: YAG wavelength range of 1.064 μ for the standard (#10) or polished finish of the material. The surface of these materials will reflect most of the laser energy incident upon them such that only a few percent of the energy will be absorbed and thereby provide heating. It should be noted that silver reflectivity degrades rapidly when exposed to air. The surface is no longer silver, it is silver oxide or silver sulfide. Before exposure to air, silver may be coated with a material that is transparent in the laser wavelength. In this manner, the high reflectivity can be maintained. Such coatings are not practical unless they can withstand the high temperature of the laser beam environment.

Material	Reflectivity % at 1.06 μ
Aluminum	95
Copper	98
Gold	98
Platinum	80
Rhodium	84
Silver	99
Table 2	

A similar degradation occurs when copper is exposed to air. However, the reaction time is considerably longer than the near instantaneous reaction of silver. If all other operational parameters are stable, a degradation in system performance provides an indicator for the condition of the copper surface. If maximum reflectivity is a requirement for your system, the surface should be re-polished on a regular basis. To maintain high surface reflectivity, spare copper units should be stored in an inert atmosphere.

Aperture substrate discs with a highly reflective surface facing the laser provide the advantage of reflecting away most of the energy so that the disc is not damaged by heat. The disadvantages of the reflective surface is the possible introduction of scattered light in the optical axis of the system. The use of optical baffles throughout the system will eliminate or significantly reduce scattered light.

Thermal Conductivity

For high power/high energy aperture applications, the ideal substrate would be the one with the highest melting point, highest thermal conductivity for heat sinking

and the highest reflectivity at the laser wavelength for the polished surface. Most aperture applications disallow this ideal condition and compromise is in order.

For reference, Table 3 outlines the thermal conductivity of a few aperture substrate or aperture coating materials. Those materials with the highest thermal conductivity value provide the best heatsinking characteristic.

Aperture Material Selection

Material	Thermal Conductivity (Cal/cm/sec/°C)
Alumina	0.060
Aluminum 6061	0.410
Brass, Naval	0.280
Chromium	0.160
Copper	0.918
Gold	0.705
Graphite	0.037
Hastaloy C	0.030
Incolen	0.036
Iridium	0.141
Molybdenum	0.346
Nickel	0.142
Platinum	0.166
Rhodium	0.210
Silver	1.006
Stainless, 410	0.057
Stainless, 304	0.036
Tantalum	0.130
Tungsten	0.476
Table 1	

In the laser field, experience and heritage are often the guidelines for selection of pinhole aperture substrate materials. There is no direct means by which one may calculate the expected temperature of a given aperture disc when irradiated with a laser beam of known power or energy. Modeling based on the first principle of Physics is not a reliable method to use for this endeavor. High resolution, detailed matrix modeling could produce useful thermal predictions. However, this is a labor, computer time and software intensive effort. As an alternative, there is empirical measurement and the trial and error method.

For empirical temperature measurement, there are some moderately accurate high temperature optical pyrometers on the market today. Pyrometer response time is generally significantly slower than the laser pulse duration and this may disallow a viable measurement. Given the relatively low cost of precision apertures, the trial and error method is practical and less labor-intensive. Select several aperture materials with melting points above that of the estimated temperature of the laser beam environment. One face of the aperture disc should be polished so as to reflect away a large percentage of the laser power. The aperture disc holder should be fabricated of a high heat conducting material. Empirical testing of the apertures in the laser environment will generate an aperture material performance data base that is unique to your system.

Footnotes:

1: Smithsonian Physical Tables; 9th Edition; Washington, D. C.

2 Handbook of Optics; Optical Society of America; 1987

3. Metals Handbook, Properties and Selection, Vol. I; ASM; 8th Edition

For further information regarding this topic, contact the author, in care of Lenox Laser, Glen Arm, MD. See advertising and listing in the Business Directory.

6

Recording Materials

This chapter discusses the major types of recording materials used by holographers: silver halide, dichromate, photoresist and photopolymer. Of special interest is the updated information we've assembled from the three main manufacturers of silver halide emulsions.

Silver Halide Holographic Recording Materials

Photographic Verses Holographic Materials

Photography has utilized silver halide recording materials for many decades. These silver halide emulsions are commonly coated onto films or glass plates for use in traditional cameras, in laboratories, and in factories. However, they were not designed for holographic applications. The important difference between photographic and holographic materials is resolving power, usually expressed in lines per millimeter. Whereas photographic films usually cannot resolve more than 50 - 100 lines per millimeter, holographic applications require between 1250 - 2500 lines per millimeter. Another difference is sensitivity, which is typically expressed as an ASA number. For example, popular photographic films are rated at ASA 120 - 400. Exposures are usually measured in hundreds of a second.

Silver halide holographic recording materials are so much less sensitive than standard films that their ASA would only be rated as fractions. Therefore, their sensitivities are usually expressed in micro-joules per square centimeters (or ergs per centimeter squared). Exposure times are typically measured in seconds, or even minutes, depending on the amount of laser light available.

Current Availability

In response to an anticipated demand, several major manufacturers of silver halide emulsions adapted their existing production techniques and formulations to provide affordable recording materials for holographic applications. This spurred the growth of the holography industry, which needed a reliable supply of basic materials. These silver halide emulsions were high quality, ready to use, and had a reasonably long shelf life.

Unfortunately, the worldwide demand for these particular photosensitive materials has slowed, due in part to the increased use of electronic imaging, especially in the field of Non-Destructive Testing. Since the current combined needs of commercial holographers, educational facilities and hobbyists do not nearly compare to other industrial and mass market customers, the production of silver halide recording materials suitable for holography has decreased, while prices have generally increased. (To prevent potential supply problems, some commercial holographers have learned to utilize silver halide recording materials with characteristics similar to holographic films but which are intended for other industrial uses, such as micro-lithography.)

Currently, the main suppliers of silver halide holographic recording materials are Agfa (Belgium) and Eastman Kodak (USA). Several distributors have begun to import previously hard-to-obtain emulsions by Slavich (Russia), but relevant product catalogs and associated technical specs are still somewhat difficult to acquire. Other distributors still have stockpiles of the popular Ilford (England) product, though production has been suspended for some time.

Basic Information

The silver halide emulsions are coated on either glass plates or film. Glass is preferred for most holographic applications, especially reflection holography, due to its rigidity. The most popular size glass plates from Kodak are 4" x 5" and 8" x 10", though display holographers often prefer to work with bigger sizes. Agfa supplies similar metric sizes, as well as larger plates. It is most economical to purchase larger sizes and cut plates to your own specs, if this is feasible.

A major advantage of film is that it is less expensive than glass, is easier to cut and curve, and is much more suitable for automated reproduction processes - as it is often available on rolls. The main difficulty faced by holographers using low powered CW lasers and film is keeping the film absolutely motionless during the exposure. Although the film can be sandwiched between clear pieces of glass, better results are obtained when the emulsion is left uncovered. Vacuum mounts and various other devices have been designed to accomplish this feat.

Some plates and films are supplied with an antihalation coating on the back, which can be useful when making transmission holograms, as it helps to cut down on unwanted internal reflections. These plates cannot, however, be used to make reflection holograms, so check product codes before ordering.

More Detailed Information

The general rules when selecting a silver halide material are: match the peak sensitivity of the material as closely as possible to the wavelength emission of the laser being used to expose the material; and, select an emulsion with the lowest possible graininess characteristics, and highest possible resolution. Let's examine why this is so.

There are five atoms which, because of their atomic similarity, are called the halides. They are chlorine, bromine, iodine, fluorine and astatine. Silver halide emulsions are made using either silver chloride, silver bromide, or silver iodide. The other two halides are not used because silver fluoride is insoluble in water and astatine is radioactive.

A typical silver halide emulsion is made by adding a solution of silver nitrate to a solution of potassium bromide and gelatin. Silver bromide crystals form in the emulsion. The emulsion is heated for a certain amount of time, which is called the ripening process.

During the ripening process, the grain size increases and the speed of the emulsion is increased. Some doping agents may be added to the emulsion at this time to foster proper crystal growth. Afterwards, the gelatin is allowed to cool. It is then shredded, and the soluble potassium nitrate is washed out of the emulsion.

The emulsion is heated again, with more gelatin added; then it is cooled and applied to a base. The thickness and hardness of the emulsion is important in holography because emulsions too thick tend to deform during development. Emulsions that are too hard can either retard chemical reactions or create vacuoles in the emulsion left by migrating atoms. These vacuoles tend to scatter light.

The Photochemical Reaction

Let's assume the emulsion is made and we now want to expose it to light. It sounds surprising, but a perfectly structured crystal of silver bromide does not react to light in any appreciable way. A crystal with defects, however, does react with light. Fortunately, most silver bromide crystals will have defects which consist of some interstitial (out of order) silver ions displaced in the crystal structure.

The process of the photochemical reaction is not known in exact detail, but it is believed that when light strikes a silver bromide crystal, enough energy is available to remove an electron from an occasional bromide ion. The electron produced is able to migrate through the crystal until it comes in contact with an interstitial silver ion. The silver ion takes the electron and becomes silver metal. Silver atoms formed by this mechanism apparently act as a nucleus for the formation of aggregates of 10 to 500 silver atoms, known as latent images because they are too small to be seen by the naked eye.

After exposure, the emulsion is developed. The developer goes to the site of any silver bromide crystal with a latent image and causes all the silver in that particular silver bromide crystal to be reduced to silver metal and deposited on the already-existing latent image of silver metal. This causes a worm-like grain of silver metal to form which is limited in size by the amount of silver available in the silver bromide crystal. This growth is considerable, amplifying the size of the latent image silver metal by a factor on the order of 10^6.

If the developer is left in contact with the emulsion long enough it eventually attacks all the silver in the emulsion. The speed of development is slow enough, though, that you can use a timer to take the emulsion from the developer just after the latent image, but not the unexposed silver bromide crystals, have been developed. At this point the developer has converted silver ions to silver metal if and only if they belong to a silver bromide crystal that was exposed to light.

The emulsion is then placed in a fixer solution which attacks all silver bromide crystals that were not exposed to light. The fixer makes these silver bromide ions soluble and removes them from the emulsion. The result is an emulsion with black spots where light has struck, and clear spots where no light struck.

An ideal silver halide emulsion depends somewhat on its use but there are three main factors to consider in any emulsion: thickness of emulsion, grain size of silver halide crystals, and sensitivity (or density of silver halide crystals) in the emulsion. We can generally state the following: It is agreed that emulsions of more than 10μm are neither practical or theoretically necessary to produce most volume holograms. Thicknesses above this size cause problems in development.

Grain size becomes an important issue in holography because it involves recording fringe patterns that are wavelengths apart. Too large a grain size may create excessive scatter, which may fog or destroy your hologram, and too small a grain size makes the emulsion have no usable sensitivity. It is generally agreed that the most ideal grain size is in the range of .01μm to .035μm.

The ideal exposure would probably be 100 - 300 mJ/cm2 to give a useful density (D=2-3). If exposures are much longer than this, the main attractic n of silver halide emulsion, its speed, comes into question and other emulsions become more attractive.

KODAK Products for Holograpby

(reprinted with permission from Eastman Kodak Company)

1. Red Sensitive Holographic Emulsions

Intended for hologram recordings m de with Helium Neon lasers (633 run) and Krypton lase s (647 run).

KODAK High Speed Holographic G ass Plates: Type 131-01, Type 131-02

(To place orders and obtain current pricing and availability information, call 1-800-823-4474.)

131-01 (with back side antihalation; dye and polymer backing):

SIZE	CATALOG #
2"x 2"	#1723485
2.5"x 3.5"	#1729656
4"x 5"	#1233139
8"x 10"	#1297431

131-02 (unbacked):

SIZE	CATALOG #
2"x 2"	#1240308
2.5"x 3.5"	#1729656
4"x 5"	#1231547
8"x 10"	#1241082

KODAK High Speed Holographic Film SO-253

(To place orders and obtain current pricing and availability information, call 1-800-248-3022.)

SO-253 (backed sheet film. Emulsion coated 9 microns on a clear ESTAR Film Base. A dyed gelatin pelloid on the base side provides antihalation protection.)

SIZE	CATALOG #
4"x5"	#1772672

Note - Rolls of film in larger sizes are available as a special order item.

General Characteristics of Red Sensitive Holographic Emulsions

- High speed, high contrast.
- Resolving power - 1250 lines + per mrn. Micro fine grain size.
- Energy required - approximately 5 - 10 ergs per sq. cm at 633 nm.
- Negative working.
- Glass plate is 0.04 inch thickness. SO-253 poly ester film base is 0.004 inch thickness.
- Processing - Process in KODAK Developer D-19.

Additional Information

"This film provides adequate speed when exposed with helium neon or krypton lasers. At the same time, its micro-fine grain structure and other emulsion characteristics combine to yield high diffraction efficiency and low noise upon reconstruction of holograms recorded at spatial frequencies as high as 1500 cycles / run. It is recommended primarily for holographic interferometry and micrography, and it is particularly useful for general holographic procedures with low power HeNe lasers."

Recommended Processing Technique

- Developing - When exposed with HeNe lasers or the red line from Krypton lasers and processed for 6 minutes in D-19 developer at 68 degrees F (20 degrees C), exposures of 5 ergs/cm sq. should be sufficient to achieve maximum reconstruction brightness. (As with other holographic materials, an increase in development time, to 8-10 minutes, will result in higher speed and diffraction efficiency at the expense of reduced exposure latitude and playback signal-to-noise ratio.)

Following development for the indicated times, processing is continued with the following steps, all at 65 - 70 degrees F (18.5 - 21 degrees C):

- Rinse - in running water or KODAK Indicator Stop Bath or KODAK Stop Bath SB-1 with agitation for 10 - 30 seconds.

• Fix - using KODAK Fixer or KODAK Fixer F-5 with agitation for 5 - 10 minutes.

• Wash - with moderate agitation for 1 minute.

• Rinse - in a solution of KODAK Hypo Clearing Agent with agitation for 4 minutes. (This rinse contributes to washing equivalent to the criterion for archival keeping as described in ANSI Standard PH 4.8 - 1971.)

• Wash - with moderate agitation for 3 minutes.

• Rinse - in a solution of three parts Methanol and one part water with agitation for 5 minutes. (The methanol rinse is required to remove a high level of residual sensitizing dye from the emulsion. The dye is distinctly blue in appearance and would greatly reduce reconstruction brightness when operating with a red-emitting laser.)

• Wash - with moderate agitation for 5 minutes. Use a wash water flow rate sufficient for one change of water every 5 minutes.

• Dry - in a dust free atmosphere. Drying marks can be minimized by treating the film in KODAK Photo-Flo Solution (prepared as directed on the bottle label) after washing. The use of Photo-Flo solution will promote uniform drying of film surfaces. For best results, dry film slowly at room temperature.

Latent Image Decay - Like most films with extremely fine grains, these emulsions exhibit significant latent image fading during the hours just following exposure. It is good practice in determining an optimum exposure level for a given holographic setup to process as soon after exposure as possible, provided that the elapsed time can be maintained for all subsequent operations with the same setup .

Storage - Unexposed film should be stored in a.cool place (70 degrees F or lower) in the original sealed package. Prevent condensation, which may result in spotting, ferrotyping, or sticking. In addition, thermal expansion during exposure will result in smearing of holographic fringes.

Safelight - Total darkness is recommended when handling this film. Green safelights at very subdued levels may be tolerable.

Note - This film can also be exposed efficiently with Helium-Cadmium (442 nm), Argon (515 nm), and frequency doubled Nd:Yag (532 nm) lasers. For exposures in the blue or green, 25 - 40 ergs/cm squared will be required. Some reduction in holographic performance is to be expected at progressively shorter wavelengths as a result of Rayleigh scattering in the emulsion during exposure, but this is characteristic of all silver-halide holographic materials.

2. Blue / Green Sensitive Holographic Emulsions

Suitable for hologram recordings made using Nd:Yag laser (532 11m) or other blue lasers below 550 nm..

KODAK High Resolution Plate, Type lA

(To place orders and obtain current pricing and availability information, call 1-800-823-4474.)

Type lA

SIZE	CATALOG #
12" x 16"	# 8943904

KODAK High Resolution Film SO-343

(To place orders and obtain current pricing and availability information, call 1-800-248-3022.)

SO-343

TYPE	CATALOG #
HOLOTEST plates	8E56 HD NAH
HOLOTEST film	8E56 T3 HD NAH
HOLOTEST plates	8E75 HD NAH
HOLOTEST film	8E75 T3 HD NAH

General Characteristics of Blue / Green Sensitive Holographic Emulsions:

• Extremely slow speed. Extremely high contrast.

• Resolving power of 2000 per mm.

• Energy requirement approximately 1000 ergs per cm squared at 532 nm.

• Negative working.

• Glass plate thickness 0.04 - 0.06 inch. Film thickness is 0.007 inch.

• Processing - Process in KODAK Developer D-19 or KODAK HPR Developer.

Additional Information

"Micro fine grain structure. Thin emulsion. Frequently used for pattern generation of fine reticles, preparation of printed circuit board artwork, television shadow masks and other masks for microelectronics."

Safelight - red

For further information, contact the KODAK information Center 1-800- 242-2424 ext. 19, or the KODAK Professional Printing and Imaging (Special Products) Division 1-716-477-7658.

AGFA Products for Holography

(reprinted with permission of Agfa-Gevaert N. V)

Introduction

Holographic emulsions must be able to record interference patterns with dimensions of the same order of magnitude as light waves. This means they have to meet extremely high requirements in terms of resolution, while possessing reasonable speed To achieve this, the silver halide crystals must be very small.

However, high resolution and high speed are opposing requirements for photographic materials. It is nevertheless possible to make silver halide crystals so small and so sensitive, and to tune their light absorption to the emission frequencies of lasers, that an acceptable compromise is achieved, in a reproducible manner.

HOLOTEST Materials

Under the name of HOLOTEST, Agfa offers a range of holographic materials for various applications. There are two different types of emulsion, available either as glass plates or as film:

TYPE	CATALOG #
HOLOTEST plates	8E56 HD NAH
HOLOTEST film	8E56 T3 HD NAH
HOLOTEST plates	8E75 HD NAH
HOLOTEST film	8E75 T3 HD NAH

Physical Characteristics

The coating thickness of holographic emulsions theoretically has to lie between 5 and 7 μm. Small silver halide crystals, in particular AgBr (refraction index n = 2.25), suspended in gelatine (n = 1.52 to 1.54) give the undeveloped holographic emulsion a refraction index of n = 1.64.

In practical terms, this means that the HOLOTEST plates have an emulsion thickness of approx. 6.0 11m. Depend- ing on the size of the plate, the thickness of the glass is 1.5 mm or 3.3 mm, with a refraction index n = 1.51. In the case of HOLOTEST 8E films, the approx. 6 11m thick emulsion is coated on a 190 J..lm thick T3 triacetate base (n = 1.485). In contrast to a polyester base, with triacetate there is no bire fringence that can cause interference when recording reflection holograms or copying holograms.

The HOLOTEST materials do not have an anti-halation layer (NAH). Thanks to their extremely small silver halide crystals, the 8E materials have very high resolution of 5000 Lines/mm, hence the designation "HD" (for "high definition").

Photographic Characteristics

Spectral Sensitivity

HOLOTEST Type 8E56 HD NAH:

This emulsion is sensitive in the range ultraviolet to yellow/green, and so it is particularly suitable for lasers which emit green light: these Include argon, krypton and frequency-doubled NdlYag lasers.

The following curve shows the absolute spectral sensitivity expressed as Log S, where S is the reciprocal of the illumination (in mJ/m2) necessary to achieve a density of D = 2 above fog level.

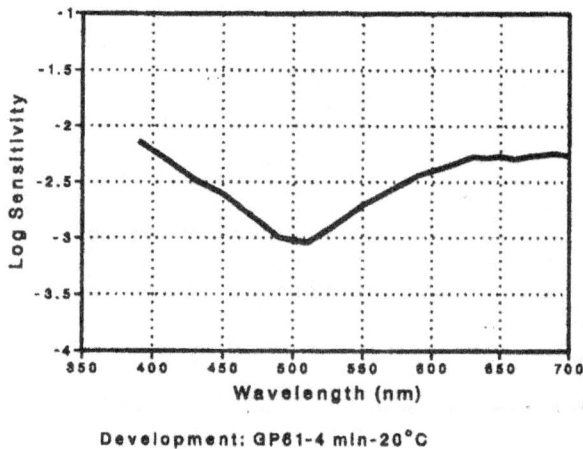

Development: GP61-4 min-20°C

Figure 5.1 Holotest film 8£56 T3 HD NAH spectral sensitivity

HOLOTEST type 8E75 T3 HD NAH:

This holographic emulsion is sensitive in the wavelength range from 600 to 750 nm, making it suitable for exposure with HeNe, krypton and ruby lasers.

The absolute spectral sensitivity ofthis emulsion is shown by the following curve.

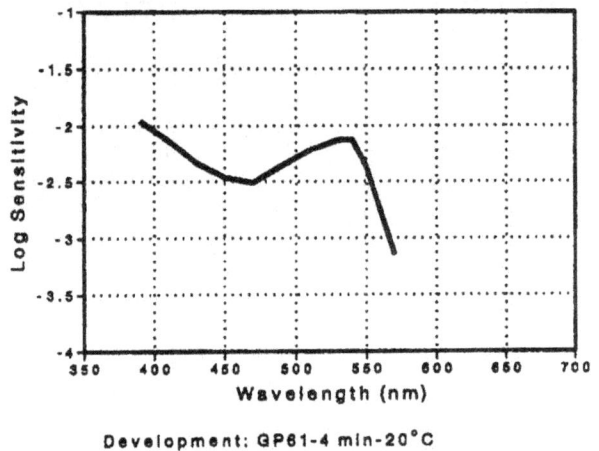

Development: GP61-4 min-20°C

Figure 5.2 Holotest film 8E75 T3 HD NAH Absolute spectral sensitivity

Sensitivity Characteristics

The sensitivity of the HOLOTEST emulsions depends . on the laser light and the development chosen. It is there- fore recommended to determine the exposure time and intensity by trial and error, referring to the absolute spectral sensitivity curve. When using Q-switch lasers, in which pulse times of 10-50 ns are common, there may be devia- tions in the reciprocity behavior of the emulsions. Deviations from the reciprocity law occur when using CW (continuous wave) lasers with low intensity, which require very long exposure times.

Diffraction Efficiency

Using the development methods described by Agfa, transmission holograms with a diffraction efficiency ofapprox. 70% can be achieved with HOLOTEST 8E56 and 8E75 HD emulsions (R.R.A. Syms and 1. Solymar, Appl. Opt. 22,14791983). Reflection holograms, processed in e.g. CW-C2 and PBQ2 bleaching bath, also achieved a diffraction efficiency of approx. 70% (DJ . Cooke and A.A. Ward, Appl. Opt. 23,934 1984).

Applications

HOLOTEST type 8E56 HD NAH:

This type of emulsion is suitable for making transmission and reflection holograms with green lasers. HOLOTEST 8E56 HD NAH is also used for holographic non-destructive testing.

HOLOTEST type 8E75 HD NAH:

This emulsion is designed for making transmission and reflection holograms with red lasers. HOLOTEST 8E75 HD NAH is suitable for making master holograms and for copying (contact copying and HeNe scans). This type is also very suitable for portrait holograms, computer holograms and research into new holographic concepts.

Processing

WARNING: always observe the necessary safety measures when handling chemicals: wear rubber gloves, goggles and dust masks, make sure the work area is well ventilated, etc. Information about the risks of the various chemicals involved can be obtained from the manufacturers in the form of MSDSs (Material Safety Data Sheets). MSDSs about the various baths can be obtained from Agfa.

Safelight - For HOLOTEST 8E56 HD NAH, a dark red filter is recommended (Agfa R4, Encapsulite R20IND.3). For HOLOTEST 8E75 HD NAH, a dark green filter can be used (Agfa V505). It is recommended to use the filter in combination with a 15 W incandescent lamp at a distance of 1.20 m from the light sensitive material (with the exception of Encapsulite filters).

Development

The development of holographic silver halide emulsions depends on the recording technique used, the type of hologram being made and the desired results.

Amplitude holograms - For simple laser transmission holograms and white light holograms in which the interferogram is built up of density differences, standard developers such as Atomal or Refinal can be used in combination- with G 328 fixer.

Phase holograms - Three basic methods are used for bleaching amplitude holograms so as to convert the original density differences into phase differences in order to obtain a phase hologram.

1.   Rehalogenating bleaching after fixing: The unexposed silver halide is dissolved out during fixing. The rehalogenating bleach bath converts the developed silver into transparent silver halide with a higher refraction index than the gelatine; i.e. the areas originally exposed have on average a higher refraction index than the unexposed areas. By means of this bleaching technique, a higher diffraction efficiency is achieved .

2.   Dissolving or reversal bleaching: In this method, the amplitude hologram is not fixed but is bleached immediately. Through the action of the silver-dissolving bleach bath, the developed silver is either converted into a water-soluble salt and washed out of the emulsion, or it is immediately dissolved and removed from the emulsion. As a result, a surface relief is additionally created from the original silver image. This results in a difference in refraction index between the originally blackened and nonblackened areas, i.e. the undeveloped silver halide crystals. Dissolving bleach baths enable holograms with very low scattering to be made.

3.   Rehalogenating bleaching or fixing: In addition to an oxidizing compound, bleach baths are used containing an alkali halide which immediately converts the developed silver into silver halide. The phase hologram is built up by transfer of material between the exposed and unexposed areas. In this bleaching technique, little or no silver salt is removed from the emulsion, so that the thickness of the emulsion alters only very slightly. These bleach baths also enable high diffraction efficiency to be obtained.

TRANSMISSION HOLOGRAMS

For laser and white light holograms, Refinal can be recommended as a standard developer. If desired, the following preparation GP 61 can be made up in the laboratory:

GP 61	
Water	700 ml
Metol	6 g
Hydonquinine	7 g
Phenidone	0.8 g
Na_2SO_3	30 g
Na_2CO_3	60 g
KBr	2 g
Na_4EDTA	1 g
+ Water up to 1000 ml	

Good results may also be achieved with GP 431, made . up as follows (Useful lifetime in closed bottle - 1 week or longer):

GP 431	
Water	600 ml
$Fe(NO_3)_3 9H_2O$	150 g
KBr	30 g
Phenosafranine	300 g dissolved in 200 ml ethanol
+ Water up to 1000 ml	

The following processing is recommended at 20°C

- Development   GP 61   2 min
- Intermediate wash   running water 2 min
- Fixing   G 328 (1+4)   2 min
- Intermediate wash   running water 2 min
- Bleaching   GP431 (1+4)   until clear
- Washing   running water 5 min
- Final wash   in distilled water

with addition of

1 part Agepon to

200 parts water 2 min

- Dry   in warm, dust free air

REFLECTION HOLOGRAMS

Most bleach baths cause the emulsion to shrink, so that the distance between the Bragg planes becomes smaller. The hologram is reconstructed with a shorter wavelength.

When shortening of the wavelength is not desired, it is necessary to ensure that the thickness of the layer is the same before and after processing.

For this purpose, developers are normally used which partially harden the gelatine (e.g. pyrogallol), in combination with rehalogenating bleach baths. In this case only minimal quantities of silver salts are removed from the emulsion.

If desired, the following processing can be used:

GP 62			
Part A		Part B	
Water	700 ml	Water	700 ml
Metol	15 g	Na_2CO_3	60 g
Pyrogallol	7 g	Distilled water up to 1000 ml	
Na_2SO_3	20 g		
KBr	4 g		
Na_4EDTA	2 g		
Water up to 1000 ml			

Mixing ratio: 1 part A + 2 parts water + 1 part B.

Useful life: Parts A and B keep have a relatively long useful life if kept separately. However, the ready-to-use developer has a useful life of only 1-2 hours.

• Intermediate washing: 2 min. in running water at 20°C

• Bleach until clear at 20°C

In **GP 432** make up the following formula (Useful life in closed bottle: 1 week or longer):

GP 432	
Water	700 ml
KBr	50 g
Boric acid	1.5 g
Water up to 1000 ml	
p-benzoquinone	2 g added immediately before use.

• Wash under running water: 5 min. 20°e.

• Final wash in distilled water to which is added 1 part Agepon to 200 parts water: 2 min. at 20°C

• Dry < 40° or room temp.

A large number of processing methods by which reflection holograms with high diffraction efficiency can be made are described in the various literature.

The following are some useful references:

L. Ioly, R. Vanhorebeek: Photogr. Sci. Eng. 24, 108 (1980)

W. Spierings: Holosphere 10. No. 7/8 (1981)

L. Ioly: J. Photogr. Sci. 31 , 143 (1983)

D.J. Cooke, A A . Ward: appl. Opt. 23, 934 (1964)

L. Ioly, R. Phelan, M. Redzikowski: SPIE 615, 66 (1986)

L. Crespo, A. Fimia, J.A Quitana: Appl. Opt. 25,1642 (1986)

P. Hariharan, eM. Chidley: Appl. Opt. 27, 3852 (1988)

P. Hariharan, eM. Chidley: Appl. Opt. 27, 3065 (1988)

N.J. Philips: Proc. ISDH 3, 35 (1988)

P. Hariharan, eM. Chidley: Appl. Opt. 28, 422 (1989)

P. Hariharan: Appl. Opt. 29, 2983 (1990)

P. HarWharan: J. Photogr. Sci. 38, 76 (1990)

R. Kostuk: Appl. Opt. 30, 1611 (1991)

SLAVICH Products for Holography

(information obtained from 3Deep Hologram Co.)

In the past few years, Russian emulsions have started to become available. These materials have long been sought by holographers worldwide, due to their characteristic high resolution and quality. In addition, Slavich is the only company currently producing silver halide plates sen- sitive to multiple wavelengths, thereby making them suitable for multi and full color imagery..

Here are the specifications for two monochromatic emulsions, PFG-03 and FPR:

Summary of Technical Specs for PFG-03

Description: Holographic High Resolution Photographic Photo Plates

Usage: Pulse holography, CW holography, to produce copies, interferential non-destructive testing; to manufacture materials for image recognition systems, to manufacture holographic optical elements.

Sensitivity:

Red Sensitive - 633/694 nm

At 633 nm : <= 40J/M2

At 694 nm : <= 70 J/M2

Diffraction Effectiveness:

At spatial frequency of 1/5000 mm

At 633 nm: >= 30%

At 694 nm: >= 45%

Grain size: 12 nm

Resolving Power: 10,000 lp/mm

Storage: (4-10) °C 12 month

Development: See manufacturer's bleaching, development procedures included, or consult the text, Silver-Halide Recording Materials for Holography and Their Processing by H.I. Bjelkhagen. Published by Springer-Verlag.

Summary of Technical Specifications for FPR

Description: High Resolution Photographic Plates

Usage: Pulse holography, CW holography, to produce copies interferential non-destructive testing, and photographic masks for integrated circuit.

Sensitivity:

Green Sensitive - 530 nm

At 543 nm: <= .6 J/M² (co-directed HeNe, bleached)

At 532 nm: <= 1.0 J/M²

Diffraction Effectiveness:

at spatial frequency of 1/1000 mm

At 488 nm: >= 45%

At 532 nm: >= 45% (pulse duration of 1.l0-10c)

Grain size: 60 nm

Resolving Power: 1200 Iplmm

Storage: (4-10) °C 12 month

Development: See manufacturer's bleaching, development procedures included, or consult the text, Silver-Halide Recording Materials for Holography and Their Processing by H.I. Bjelkhagen. Pages: 44-47, 108-109,36 1 Pub. by Springer-Verlag.

Additional Information

FPR High Resolution Photographic Plates

FPR High Resolution Photographic plates are intended for use in manufacturing of photographic masks for integrated circuits. High resolution, high durability, low presence of microdefects, and excellent adhesion of emulsion layer to base provide high quality of FPR photo plates, allowing to receive photomasks having elements with size not more than 2 to 3 micrometers.

Holographic properties of the FPR photo plates allow to use them for recording and manufacturing of holograms and for testing of important parts of different kinds of equipment by non-destructive methods.

As concerns properties FPR photo plates are analogous to photo plates used in microcircuit production of other companies.

Holographic Data

Diffraction Efficiency at spatial freq. of 1000 mm-1:

- exposed by argon laser, %: more than or equal to 45.

- exposed by neodymium laser with pulse duration of 1.10-10c: more than or equal to 45.

Holographic Sensitivity:

- when exposed by helium-neon laser according to codirected scheme and processed with bleaching, J/m2: less than or equal to .6

- when exposed by rteodymium laser, J/m2: less than or equal to 1.0

Average Diameter of Microcrystals, nm: 60

Instructions for Application of FPR Photographic Plates

Purpose - FPR photo plates are used to produce masters of holographic three-dimensional portraits of people and animals with the use of impulse laser.

Terms of storage - The photoplates must be kept in trans- portation boxes in a warehouse at the temperature 12-25 degrees C and relative humidity 40-60%. The photoplates must not be kept jointly with X-ray matters and luminous paints. Guarantee storage period is 12 months.

Chemical and Photographic Processing of FPR Photoplates

(Ed. Note - Holographers reported that they have obtained good results using their standard chemistries, rather than the following Russian formulations)

D e v e l o p e r	
HYDROQUINONE	25 g.
4 - METHYL - 1 PHENYL 3 PIROSOLYDON (4 METHYLFENIDON) $C_{10}H_{12}N_{20}$	1.5 g.
SODIUM SULPHITE, ANHYDROUS Na_2SO_3	193.8 g.
POTASSIUM HYDROACID KOH	22 g.
POTASSIUM METOBORAT $2KBO_2$ x $2.5H_2O$	140 g.
POTASSIUM BROMIDE KBr	20 g.
BENZOTRIASOL $C_6H_5N_3$	0.1 g.
WATER UP TO 1,000 ml.	

F i x e r	
SODIUM THIOSULPHATE $NA_2S_2O_3$ x 5 H_2O	300 g.
SODIUM SULPHITE NA_2SO_3	26 g.
WATER UP TO 1,000 ml.	

BLEACHES

BLEACH 1:

B l e a c h 1			
A		B	
$K_2Cr_2O_7$	30 g	$CuSO_4$ x 5 H_2O	100 g
H_2SO_4	30 ml	KBr	100 g
WATER UP TO 1,000 ml.		WATER UP TO 1,000 ml.	

Before use, 2 parts solution A are mixed with 1 part of solution B and the mixture obtained is diluted with water 1:6 ratio.

B l e a c h 2	
$K_3Fe(CN)_6$	8 g
KBr	7 g
WATER up to 1,000 m	

Bleach 2 is used without dilution. Usually we use VRP Developer and Bleach 1, but the best holograms were obtained with the use of bleach by holographer and author N. PHILLIPS.

PFG-03, PFG-03-M Holographic High Resolution Photographic Plates

PFG-03 HOLOGRAPHIC HIGH RESOLUTION PHOTOGRAPHIC PLATES are designed to record holograms in contrary beams by Denisyuk method using continu- ous or pulsed laser emission in red spectrum (for instance, helium-neon, argon or ruby laser).

PFG-03 photographic plates are used to produce copies of holographic three-dimensional portraits of people and animals; to record artistic holograms of different subjects for demonstrating them in usual light; to register three-dimensional images of moving subj ects and rapid processes; to do interferential non-destructive testing of important parts of machines, motors, and other equipment; to manufacture materials for image recognition systems; to record information holographically frame by frame; to manufacture holographic optical elements.

Holographic Data PFG-03

Diffraction Effectiveness at spatial freq. of 5000 mm-l

- when exposed by helium-neon laser (633 nm), %: more than or equal to 30

- when exposed by ruby laser (694 nm) in free-running mode (with pulse duration of 3*10-6 c), % : more than or equal to 45

Holographic Sensitivity:

- when exposed by helium-neon laser according to counter directed scheme, 11m2: less than or equal to 40

- exposed by ruby laser, 11m2: less than or equal to 70

Average Size of Microcrystalls, nm 10 - 15

Processing of PFG-03 Photoplates

Process	Time (min.)	Temp (C)
Hardening	2-3	18 + 2
Developing	10 -15	20 + 0.5
Washing in running water	0.5 - 1.0	15 - 20
Fixing	3.0 + 1	18 - 21
Washing in running water	10.0 + 1	15 - 20
Processing in 0.8% solution of moistener OP-7 (1 g/l)	2.0 + 0.1	17 + 3

Notes:

Drying in aqueous solutions of ethyl alcohol with increasing concentration 50, 75, 96% is used to provide homogeneity.

It is admissible:

a) to process photoplates without fixing

b) to carry out hardening after developing.

Chemical and photographic processing is carried out in the solutions with composition shown in the following tables:

Hardening solution

Formalin 40% (Formaldehyde)	10.2 + 0.5 g
Potassium bromide	2.0 + 0.005 g
Sodium carbonate anhydrous	5.0 + 0.1 g
Distilled water	up to 1 l

Developing solution

Sodium sulfite	100 + 1
Hydroquinone	5.00 + 0.05
Potassium hydroxide	5.00 + 0.005
Ammonium rhodanate	4.5 + 0.5
Distilled water. ml	up to 1000

Fixing solution

Sodium thiosulfate crystalline	250 + 1
Sodium sulfite pyro	30.0 + 0.1
Distilled water, ml	up to 1000

For preparation of hardening solution potassium bromide and sodium carbonate are dissolved in 600 ml of distilled water, then formalin is added, and the total volume is in- creased to 100 mL 400 ml of distilled water are added to 15 ml of the concentrated developer before use of developing solution,

PFG -03 C Photographic Plates For Color Holography

These plates are designed to record holograms in contrary beams by the Denisiuk method in the range of wave lengths in 457 nm, 514 nm and 633 nm, The holograms are restored by white light.

Diffraction Effectiveness, %

(lambda + 457 and 514 nm): not less than 30

(lambda + 633 nm) : not less than 30

Holographic Sensitivity when exposed by laser ac-

cording to counter directed scheme, providing diffraction effectiveness 'not less than 30% 11m^2:

- when exposed by argon laser (lambda = 457 and 514 nm): less than or equal to 50

- when exposed by helium neon laser (lambda + 633 nm): less than or equal to 50

Dichromated Gelatin (DCG) Recording Materials for Holography

(information provided by Lasart Ltd. and Holocrafis)

DCG has the highest index of refraction of any emulsion used in holography. Therefore, holograms recorded on this emulsion create the brightest and most easily viewable images under a variety of lighting conditions. This makes it an ideal material to use for commercial displays, art and giftware. In addition, DCG produces little scatter in blue light, making it valuable material to use when manufacturing precision optical components, such as HOEs.

Making Plates

This emulsion consists of Ammonium Dichromate or Potassium Dichromate, gelatin and water. Dichromates are available through chemical supply houses. Gelatin is easily obtained from gelatin manufacturers, by the balTel. Recording plates can be made by coating a piece of glass with a uniform thickness of the liquid emulsion using standard application methods, such as spin coating. To obtain the best results, this should take place in a dust free environment.

DCG is one of the easiest emulsions to work with and high quality holograms can be consistently produced once manufacturing variables are identified and controlled. The major variables to be aware of are: the concentration of dichromate used in the emulsion, the "hardness" of the gelatin used in the emulsion, and the temperature and amount of humidity that the emulsion experiences once it is coated onto the recording plate.

Mastering and Processing

Mastering set-ups depend on the size and complexity of the particular job. A single beam "Denisyuk" type of set up is most commonly employed when recording small objects, such as miniature models. Due to the ease of preparing a shot, mastering charges for dichromate production runs are often comparable to, or lower than, mastering charges for other types of holograms.

The laser power necessary to expose DCG emulsions varies according to the formula used, but a rule of thumb is that the more dichromate in the emulsion, the shorter the exposure. DCG is blue green sensitive -- the shorter the wavelength, the more sensitive the emulsion becomes. Most holographers working with DCG use 5 watt Argon lasers, but small holograms can be made with smaller Argon (40 mw) or Helium Cadmium lasers.

Color control is somewhat limited in DCG since exposures are made using the shorter wavelengths of light. Some holographers shoot wide band, which results in a white or silver tone image. Others shoot narrowband, which often produces a gold or bronze colored image. However, some laboratories have perfected multi color production techniques by selecting certain wavelengths and painting the subject matter to match.

Processing is quite simple - starting with a fixer, then a water rinse, then drying with isopropyl alcohol. After processing, one must isolate the emulsion from atmospheric moisture and direct contact from water. This is done by laminating another piece of glass over the emulsion and sealing the sandwich with an appropriate glue. Otherwise, the emulsion will dissolve and the holographic image will disappear.

Mass Production

It is very cost effective to produce short runs of DCGs, however large runs are more expensive due to the amount of time consuming, hand labor used throughout the process; especially in the sealing and finishing stages. Many production facilities mass-produce DCG holograms by repeatedly "remastering" each shot using the original model or subject matter. Higher quality results can usually be obtained by copying from a master hologram.

The most popular finished product is glass discs (they can be more easily sealed) which are used as watchfaces and as jewelry items. Other companies make produce ready to frame plates for wall decor and executive gifts. In the past, manufacturers have attempted to mass pro- duce DCGs by laminating them in plastic, but holograms sealed in this way tend to fade rather quickly.

Limitations

The major drawbacks to using DCGs for a wide range of commercial applications is the fact that they are thick and fragile, since they are sealed in glass. Some manufacturers have experimented with using plastics to seal their dichromate products, but since most plastics are permeable, these tend to fade over time. This usually makes them impractical to use in product packaging and publishing. DCG holograms display more image depth than embossed holograms, but less than those produced on silver halide.

Photoresist Recording Materials for Holography

As mentioned throughout this publication, most commercial holographic applications utilize embossed holograms. A crucial step in the production of embossed holograms is transforming the microscopic optical recording into a useful production tool. To accomplish this, holographers record their images onto a high resolution photosensitive material called photoresist. Once an image is recorded onto a photoresist emulsion, it can be then be processed in a manner suitable for mechanical mass replication.

Although most holographers shooting holograms on photoresist materials prefer to use pre coated, ready-made recording plates, some holographers might want to (or need to) coat and process their own plates. The first part of the following article outlines the procedure for making plates 'from scratch" using Shipley photoresist emulsion. Shipley is the main manufacturer of these emulsions and sells their product in liquid form mainly to high volume users. Although it is mostly used by the microelectronic and semiconductor industries, their photosensitive formula has proven suitable for holographic recordings.

After the aforementioned information Fom Shipley, we - will provide a description of the pre-made products available from Towne Technologies - a major US manufacturer and distributor of photoresist recording plates. Towne is not the world's only supplier of these products, however, they have extensive experience serving the holographic industry and their products have been used successfitly by many embossed hologram production facilities.

Shipley Products for Holography

(reprinted with permission of Shipley Co. Inc.)

Intended Uses

The following instructions cover the use of MICROPOSIT S1800 SERIES PHOTO RESIST for ho- lographers interested in coating and processing their own plates. MICROPOSIT S 1800 SERIES P H O T O RESIST is a positive working photoresist system optimized to sat- isfy industry requirements in advanced optical lithogra- phy and related holographic applications. MICROPOSIT S1800 SERIES PHOTO RESIST is a replacement for 1300/ 1400 type resists providing an alternative to Cellosolve acetate as the casting solvent.

Step 1. Dehydration Bake to Prime Substrate

To obtain maximum process reliability, bake all substrates immediately prior to coating at 200°C for 30 minutes. Cool to 18°-25°C ambient before coating.

For maximum resist adhesion to all semiconductor surfaces, vapor phase priming with MICROPOSIT PRIMER is recommended. For liquid phase priming use MICROPOSIT PRIMER TYPE P.

Figure 1

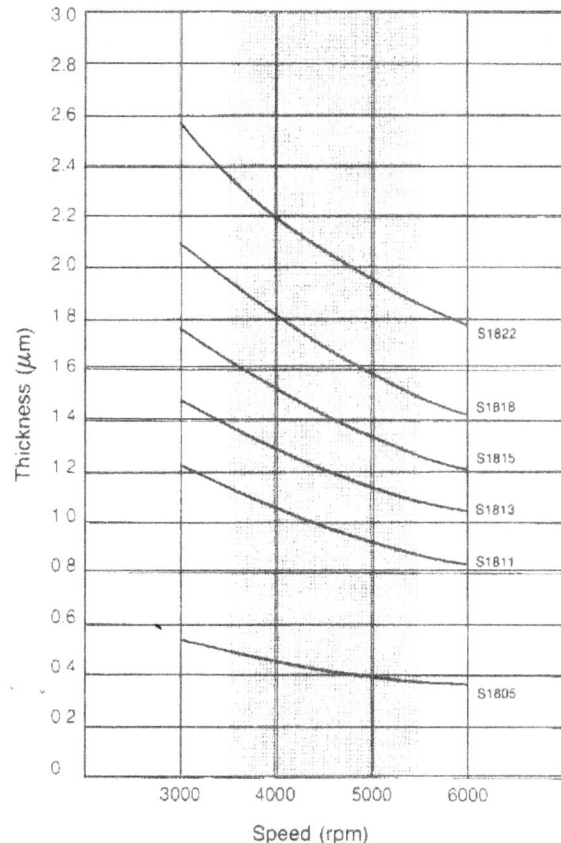

Figure 1 is a graph showing coating Ihickness for S1800 products from 3000 to 600 'pm. The shaded area is optimum spin range Step 2. Spin Coat

MICROPOSIT SI800 SERIES PHOTO RESIST is a resist designed to produce low defect coatings. The improved coating characteristics include enhancements to thickness uniformity for optimized critical dimension control across a wafer as well as wafer to wafer.

Select the appropriate S1800 SERIES PHOTO RESIST to give the optimum coating thickness at the optimum spin speed. The chart above shows typical coating thickness vs. spin speed for the standard MICROPOSIT S1800 SERIES PHOTO RESISTS.' Spin speeds between 3500 and 5500 rpm are recommended for maximum coating uniformity.

Use the following parameters to obtain maximum resist coating uniformity:

Dispense .... Static

Spread

      Static............... 2 seconds recommended

      or

      Dynamic .... 500 rpm, 2 seconds maximum

Ramp............................ Maximum acceleration

Spin .... 3500 - 5500 rpm

Spin time........................ 25 seconds minimum

MICROPOSIT S1800 SERIES PHOTO RESISTS are manufactured to give reproducible coating thicknesses. The nominal thickness your process produces may vary slightly from our published nominal thickness due to pro- cess and ambient variables. Reproducibility will be achieved when processing parameters are held constant.

STANDARD MICROPOSIT S1800 SERIES PHOTO RESISTS					
Product Type	Approximate Viscosity (cSt)	Resist Thickness (µm)			
		3000 rpm	4000 rpm	5000 rpm	6000 rpm
S1805	5	0.54	0.46	0.41	0.36
S1811	15	1.23	1.06	0.93	0.83
S1813	20	1.51	1.28	1.14	1.02
S1815	26	1.78	1.53	1.34	1.22
S1818	35	2.10	1.82	1.59	1.45
S1822	48	2.58	2.20	1.95	1.77

Table 1 (above) relates specific S1800 SERIES products to viscosity and corresponding film thickness at 3000 to 6000 rpm. The standard product name is derived from the coating thickness obtained at 4000 rpm spin speed This should be used as a guide to selection of a particular product.

Step 3. Edge Bead Removal

MICROPOSIT EBR-10 is recommended for removing resist build-up occurring at the edge of wafers during spin- ning. The formulation has been specified to be free of Cel- losolve acetate, acetone, and xylene. EBR-10 can be used with most coating equipment designed to include an edge bead removal process.

Step 4. Soft Bake

The following baking parameters are optimum:

Oven Forced air convection (do not use nitrogen)

Temperature

Time 30 minutes (after recovery to operating temperature)

Cool To ambient

In-line track baking equipment should be adjusted (speed! temperature) to yield photospeed, contrast and unexposed resist loss through developing equivalent to or better than those obtained using the above forced air con- vection conditions.

Step 5. Expose Plate

MICROPOSIT S1800 SERIES PHOTO RESISTS can be exposed with light sources in the spectral output range of 350 - 450 nm contained in commercially available ex- posure equipment.

Standing Waves and Exposure Requirements

The resist film thickness on a reflecting surface will de- termine the intensity and location of the standing waves in the film. Exposure time is minimized and line size con- trol in the photoresist will be optimum in a resist film thickness that is an odd mUltiple of one quarter the expos- ing wave- length By choosing this thickness a valid com- parison of relative exposure energy required versus wave- length can be made. The exposure energy threshold (Et) is the mini- mum dose required to remove lOO% of the pho- toresist in a large open area on a wafer. Contact the manu- facturer for additional data regarding exposure times.

Step 6. Develop

The following developers are designed for use with MICROPOSIT S1800 SERIES PHOTO RESISTS.

Ready to Use Developers:

MICROPOSIT 352, MICROPOSIT 353, MICRO- POSIT 354, MICROPOSIT 355, MICROPOSIT MF-312 DEVELOPER CD-27, MICROPOSIT MF-314, MICRO- POSIT MF-316, MICROPOSIT 452 , MICROPOSIT 453, MICROPOSIT 454, MICROPOSIT 455, MICROPOSIT DEVELOPER CD-30

Custom solutions and ready to use developer formula- tions are manufactured to tight analytical and functional specifications, and are therefore recommended for maxi- mum process control. These developers will also avoid costly production line downtime caused by in-house dilu- tion errors.

Note - Contrast data is available from the manufacturer.

Step 7. Hard bake

A hard bake is recommended for all photo levels to op- timize process reliability in wet etching and maximize

Oven	Forced air convection (do not use nitrogen)
Temperature	100°-1 10°C
Time	30 minutes

selectivity in dry processing steps. The following baking parameters, or equivalent, are recommended:

For maximum thermal stability, blanket deep UV ex- posure or plasma treatment prior to hardbake is recom- mended Either of these procedures will assure image in- tegrity through plasma etching, high dose implanting and other high temperature process steps.

Where taper etching is desired, using dry processing,

the thermal properties of Sl80C allow for the appropriate resist profile adjustment through controlled hard bake.

Step 8. Etch / Ion Implant

MICROPOSIT S1800 SERIES PHOTO RESIST can be used with all common semiconductor wet etchants as well as with plasma and ion implant processes.

Step 9. Strip

MICROPOSIT S1800 SERIES PHOTO RESISTS can be removed using MICROPOSIT REMOVER 140. MICROPOSIT REMOVER 1112lA, MICROPOSIT REMOVER 1165, or oxygen plasma. Refer to the individual remover data sheets for specific processing instructions, specifications, and other product information.

Properties as Delivered

MICROPOSIT S1800 SERIES PHOTO RESIST is filtered to 0.2 ,urn absolute. Each container is date coded.

MICROPOSIT 51800 SERIES PHOTO RESISTPROPERTIES (For S1805, S1811, S1813, S1815, S1818, S1822:

Film thickness reproducibility for S1805, S1811, S1813, S1815:	+ 200Å with respect to a reference
S1818:	+ 300Å with respect to a reference
S1822	+ 500Å with respect to a reference
Filterability constant, n/nO	0.0075 maximum
Water content	0.5% maximum
Index of refraction	1.64 @ 6328 Å 1.68 @ 4360 Å
Na content	< 1 ppm
Fe content	< 1 ppm
Type of solution	solvent base propylene glycol monomethyl ether acetate
Flash point (closed cup).approximate	46°C
TLV rating*	100 ppm

* Rating is for propylene glycol monomethyl ether

Handling Precautions

CAUTION: Combustible solvent mixture. Harmful if swallowed. Use adequate ventilation. Avoid breathing of vapors. Keep away from heat. sparks. Or open flame Avoid contact with skin and eyes. Handle with care. Wear chemical goggles, rubber gloves and protective clothing.

Toxicological and Health Advantages

Ethylene glycol monoethyl ether acetate (also known as 2 ethoxyethyl acetate or Celloso ve-R- acetate) is used as a diluent solvent for most conventional positive photore-

sists. The solvent used in MICROPOSIT S1800 SERIES PHOTO RESIST is propylene glycol monomethyl ether acetate. It has been demonstrated in toxicological studies reported in the NIOSH Current Intelligence Bulletin 9, (512183) that the propylene glycol derivatives contained in MICROPOSIT S1800 SERIES PHOTO RESIST do not demonstrate the adverse blood effects and reproductive effects that the ethylene glycol derived ether acetates do. These significantly lower health risks are reflected in the most current American Conference of Governmental Industrial Hygienists (ACGIH) TLV values which lists: TLV ethylene glycol monoethyl ether acetate 5 ppm propylene glycol monomethyl ether - 100 ppm. Material Safety Data Sheet is available upon request.

Equipment

MICROPOSIT S1800 SERIES PHOTO RESIST is com- patible with most commercially available photoresist pro- cessing equipment. Recommended compatible materials include stainless steel, glass, ceramic, unfilled polypro- pylene, high density polyethylene, polytetrafluoroethylene, or equivalent materials.

Disposal

• MICROPOSIT S1800 SERIES PHOTO RESIST should be disposed of according to Shipley Waste Treatment Procedure WT 78-13 (include with other solvent wastes). Contact your Shipley Technical Sales Representative for details.

Storage

Store in dry area at 50°-70°F (10°-21 0c) in closed original containers away from light, oxidants, heat. sparks. and open flame,

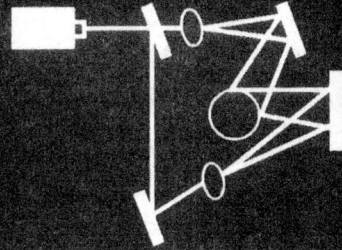

Towne Products for Holography

Towne Technologies of Somerville, NJ (USA) is a producer of photoresist plates for use in holography and other fields. Towne supplied the following description of how its plates are manufactured.

The materials that are used to produce the large Iron-Oxide coated holographic plates are purchased to specification requirements of the microelectronic, semiconductor and printed circuit board industries for which it was originally designed.

For example, an optical grade polished (both sides) soda lime, float glass substrate 24" x 32" x 0.190" (609.6 x 812.8 x 4.83 mm) has a flatness tolerance of 150 x 10-6 inches per linear inch (flat to within 150 microinches per linear inch.)Before it is acceptable for FeO_2 coating, each piece of glass is cleaned and surface-polished to ensure that the slightest surface imperfections and even microdust particles are removed.

The pure Iron-Pentacarbonyl used has a controlled specific gravity of 1.44-1.47 @ 20°C and its deposition is carried out in Class 100 clean room conditions. After the FeO_2 deposition, 100 Angstons thick, the plate is inspected, for integrity of coating. Pinholes are marked, and when the 24" x 32" plate is cut into final working plates, the pinholes are avoided.

The plates are dried in a thermostatically-controlled Class 100 environment, then cleaned and inspected again. From that time to the deposition of the photosensitive coating, nothing is permitted to contact the surface of the plate. The Microposit-S-1400-30 highly sensitive photoresist is used for coating because it is specifically formulated to be striation-free. On plates up to 15" X 15", (381 mm square) the photoresist is applied by a spinning process to a final standard thickness of 1.5 +/- 10% micrometers subsequent to a 0.2 micrometer filtration process.

The success of the iron-oxide coating is owed primarily to two inherent characteristics of FeO_2 coating, e.g. the iron oxide coating effectively absorbs any laser light that may be transmitted through the photosensitive coating. This virtually eliminates light backscatter and the possibility of damage to the primary image. Second, and possibly more important, the iron-oxide coating greatly increases the adhesive quality of the photosensitive coating, thus ensuring the integrity of the imaging and electroplating processes to follow.

Sal LoSardo, the sales representative from Towne, says that of the people who buy their plates, about 30% use them for holography, and "the number is growing." A par- ticular area of expansion has been the Far East and China.

Towne has not needed to make any specific modifications to its plates to accommodate holographers, except for the need to install two "oversize" spinners. Plates for the electronic industry are usually dip coated when the size is above 7" x 7". Holographers require the smoother finish of spin coated plates. Prior to 1984, Towne could only spin coat up to 7" x7"s, but now they can spin up to 15 inch square or 18 inch octagonal plates.

Photopolymer Recording Materials for Holography

There are two major producers of photopoly mer materials, DuPont and Polaroid. (Note that the former's formal name is E.I. DuPont De Nemours & Co.) Following is a description oftheirproducts.

Polaroid Products for Holography

Polaroid continues to supply the industry with photopolymer holograms. Although Polaroid's photopolymer recording material is not for sale, the company offers a full range of in house production services for its clients. As of Fall 1996, the company was heavily involved in producing HOEs for industrial clients with long term sup- ply requirements, such as Motarola. These particular HOEs, called ImagixTM Holographic Reflectors, are intended to improve the brightness and contrast of reflective and transflective LCDs by concentrating the emerging light from the LCD at a narrow viewing angle, while redirecting glare. Motarola currently uses these particular HOEs to enhance the viewability of its Envoy TM line of wireless communicators.

Polaroid is still producing its Mirage TM Holograms for commercial users of display holograms who prefer photopolymer holograms over embossed. The company offers a line of stock images and welcomes all inquires for custom work.

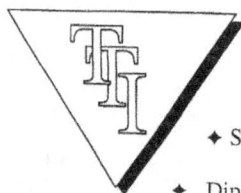

Stock holograms: Choose from selection of already-created holograms. These holograms are available on panels of pressure-sensitive (self-adhesive) labels.

Custom holograms: Hologram images that are determined by the customer. Choose the image or model you would like to capture in a hologram, and Polaroid will create it.

For more information call 1-800-237-5519 or fax 617-386-8857.

DuPont Products for Holography

DuPont manufacturers photopolymer recording films and replication equipment which it sells directly to authorized replication facilities. Currently, there are only a handful of such operations worldwide, however they are capable of high volume production. DuPont also provides selected holographers materials for research and development work.

DuPont's line of holographic photopolymer films features OmniDex-706, a blue-green reflection material, currently available in 500' x 12" rolls. Two other OmniDex films are also available: HRF-700 and HRF- 600. HRF-700 is a high Dn blue-green sensitive reflection film and HRF-600 is a blue-green sensitive high Dn film for recording transmission or reflection holograms. In addition, DuPont also manufactures full-color mastering and replication films (see chart for details).

Processing of DuPont's photopolymer consists of a few steps:

- holographic exposure

- followed by UV or white-light exposure

- optional lamination with a color-tuning film

- and heating to brighten the image. The finished hologram is stable to environmental conditions such as heat and moisture.

High-volume replication and processing equipment is also available from DuPont for use with OmniDex films. The DuPont OmniDex Replicator operates much like a printing press, using a cylindrical scanning geometry for continuous production of holographic images typically at five square feet per minute.The DuPont Omnidex Laminator is available with the replicator for color-tuning of rolls or sheets of images, and for downstream conversion operations such as application of pressure-sensitive adhesives or anti-scratch cover films to rolls or individual sheets of holograms. Equipment is also available for heating of large rolls of holograms.

For further information on DuPont materials or replication equipment, contact DuPont Holographic Materials, Experimental Station, Po. lJox '803"52, Wilmington, DE 19880-0352 USA. Tel. 302-6954893. Fax: 302-695 4635.

Table 1: DuPont Photopolymer Products			
Film	Thickness (μm)	Type of Film	Film Speed* (mJ/cm²)
HRF-150X001	38	Blue/Green Transmission	514nm: 100
HRF-600X001	10 or 20	Blue/Green Transimiison high Δn	514nm: 50
HRF-700X001	10 or 20	Blue/Green Reflection high Δn	514nm: 15
HRF-800X	20	Full Color Reflection high Δn	476nm: 11 514nm; 10 532nm: 7 633nm: 10 647nm: 12
HEF-850X	20	Full Color Mastering Film** (removable Mylar base)	476nm: 13 532nm: 9 647nm: 17
OmniDex 706	20	Blue/Green Reflection	514nm: 25

OmniDex 706 comes with equal amount of Color Tuning Film (CTF)
For Quantity purchases we also offer the following 500 ft rolls :(500 ft roll is 72,000 sq. inches with 12 inch wide imaging area):
500 ft. roll OmniDex 706 with 500 ft roll CTF

* Single beam exposure with front surface mirror for reflection films, or single beam copy of 50% Diffraction Efficiency master for transmission films
** Film has enhanced glass adhesion premitting removal of cover sheet and Mylar base, leaving only photopolymer layer and nonbirefringent protective layer

HUGHES POWER PRODUCTS
Hologram Replication Services

Exposed

[Image copyright © by Ron and Bernadette Olson. Copied by HPP into Du Pont OmniDex® Hologram Recording Film.]

Custom Production Services for Commercial, Industrial, or Artistic Applications - We can mass-produce high quality reflection holograms using your master image or our origination. Cost effective high volume or short runs; as low as $0.18 per image! Standard sizes range from 1"x1" to 12" x12".

Open Edition Holographic Prints - We produce and distribute a line of affordable wall decor hologram images by world-renown artists. Ready-to-frame or pre-framed. Wholesale catalog available.

Call us today (310) 414-7086 or visit our website (powerholos.com) for more information.

Join SPIE's International Technical Holography Working Group

...and be part of a worldwide communication network

Benefits include:

• A semi-annual copy of the Holography newsletter.

The publication offers a networking opportunity for holographers working in different fields and different parts of the world. Articles focus on the latest work from academia and industry, news about the holographic community, and announcements of meetings, courses, and other events of interest.

• Holography Online `info-holo-@spie.org`

Now there's an easy way to reach your colleagues around the world—instantly. SPIE's Holography Listserver is an automated e-mail server that connects you to a network of engineers, scientists, vendors, entrepreneurs, and service providers.

We make it possible for you to stay informed...

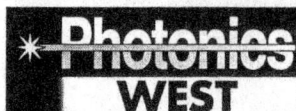

Photonics WEST

February 8–14, 1997 • San Jose Convention Center • San Jose, CA

offers conferences on:

3D Displays
- Practical Holography XI
- Holographic Materials III
- Stereoscopic Displays ansd Applications VIII
- The Engineering Reality of Virtual Reality 1997

For membership information and application, or related program and proceedings information contact:

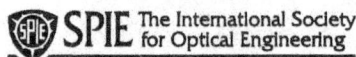

SPIE The International Society for Optical Engineering

SPIE is an international technical society dedicated to advancing engineering, scientific, and commercial applications of optical, photonic, imaging, electronic, and optoelectronic technologies. Its members are engineers, scientists, and users interested in the development and reduction to practice of these technologies. SPIE provides the means for communicating new developments and applications information to the engineering, scientific, and user communities through its publications, symposia, education programs, and online electronic information services.

SPIE International Headquarters • P.O. Box 10 • Bellingham, WA 98227-0010 USA • Phone 360/676-3290 • Fax 360/647-1445
E-mail spie@spie.org Telnet/FTP spie.org • Dialup 360/733-2998 • World Wide Web http://www.spie.org/

Shipping Address • 1000 20th Street • Bellingham, WA 98225-6705 USA

SPIE in Europe • Box 223 • N-5095 Ulset, Norway • Phone 47 55 18 94 20 • Fax 47 55 19 27 80 • E-mail spie@spie.no

SPIE in Japan • c/o O.T.O. Research Corporation • Takeuchi Building • 1-34-12 Takatanobaba • Shinjuku-ku, Tokyo 160 • Japan
Phone (81 3) 3208-7821 • Fax (81 3) 3200-2889 • E-mail otoresco@gol.com

SPIE in Russia/FSU • 12, Mokhovaja str. • 121019, Moscow, Russia • Phone/Fax (095) 202-1079 • E-mail edmund@spierus.msk.su

7

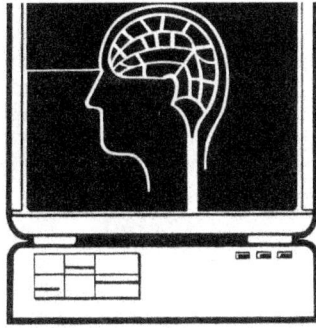

Industrial Holography

In this chapter we give an overview of two areas of industrial holography: Holographic Interferometry, a major tool for NDT (Non-Destructive Testing) and direct-write laser ablation, for the quick fabrication of CGHs (Computer Generated Holograms) and DOEs (Diffractive Optical Elements).

Holographic Interferometry: A Brief Overview of Current Applications and Technology

by Howard Fein

New methods of Non-Destructive analysis are becom- ing important and critical tools in the characterization of materials and sophisticated mechanical configurations. The unique properties of laser light have offered new ways for technologists to look at structures and materials, and physical phenomena and processes. Applied laser-based nondestructive test techniques can reveal significant and hitherto unachievable data concerning the actual dynamics and structure of a body under study. There is a virtually unlimited scope as new applications are discovered.

Primary among these technologies are methods of Holographic Interferometry. Derived from well known holographic imaging techniques, this analytical measurement ability has found wide- spread use in addressing a broad range of appli- cations during the last thirty years.

There is a broad spectrum of applications for holographic methods covering virtually every aspect of static and dynamic structural analysis.

This includes, but is not limited to applications in Vibration and modal analysis; Structural analysis, composite materials and adhesive testing; stress and strain evaluation; and flow, volume, and thermal analysis.

The mechanisms of holographic interferometry are well understood and various methods have been developed, but all are based on simple holographic optical configurations such as that illustrated in Figure 1. The methods of recording the holographic data vary: from silver halide photographic emulsion films and plates to

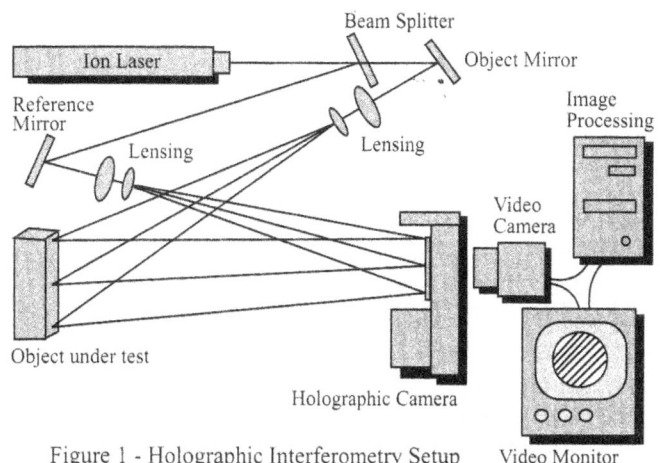

Figure 1 - Holographic Interferometry Setup

advanced photopolymer plates, and even non-photographic CCD based "electronic" or "TV" holography. Photorefractive crystals are now being used to record and store multiple high volume holographic images and data as well.

Each of the recording methods is uniquely applicable to a range of analytical applications and the type of testing or data generation determines which would be most useful and efficient. Diffraction efficiency, contrast, resolution, and fringe precision are all considerations in the applicability and suitability of holographic interferometry to any given analytical problem.

There are three basic methods of applied holographic interferometry. Real Time., Multi-Exposure, and Time Av- erage. The common element in all three cases is the laser which enables the production of holographic information. Interferometric nondestructive testing can be accomplished employing both continuous wave and pulsed lasers of virtually all recordable wavelengths. Pulsed lasers can be synchronized with motion and can also be used to record holograms of extremely fast transient phenomena. CW lasers are ideal for real time studies of displacement and motion. In some cases the frequency of the laser light is purposely varied during or between exposures to make comparative interferograms.

Real Time holography describes the superposition of a hologram of an object over the object itself while it is being subjected to some small stress. This is done in order to observe the effects of minute changes in displacement on, or in, the structure or volume of the object as stress affects it in real time. Multi-Exposme Holography describes a hologram created as a result of two or more exposures. The first exposure is a hologram of an object in an undisturbed or base state. Subsequent exposures on the same image recording medium are made after the object is subjected to some finite stress. The resulting image records the difference between the two states. The third technique, Time Average Holographic Interferometry involves the creation of a hologram while the object is subjected to some periodic displacement or stress. The recorded image shows the displacement effects of the average motion or stress. All these techniques reveal aspects of the geometry and magnitude of the stress induced displacements in the structure being studied. An important key is that the stresses are very low level and nondestructive in nature.

Electronic holographic analysis uses the capabilities of advanced image processing techniques in a computer captured image map which compares and calculates the phase differences between successive images on the CCD of the recording camera. The result is a computerized image of the holographic interferogram derived like a standard "photographic" hologram.

The basis of applied holographic interferometry is the fringe patterns of bright and dark lines generated on the holograms which appear superimposed over the object under study. An example is shown in Figure 2. These interference fringes define isobars of stress or vibration induced displacement and create a map of the resulting to-

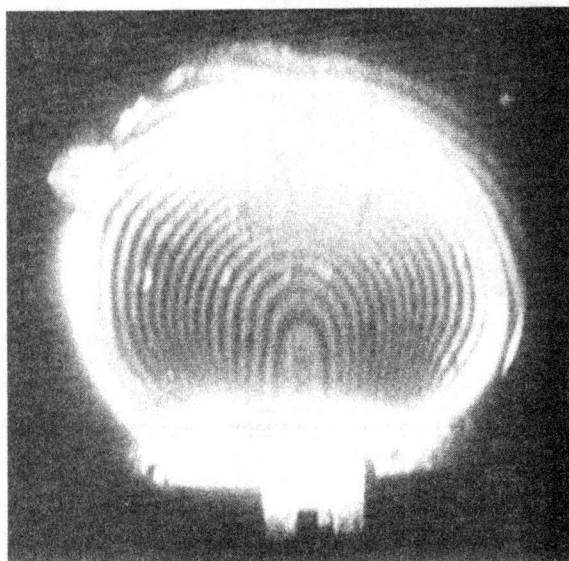

Figure 2

pography of the surface of the object. Specific geometries and symmetries in the holographic fringe patterns map the behavior of the structure by defining characteristics of motion and displacement. In the case of fiow and volume thermal analysis the fringes can also define changes in the index of refraction or the pathlength of the transparent volume being investigated.

The magnitude of the displacements represented by the holographic fringes is very small and in virtually all cases non-destructive to the object under test. Each fringe in a real-time and multi-exposure hologram denotes an isobar of displacement of one half wavelength of the illuminating laser light. For time-average holograms, displacements can be related to the coefficients ofa Bessel function which generalizes the sinusoidal excitation. Both qualitative and guantitative information can be derived from these fringe patterns.

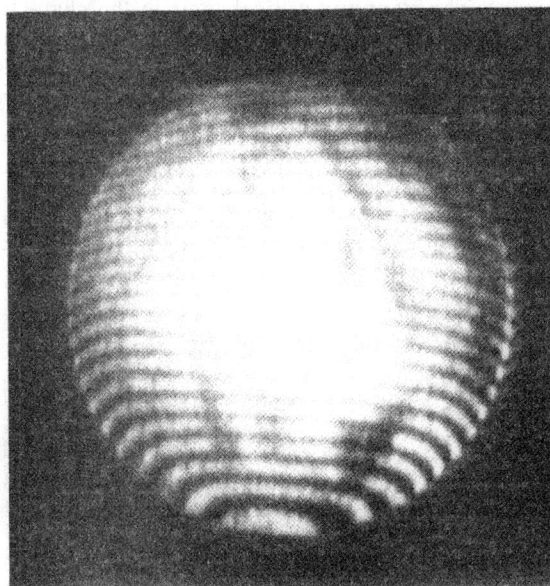

Figure 3

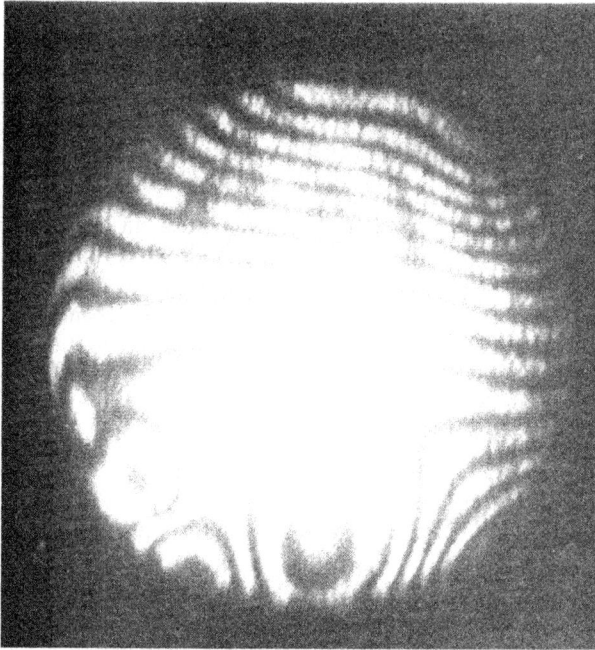

Figure 4

Some general areas of application are illustrated by the following examples of real analytical data.

Analysis of bonded structures is applicable in a wide range of considerations. This includes complex composite materials studies, bonding and adhesive analysis, thermal studies, rigid and compliant material examination, as well as other applications.

Holographic data for a bonded structure is shown in Figure 3. Defects in the structure are seen as irregularities in the fringe patterns.

Studies of structure borne noise, vibration, and dynamic modal analysis are particularly suited to holographic applications. Figure 4 shows holographic images of the modal patterns on a resonant structure. The methods are straight-forward and can offer data generally very difficult or impossible to derive from other more conventional test methods. Information pertaining to the structural transmission, construction, and configuration is derivable from this sort of data.

There are many current and potential applications for this technology as new methods of structural and materials analysis are sought. Nondestructive test and evaluation is becoming increasingly more important in a wide range of technology areas. It can augment and aid advanced computer modelling techniques for structural analysis by showing the true behavior of anything from the smallest micro-circuit component to highway bridge structures.

There are a number of ways for this technology to be exploited when it is needed. Certainly full capital acquisition of the necessary laboratory technology and personnel is a possibility but it has an admittedly high initial cost. There is also contract consultation with organizations which already have the facilities and offer services with specialized capabilities for R&D, engineering, and even production testing.

In order to find out how to solve problems using holographic interferometry one can make use of industry guides like Holography MarketPlace, Photonics Spectra, SPIE, etc. Also the Internet and the World Wide Web are becoming the newest and most comprehensive real-time way to find information.

For further information about this topic, contact the author:

Mr. Howard Fein President/Chief Scientist in care of:

Polaris Research Group, 24400 Highland Road, Richmond Heights, Ohio 44143 Phone: 216-383-9480 FAX: 216-383-9488 Email: PolarisRG@aol.com

Howard Fein is the President and Chief Scientist for Polaris Research Group in Cleveland, Ohio. He formerly managed the Westinghouse Naval Systems Division Electro-Optics Laboratory in Cleveland, Ohio where, for fourteen years, his work had centered on Holographic Interferometry and NDT methods applied to various underwater weapon and sensor systems. He has authored over 20 other articles and papers on these subjects, holds several patents, and is a member of the LIA, the SPIE and Holography Working Group, as well as the OSA. Mr. Fein directs a continuing program of consultation and development of industrial, commercial, and defense related applications, as well as commercial services of holographic analytical methods for vibration, motion, and structural analysis.

Using Direct Write Laser Ablation Techniques to Fabricate CGHs and DOEs

by G. Behrmann and M. Duignan

Holograms can be designed and fabricated without the actual interference of an object wave and reference wave. Computer programs have been used for more than thirty years to calculate hologram patterns based on diffraction theory and Fourier optics. Originally, these computer generated holograms (CGH) were produced by pen plotters and photo-reduced to create transparencies suitable for image reconstruction. Early CGHs were referred to as binary or binary amphrude due to the fact that the hologram consisted of only transparent and opaque regions. One of the most successful applications ofthese holograms was the testing of optical surfaces.

However, advances in semi-conductor fabrication techniques resulted in a new class of CGHs known as kineforms, now more commonly referred to as diffractive optical elements (DOEs). DOEs produce interference patterns through phase variations across the element. These phase variations are the result of precisely fabricated surface relief structures in a transparent material. DOEs of- fer higher diffraction efficiencies than binary amplitude CGHs and have found application in laser systems, traditional imaging applications, and optical interconnects.

The most common method for fabricating surface relief DOEs requires the generation of a mask set which is used to sequentially expose and develop photoresist that is spun on a glass substrate until the desired structure is formed. This process, though well developed, is costly and time consuming, requiring weeks or more to complete the fabrication cycle.

An alternative to multiple mask and exposure makes use of lasers to directly expose the resist. Control of laser intensity allows for controlled exposure which results in a surface relief profile when the resist is developed. In both cases, the pattern can be transferred into the glass substrate by etching techniques. In addition, the photoresist patterns can be used as masters for polymer replication if mass production is required.

One company, Potomac Photonics, has developed a new technique for fabricating surface relief DOEs. By utiliz- ing a compact, high repetition rate ultra violet laser and a process known as laser ablation, a DOE surface relief is laser micromachined directly into polymer films without the need for masks sets or chemical processing of photo- resist materials. The company claims this process is easier to use than existing methods and is capable of producing functional DOEs in minutes to hours.

Potomac Photonics, Inc. primarily manufactures compact, high repetition rate (2000 Hz) ultraviolet lasers. These lasers operate at 248 nm with pulse energies oftens of microjoules and average powers of tens of millliwatts. They are used in the micromachining workstation the company has designed for DOE and CGH production. The workstation allows a user to quickly sketch the shape of an exposure area, set exposure levels and view the exposure process. A small aperture is imaged onto the work surface with uv transmitting microscope objectives and is capable of producing spot sizes that are adjustable between 1 and 100 microns.

Laser energy levels reaching the work surface can be adjusted from 0 - 15 Jcm^2 to accommodate processes ranging from exposure of sensitive photoresist to ablation of very hard materials. Polymers such as polyimide and polycarbonate are well suited for the fabrication of DOEs by uv laser ablation as they absorb strongly, allowing for precise depth control.

By adjusting laser energy and the number of pulses, it is possible to directly produce multi-level DOEs. The number of phase levels determines the overall diffraction efficiency of the optical element. Currently, optical elements with 4 phase levels produced with this system exhibit overall efficiencies of 45 - 50 %, with most of the loss due to wide angle scatter.

To fabricate DOEs or CGH with this system, a "text (.txt)" file is generated which represents the phase height at each x and y location of the element. These files can be received at the fabrication site electronically for users who are off site. The text file is read into a bit map translating program which displays the hologram and generates motion and laser firing programs for the fabrication step. The laser spot and energy is adjusted for so that the correct amount of material is removed for each phase level. Machined depth is verified by use of a white light interferometric microscope which can measure three dimensional profiles. Currently the system is capable of pixel sizes from 2 to 15μm in a round or square shape.

Once the machining is complete, the part is a functional element with no additional processing steps required. In addition the patterned material can be used as a resist for etching in glass substrates or as a master for electroplating reproduction. To date, this method has been most useful in rapid prototyping.

For further information, contact the authors in care of Potomac Photonics, Inc., 4445 Nicole Drive, Lanham, Mwyland 20706 Phone 301-459-3031 Fax 301-459-3034

8

Sales & Distribution of Stock Holograms

This chapter discusses the commercial development of artistic holography, especially the sale and distribution of stock holograms and related products by the giftware industry. Tips on opening your own hologram store are included.

It also includes information about sourcing stock holograms for other commercial applications.

The Commercialization of Artistic Holography

In its early stages holography remained unseen by the general public. Only scientists and researchers had access to the lasers and other specialized equipment that were needed to create and view a hologram . When methods were developed in the late 1960s that enabled a hologram to be created and viewed in more practical ways, holography" slowly left the laboratory and began a journey that has resulted in a multi-million-dollar, worldwide industry. A great portion of this industry deals with "artistic" holography, (i.e. three-dimensional images of things) which is also commonly referred to as "pictorial" or "display" holography.

During the 1970s and early '80s holograms were made by individual holographic "artists" on a one-by-one basis. The process was labor-intensive and time-consuming. Production techniques were developed through trial and error. Raw materials such as film emulsions were scarce, equipment was often homemade, and production quality often inconsistent. Unfortunately, the individuals and small companies that were capable of-making high-quality holograms generally did not have the money or marketing expertise needed to get their work into widespread distribution, so holograms were still out of view of the public-at-large.

The handful of galleries and stores that did show holo- grams proved that the public was fascinated by this emerg- ing medium. Although most holograms were treated as futuristic artworks or novelty items and were relatively expensive - the public constantly asked for more affordable ones. Enterprising gallery owners,

retailers and holographers recognized this demand for holographic merchandise, and a small industry slowly evolved. Holographers and their entrepreneurial partners began to create products rather than artworks; well connected retailers began to distribute holograms to other retailers; and hologram aficionados became customers. Everyone recognized the potential of this new industry, yet manufacturing and display/lighting problems still needed to be overcome before holograms and related products could enter the mainstream marketplace. Limited production runs kept prices high.

Over the next decade technological advances enabled holograms to be mass-produced in a variety of ways. This made it feasible for artists, technicians and businessmen to join together to create facilities dedicated solely to producing large runs of affordable, high-quality holograms. These holograms were intended for a variety of commercial applications including security, packaging and advertising - as well as products for fhe giftware industry.

Art holographers copied their most popular images onto film (which was less expensive and easier to handle than holograms produced on sheets of glass) and began to use assembly-line production methods in their labs. Whole catalogues of images soon became available, intended for sale as wall decor. Retail price points dropped considerably. Other holographers perfected methods of mass producing very bright dichromate holograms for use as jewelry. Still others concentrated on developing high-speed automated replication technologies capable of embossing holograms on very inexpensive foils and plastics - perfect for use on toys, optical novelties and paper products.

Once reliable supply lines were established, it became feasible for other companies to package and market these holograms in a variety of ways and integrate them into the nom1al chain of giftware distribution. Businesses in the United States and England quickly grew into major distributors. Film holograms were matted and/or framed and marketed as high-tech art. Holographic fashion accessories (including watchfaces, pendants and earrings) were developed. Executive gifts and desktop accessories were created. Rolls and rolls of kids' stickers were produced. New toys were invented. Holograms started to appear at national gift shows, in giftware catalogs, and in the media.

Savvy retailers soon realized that holograms and re- lated products were very popular with the buying public, and if displayed correctly, could prove quite profitable. A good display of hologrqms drew a crowd, generated customer excitement and more importantly, generated dollars! (Most giftware items sold in stores are priced at twice their cost.) Holographic merchandise spread from science museum gift shops to mainstream outlets, and even included a number of specialty stores set up to sell only hologram products.

Increased visibility created greater public awareness of the product and demand for new and better holograms. Artists added color and motion to their images. Manufacturers automated further and invented materials especially suited for holographic applications. Holograms became brighter and easier to see under typical viewing conditions. Distributors created new product lines by integrating holograms into existing merchandise. Packaging was brought up to commercial standards. Wholesalers adopted more sophisticated marketing techniques, while retailers offered a wider selection of goods.

Today, a variety of holograms are manufactured around the world for the giftware industry, with the highest concentration of factories located in North America and Europe. English holographers have traditionally domi- nated the silver halide film replication business. American holographers are actively developing photopolymer replication factories. The production of dichromate holograms seems to have slowed, while facilities capable of producing embossed holograms have multiplied significantly, especially throughout Asia. Surprisingly, very few holograms are exported from Japan.

The number of distributors and wholesalers dealing exclusively in holograms has dwindled as the market has diversified. To stay profitable several major distributors have developed their own custom images in order to target specific consumer groups. One major US distributor has developed a very successful product line based on the ever popular hologram eyeglasses with stock and custom photopolymer images as lenses. Another has developed close working relationships with the product development departments at several major retail chains, thereby ensuring longterm sales. The sale of "licensed" holographic images featuring popular sports figures, cartoon characters and movie scenes has grown steadily, while the sale of more mundane images has stagnated.

There are fewer holography specialty shops in business now than a few years ago. Those that continue to do well have increased the variety of goods that they carry and often include related optical novelties in their product mix. Sales of holographic artwork are practically nonexistent. However, more stores than ever before are carrying some sort of hologram-related product. It is not uncommon to find an inexpensive hologram item at the comer store.

The Chain of Distribution

Let's examine the chain of distribution as it typically exists for a holographic product.

The Copyright Holder

The distribution process starts with the copyright holder. Any unique work of art, including a painting, photograph, or computer-generated graphic, can be protected from unauthorized duplication (in most countries) by registering the image in the appropriate manner. In the case of holography, the original work of art is either a model, a graphic, or a computer program designed to generate a holographic image. Whoever creates the unique work of art that later becomes a hologram is considered the copyright holder of the image. It is also possible to copyright the finished hologram itself as a unique work of art, provided that none of the components that appear in the holographic image belong to another party. There are, however, statutory limits stating that after a number of years, a piece of art can become public domain and may be used freely.

Every hologram, if properly copyrighted, has only one owner, the copyright holder. The copyright holder therefore controls all subsequent distribution and is positioned at the top of the distribution chain. The copyright holder can be an individual or a group of people such as a business. Most commonly, a business commissions an artist to make a model or to design graphics and the artist turns over all copyright privileges to the business as part of the arrangement. This is legally known as "work for hire" and each party's responsibilities and rights must be documented to avoid problems concerning ownership. Holograms that are not copyrighted can be copied by whomever owns the "master" hologram.

The Manufacturer

Different companies specialize in manufacturing specific types of holograms. The manufacturing process generally involves three processes - mastering (creat- ing the original hologram), reproduction (producing some quantity of copies), and finishing (lamination, cutting, sorting, etc.). Some companies do everything - others subcontract out some part of the job. Often a company that manufacturers holograms also owns copyrights in order to have a selection of stock images to offer their customers. A few manufacturers bypass the normal chain of distribution and sell directly to retailers.

The copyright holder needs to know the exact cost of each unit produced, since the manufactmer's charge will obviously influence the final price billed to the end user. To figure the unit cost, one would take the total bill from the manufacturer (including any additional shipping and handling charges) and divide it by the number of usable copies actually delivered. As in most manufacturing businesses, prices decrease as quantity increases. In order to figure the suggested retail price of their product, it is very common for a copyright holder to mUltiply the manufacturer's unit cost five times. For instance, a product that costs the copyright holder $4.00 will be resold to a distributor for $6.00, and will end up selling in a store for $20.00.

If you are having holograms made to your specifications, choose a company that produces holograms appropriate to your final application. Be aware that manufacturers have not yet standardized their pricing - some itemize production processes, others quote a finished price. Some quote by the square inch, others according to a sliding scale based on quantities ordered.

The Distributor

The distributor is a business that specializes in buying large quantities of product from a copyright holder and distributing it to other businesses that cannot afford to, or are not interested in, stocking inventory. Distributors are often contractually obligated to order large amounts of merchandise, carry an entire line of their supplier's prod-

ucts, and not sell competing products. This alleviates many problems for the copyright holder and the manufacturer who are not usually set up to market their own products to numerous customers.

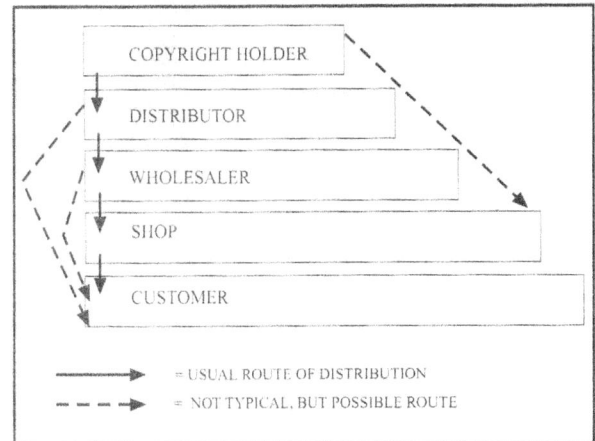

Chain of Holography Distribution

In return, the distributor commonly receives the sole rights to sell the product in a pallicular geographic region or to a pallicular group of customers and pays less than any other customer down the line. Distributors commonly pay 70%

below the suggested retail price, resell these products for 60% below suggested retail, and depend upon a large volume of sales to make their profit. For example, if they pay $6.00 for an item from a copyright holder, they would resell it to a wholesaler for $8.00.

Many distributors also repackage goods under their own names, deal with import/export procedures and constantly work to expand the marketplace. A popular product can make a distributor a lot of money, due to the fact that potential customers have no alternative supplier. The distributor sells mostly to wholesalers.

Typical costs, margins, and profits for hologram with retail price of $20				
Business	Buys For	Resells For	Discount Off Retail	Markup (%of Cost)
Copyright Holder	$4	$6	80%	50%
Distributor	$6	$8	70%	33%
Wholesaler	$8	$10	60%	25%
Retailer	$10	$20	50%	100%
Customer	$20	N/A	N/A	N/A

The Wholesaler

The wholesaler connects distributors to retailers. The essential function of a wholesaler is to get the product into shops. Most wholesalers use a combination of in-house salespersons or independent sales reps working on commission to persuade buyers to try a product. Many rely on telemarketing departments, catalogue mailings and trade shows to establish new accounts and service existing ones. A good wholesaler will teach a shop owner how to best merchandise a product, provide point-of-purchase materials, restock displays, update product selection, and generally keep the customer happy.

A wholesaler normally carries many different product lines, which is a convenience to retailers that want to consolidate the number of their suppliers. Also, wholesalers stock far less merchandise than a distributor, which allows them to react quickly to changes in the marketplace.

Wholesalers do not generally have exclusive rights to a product. They often extend payment terms to their customers (net 30 days is common) after a probationary period or credit check. For their efforts, wholesalers receive special pricing that allows them to make a profit when they resell the goods to retailers. Wholesalers typically work off 25% profit margins, i.e.; if they buy an item for $8.00, they will resell it for $10.00. Some distributors offer additional discounts to wholesalers for larger orders.

The Retailer

Retailers are the point of contact with the public, the place where merchandise is displayed and purchased by the customer. Holograms and related products have been sold in a variety o f settings, ranging from temporary table-top setups (flea markets, trade show booths) to art galleries and department stores. Many entrepreneurial businessmen start with a small cart or kiosk in a busy mall during holiday season and graduate to a bigger store that runs year round. Several single-store operations have expanded to multi-store chains.

Wholesalers have placed merchandise in obvious locales like museum gift shops, technology stores, and poster shops; less obvious locations include airport shops, nature stores and stationery stores. Other dealers have targeted specific interest groups such as hobbyists and collectors, and sell licensed products to comic book shops, trading card stores, and the like. Larger suppliers have cut deals with amusement parks, national chains, and event promotion agencies. Holograms have even been sold by mail order using catalogs and classified ads, even though a written description or photograph does not adequately capture the wonder of a three dimensional image.

All successful hologram retail businesses have several things in common - they are located in high-traffic (pedestrian) areas in places where people go to buy interesting items (often tourist destinations), they display the merchandise correctly and with flair, and they offer a high level of friendly customer service. Although the retail price suggested by the copyright holder is only a guideline, most retailers double their costs to establish the final price a customer sees in the store. Therefore, an item that costs them $10.00 will end up on the shelf for $20.00.

Categories of Merchandise

Most successful hologram stores sell a selection of holographic merchandise, including pictures (stock wall decor images and limited-edition fine art), jewelry (and related fashion accessories), executive gifts (desktop accessories), toys and optical novelties. There are several ways to categorize this merchandise - by price, by size, by manufacturer, and so on. For now, we'll discuss selection by "format" (which refers to the type of material the hologram is produced on).

Silver Halide Glass Plate

Traditionally, most holographers have produced their holograms on sheets of glass (glass plates) coated with a high resolution light-sensitive emulsion called silver halide. This is similar, but not identical to, the emulsions used in conventional photographic films.

These glass plates are rather costly, but can be used to make the highest quality holograms - mostly because the glass plates are quite rigid and will not move during the exposure period (movement will ruin the hologram). Working with these glass plates is quite time-consuming, as each plate has to be handled with great care during each production step.

Due to the time and the cost involved in making holograms on silver halide glass plates, they are mainly used for limited-edition holographic wall art, archival images or custom work. A finished 8" x 10" glass plate hologram typically retails at $300.00 or more for an open edition work.

It is possible to produce striking "deep image" holograms

on silver halide glass plates - that is, images which display considerable projection in front of the hologram's surface, and/or appear to be a considerable distance behind the hologram's surface. To achieve such dramatic effects, glass plate holograms require the best possible illumination, which usually takes some foresight.

Glass plate holograms are extremely fragile and should always be securely wrapped and handled with care. When the hologram is framed, the actual emulsion is facing toward the back and is protected. The surface which we can see (and touch) is ordinary glass and can be cleaned by gentle polishing with a soft cloth.

Silver Halide Film

Since glass plates were too expensive and impractical to mass produce, holographers developed methods to produce holograms on silver halide holographic film, which is cheaper and easier to handle. It is thin and flexible, similar to the film used in an ordinary camera. After exposure and processing, each piece of film is usually sandwiched in a cardboard matte, ready to package and/or frame as wall decor.

An 8" x 10" silver halide film hologram typically retails for $150.00 (less than half the cost of a comparable hologram produced on glass), making this format much more suitable for the giftware market. The best-selling size has traditionally been a 4" x 5" film mounted in an 8" x 10" matte due to the retail price (which has hovered around $35.00). Other standard sizes include a 2" x 2.5" film mounted in a 5" x 7" matte (which typically sells for $15.00 - $20.00), and a 5" x 8" film mounted in an 8" x 10" matte (which typically sells for $70.00).

Silver halide film holograms do not display quite the resolution, projection and depth of their glass plate counterparts. However, under proper illumination they are quite striking.

The emulsion of a film hologram is usually covered by a cardboard matte; however, the front surface can scratch easily. Fingerprints and dust can be cleaned by gently polishing with a soft cloth. Framing these pieces behind clear glass is recommended. Like all photographic films, they should be kept away from excessive heat and moisture.

Photopolymer Film

Several companies (notably Polaroid and DuPont) have developed a plastic film material that is especially suited for reproducing holograms. These photopolymers are extremely bright and durable, which makes them appropriate for a wide variety of commercial applications. The automated production process used to reproduce these holograms cuts costs considerably. An 8" x 10" matted photopolymer film hologram can retail for well under $100.00.

Image depth is a bit shallower and projection distance is a bit less than in silver halide films. However, photopolymer holograms are a magnitude brighter and generally have a much wider viewing angle. Although this allows for more latitude when using a less-than-ideal light source, a single overhead light is still recommended for illuminating wall decor. The material works exceedingly well for items that will be illuminated by normal outdoor lighting, such as jewelry, bookmarks and postcards. The plastic surface of the photopolymer films can be cleaned with a soft cloth.

Dichromate Glass

Another photosensitive material used to make holograms is "dichromate." Unlike silver halide glass plates or rolls of film, this material is actually a chemical/gelatin mix that is coated onto a piece of glass. After exposure, a cover glass is securely attached to the base plate, creating a permanent seal - since moisture or excessive heat can dissolve the dichromate gel and the hologram will eventually disappear. Sometimes plastic is used instead of glass, lowering production costs.

Dichromate holograms are very bright and can be viewed under less-than-ideal lighting conditions, e.g., outdoors. Therefore, they are most frequently used as jewelry items and fashion accessories. The glass can be cut into any shape, but it is often cut into small discs. These hologram discs have traditionally been used in watchfaces, broaches, key-rings and belt buckles. For years, the $20.00 dichromate glass pendant has been a staple for retailers. An 8" x 10" framed dichromate plate usually retails for over $100.00, but they are seldom made anymore.

Embossed Foils

These holograms are reproduced by a process that is not optical and does not use costly photosensitive materials. In this case, the holographic image is transferred onto a mechanical stamping die, which is then used to emboss the image on rolls of very thin plastic films or metal foils. These holograms can be reproduced for cents per square inch, making it the least expensive way to copy an image. These foils are commonly hot-stamped on paper or plastic, or backed with an adhesive layer. The material is very durable .

The embossing process creates very shallow images that can be viewed under poor illumination. They are usually silver-backed and reflect a rainbow of colors. Some of the best-selling holographic products are animated 3D embossed images which display fluid motion and realistic colors. These commonly retail for $15.00 - $20.00, matted and packaged, ready to frame . Most o f the embossed holograms on the market are called 2D/3D. They display multilevel graphics and are usually used as inexpensive stickers, or part of keychains, pins, and magnets.

Often a repeating prismatic pattern is embossed onto the plastic or foil, which creates a ever-changing rainbow effect. This colorful material is used in toys and optical novelties.

Opening a Shop - Basic Guidelines

1. Choose the best location you can afford. Tourist destinations in "festival" type shopping districts located near dining and drinking establishments traditionally do very well. High-traffic locations ensure a steady stream of curious shoppers. Malls do very well around the holidays, but

can be slow during the off-season. Do extensive market research.

2. Shop for a reliable supplier that can stock you in a timely manner. A good supplier will be able to replace the merchandise you have sold quickly, thereby reducing the number of units you need to backstock. A good supplier will also assist you with displays, inventory selection, and point-of-purchase materials.

3. Stock a wide selection of merchandise in various price ranges. Many hologram stores cover their overhead with the sales of inexpensive items in the $1.00 - $10.00 range. Most retail sales average $5.00 - $25.00. Wall art typically sells well at the $35.00 price point, as do $65.00 watches. Sales of more expensive pieces can be icing on the cake.

4. Plan your displays carefully. Due to lighting considerations and limited viewing angles, holograms need to be merchandised more carefully than many other products. Installing adjustable track lighting is more practical than using stationary fixtures. Halogen spot lights (50 - 100 watts) work very well. Avoid fluorescent and recessed incandescent lights.

5. Create an entertaining atmosphere. No one needs to buy a hologram. A friendly, informative staff will boost sales dramatically.

Start-up Costs

The amount of capital you need depends on how large an operation you want to have. On the low end, you can stock a "cart" operation, as opposed to an actual shop, for as little as $5,000. Buying starting stock for a small shop (several hundred square feet) will cost several times that. A larger space of 800 - 1,000 square feet can easily hold $50,000.00 worth of goods. Plan on spending $10,000 - $15,000.00 for high-quality lighting and displays for" a store of this size.

Based on a "double markup for inventory," $200,000 worth of annual gross sales should support an owner/manager making $30,000.00, a small staff (dividing another $30,000.00), rent/overhead in a high-traffic location ($25,000.00), and first year build-out ($15,000.00).

Embossed holograms are the most economical way to mass produce a holographic image but, until recently, there has been resistance among many buyers because they feel that there is a large "origination fee" that they will have to face which will drive up the unit price of short runs enough to make the projects uneconomical. Fortunately, in recent years there has been significant improvement with this problem, making short run embossed holograms much more affordable. In the following article, Peter Scheir of AD 2000 reports on this topic.

Embossed Holography Made Affordable in Low Production Runs

by Peter Scheir

Over 15 years' experience in the holographic industry has helped me to identify a number of recurring themes in potential customers' objections. I have made it my goal to create a path of lesser resistance- to make holography "a solution, rather than a new set of problems." The aim of this article is to outline what many customers perceive as obstacles to using holography, and to describe at least one solution. That solution lies in the customizing of stock image embossed holograms.

A Powerful Promotional Tool

Impact is the word, in my mind, which best befits graphic holography. It is the visual impact of the medium which gives it such great power and generates so much interest. Companies of all sizes are constantly searching for means of promotion which will set them apart from competitors and get them noticed. Holography adds three distinct advantages to any promotion:

1) **Impact**: Holograms are eye-catching. People will look at holograms for significantly longer periods of time than other graphic mediums. If a customer's name or slogan is on the hologram, that information is enforced!

2) **Pass-Around Value**: people are often impressed with good holograms, and they are likely to bring them to the attention of colleagues and associates.

3) **Retention**: people tend to keep holograms. We receive new leads from potential customers who have holograms on their desk which were created more than 10 years ago!

Holographic Misperceptions

Given these three factors, it may be hard to understand why holography is not found in much wider use in the promotional marketplace. I feel that it is largely due to false perceptions about holography and its costs.

Unfortunately, in the minds of many, affordability has never been a word associated with holography. It would be impossible for me to count the times that I have heard potential customers say "holography is great but too expensive," "we can't afford to use holography for our project" (can they really afford to ignore the impact of holography?) or, "holograms are only for large runs." These decisions are made before a price quote has even been made!

Objections such as these are sometimes based upon past experience; but more often than not, they come from an incomplete or incorrect conception of the holography industry and what we have to offer. Mainly, objections of this sort stem from experiences in which an inappropriate solution was applied to a project.

For instance, a customer may have been looking for 5,000 hologram labels to give away at a trade show. Calling a company found in the phone directory, they were told that there would be a set-up cost of $8,000; or, that the minimum run had to be 50,000 pieces. This type of incomplete information creates obstacles in customers' minds. After such an experience, further exploration into the use of holography would be limited due to the customer's presumption that holography could not be cost-effective.

It is true, however, that fully-custom embossed holography carries set-up (origination|tooling) costs which are many thousands of dollars. Making embossed holograms affordable in low production runs (runs too short to amortize the tooling costs associated with fully custom-embossed holography) is the key to tapping this marketplace. I have found that the best solution to this problem is the use of stock images.

There are a great number of projects which are defined as "not cost effective", given fully custom holography's high set-up costs. Stock image holograms have just as much impact as fully custom holograms, without the high set-up costs and stock image embossed holograms can, with great ease, add huge impact to any promotion.

Direct mail can be enhanced with a stock hologram label. Memorable premium and specialty items such as magnets, calculators, and t-shirts can be created from stock holograms. Powerful brochure covers, pocket folders, annual reports, magazine covers, advertisements, etc. can be produced and packaging can benefit from the draw and appeal of a stock hologram. New life can, in fact, be breathed into almost any promotion with a stock image hologram.

Many hologram production companies have existing stock images. Sometimes images are created 'on spec' for projects which do not come to fruition, and the originator retains the rights of 'replication' (embossing the image). In other circumstances, originators create holograms for self-promotion or for retail markets that can then be utilized as stock images for advertising and promotional projects. These images can often be customized to suit individual customers' needs.

Customizing a stock image can take a number of forms. A simple post-printing stage adds a customer's logo, ad copy, or slogan to a stock image, making it (for all practical purposes) a custom image! Stock images can also be die-cut (cropped) to specific shapes and/or sizes which give them a custom appearance. Placing ad copy around the image on the literature or product can further enhance the design and make a stock image unique.

Generally, stock images are more expensive per piece than custom images due to 'short run costs' and/or royalties (paid to the owner of the image) which are included in the price per piece. This higher cost per piece is deceptive. Since no custom tooling costs need to be amortized, the total cost for the production run is low enough to make the project viable where a fully custom hologram would have been out of the question.

Though the average "total dollar value" of stock image jobs is lower than that for fully custom projects, the lack of high set-up costs and the availability of low minimum runs create a much greater conversion rate, so there are more jobs. Just as important, customers see the image before the project begins, eliminating the long, potentially problematic proof stages and chances for image rejection.

Origination is (conservatively) based upon a fully 3D image, including basic model, photoresist master, electroforming, recombination, and die work.

Qualification of customers is often complicated in fully custom holography. It can take a long conversation and an expensive packet of samples just to determine that a customer does not have the budget for a custom image. Stock images offer an excellent back-up option. If the budget does not allow for custom work, the project may often be salvaged with a stock image hologram. Furthermore, some of our sales reps have found that offering an affordable alternative in stock imagery can remove the obstacles in a customer's mind. As the customer looks further into the project, he or she finds that a custom piece is really more attractive, and a budget is 'created' which includes custom tooling!

Stock Images Work in Other Industries

In the photographic industry, stock images have long been a well-used alternative to custom photography. Various collections of stock photos offer a dazzling array of images, so that one can be found to suit even the most obscure project parameters.

Holography has not yet developed into such a serviceable industry as photography. If you believe that a stock image may work for your particular project, request a list of available images from your favorite hologram supplier; or, look to one of the recently-developing compilations of stock images [NOTE: If you are shopping for a stock embossed hologram, be sure to see if it is available in the form you require (i.e.- hot stamp foil y.s..pressure sensitive stickers, logo add-ons, custom cut sizing or cropping, roll vs. sheet, etc.).

Some Stock Image Suppliers

AD 2000 INC. of New Haven, CT has created a stock hologram collection named HOLOBANK. We have assembled a large selection of stock images by licensing available images from many of the world's finest origination facilities. This growing compendium has over 400 stock images ranging in usable size from 3/4" to 6" square, and customizing is supported in the fOlm of add-on printing as well as custom die shapes, all for runs as low as 500 pieces. You can also find limited parts of the HOLOBANK collection available in various forms from companies ineluding Applied Holographics, Astor Universal, Bridgestone Graphic Technologies, and Light Impressions.

Smith and McKay Printing in San Jose, CA is one of the country's leading hot stampers of holograms. They have taken a unique approach to the sale of stock image holograms. Purchasing large rolls of stock images in hot stamp foil and producing a number of promotional products using these images allow them to keep products on the shelf ready to customize as required. Their line of business cards, note cards, stationery, and presentation folders is available in volumes as low as 100 pieces, and can be customized through hot stamping.

American Banknote Holographics offers a large variety of stock images ranging"in'size from under an inch to postcard size. Many of their images are available in both hot-stamp foil and pressure-sensitive formats, though custom die shapes and add-on printing are only available in larger runs.

Summary

It is important to recognize that short runs and low budgets are not restricted to small companies. Many large corporations make use o f short runs for specific projects which have relatively small budgets but are extremely important nevertheless. These projects can be high profile; and, they can create an 'in' because your company was there for them on the smaller project. The customizing of stock images ultimately makes them custom images. The majority of those viewing the images will never know that they were not created specifically for the company using them, particularly if holographic diffraction imprinting is employed.

In summary, customized stock embossed holograms add a powerful new dimension to the sales of holograms and related products. Consider the benefits:

- The visual impact of custom holography without the high tooling costs.

- Affordability in low production runs.

- Multiple applications, from print advertising to premium items.

- Additional sales to customers who are presently turned away.

- Ease of presentation and higher lead conversion.

All in all, the best solution I have found to the issue of "affordable impact embossed holography in low production runs" is the customizing of stock images. Companies dedicated to the promotion of stock holography will continually increase their image selection, and begin targeting images to particular industries. Eventually, stock image holography could become as inexpensive and common a graphic element as stock image photography is today.

[1]Comment made by Doug Miller while president of Holographic Design.

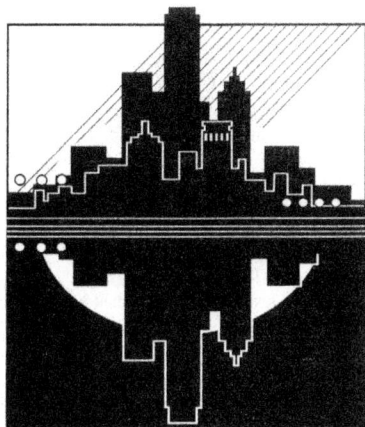

9

Business Listings

In this chapter we list businesses and people involved in the field of holography. First, there is a complete corporate directory, with companies listed in alphabetical order. Next, there is a listing of corporate names, sorted by country. At the end of the chapter there is a list of individuals, also listed alphabetically.

Please note that several countries are in the process of changing telephone prefixes. Also, please assume that all website addresses begin with "http://www."

We would appreciate if you notify us of any changes, omissions or mistakes by fax (510-841-2695) or by email (stafJ@rossbooks.com).

Corporate Listings

21 st Century Finishing Inc.
215 Pennsylvania Avenue
City: Paterson
State/Province: NJ
Postal Code: 07503
Country: United States of America
Voice Phone: 1 20 1 279 2100
Fax Phone: 1 201 2795659
Contact # 1 : Anthony Olmo
Business Description: Multi-faceted converting specialists. 10 years experience working with trade and corporate clients. Full range of web or sheet finishing services offered including: cutting, hot stamping, labels, lamination, overprinting, etc. Capable of applying holograms to a variety of substrates.
****

3 Deep Hologram Company
609 California SI.
City: Huntington Beach
StatelProvince: CA
Postal Code: 92648
Country: United States of America
Voice Phone: 1 7149695354
Fax Phone: 1 7149695354
email: acheimets@aol.com
Web Address: member.aol.comlthe3d1home.htm
Contact #1: Alex Cheimets
Business Description: Supplies Russian silver halide emulsions on plates. Green sensitive, red sensitive, and new full color emulsions available. Can also recommend appropriate developing and processing procedures.
SEE OURADVERTISEMENT
****

3-D Systems
PO. Box 145
City: Pl. Arena
State/Prov ince: CA
Postal Code: 95468
Country: United States of America
Voice Phone: 1 707 882 2066
email: Igcross@intercoastal.com
Contact # I: Lloyd Cr~ss
Business Description: Currently produces virtual optics laboratory software for holographers and optics research. Pioneered integral holograms and other holographic techniques.
****

3-D Worldwide Holograms, Inc.
7503 N.W. 36 Street
City: Miami
StatelProvince: FL
Postal Code: 33166
Country: United States of America
Voice Phone: 1 305 9947577
Fax Phone: 1 305 994 7702
email: hologram@shadow.net
Web Address: www.3dworldwide.com
Contact #1: Mark Diamond
Business Descript ion: We are specialists in full color animated holographic stereo grams. Our fully controlled laser labs and origination studios are supported by our 20 million dollar S.G. I. digital production facility.
****

3D Holograms Inc.
PO. Box 2
Ci ty: Freeport
State/Province: NY
Postal Code: 11520
Country: United States of America
Voice Phone: 1 516 872 8939
Fax Phone: 1 516872 1307
Contact # 1: Marcy Scher
Business Description: Distributors and innovators of 3D and holographic gifts, novelties, advertis ing premiums and specialty products.
****

3D Holographics
31 Willow Park Ave
Glasvenin
City: Dublin II
Country: Portugal
Voice Phone: 353 1 843 6200
****

3D Images
31 The Chine
Grange Park ...
City: London
State/Province: England
Postal Code: N21 2EA
Country: United Kingdom
Voice Phone: 44 181 3640022
Fax Phone: 44 181 364 1828
email: burder3d@aol.com
Contact # I: David Burder
Business Description: Manufacturer and distributor of 3D images. Also supplies lenticul ar 3D products. Producers ofVirtualVideo. 3D supplies and glasses.
****

3D Optical Illusions
P.O. Box 765
City: Bayswater
Postal Code: 3153

CORPORATE LISTINGS

Country: Australia
Voice Phone: 61 397296337
Fax Phone: 61 39 729 6020
Contact # 1: Terry Roberts
Business Description: Specializing in lenticular and holographic movement illusions with up to 20 different motion images.

****

3DVision
Hologramme-Laserprodukte
Ostertorsteinweg 1-2
City: Bremen
Postal Code: 28203
Country: Germany
Voice Phone: +49 (0)421 767 97
Fax Phone: +49 (0)421 767 97
Contact # 1: Uwe Reichert
Business Description: Holograms,Holographic projects. General commercial holograms for sale and distribution.

****

3D-4D Holographies
Panther House
38 Mt. Pleasant
City: London
State/Province: England
Postal Code: WC1X OAP
Counl1y: United Kingdom
Voice Phone: 44 171 8375767
Fax Phone: 44 171 837 713 1994
email: hologram@demon.co.uk
Web Address: hologram. demon
Contact # 1: Graham Tunnadine
Business Description: Production house for all types of 3D visual displays. Large fomlat, silver halide mastering facility on premises.

****

3M - Safety and Security Systems
3M Center, Bldg 225-4N-14
City: St. Paul
State/Province: MN
Postal Code: 55144-1000
Country: United States of America
Voice Phone: 1 800 328 7098 ext. 5
Fax Phone: 1 800 223 5563
Web Address: www.lmnm.com
Contact #1: Maureen Tholen
Business Description: Supplier of authenticating labels, including some holographic applications. Full range of origination services offered.

****

A.D. Tech (Advanced Deposition Technologies)
580 Myles Standish Blvd.
Myles Standish Industrial Park
City: Taunton
StatelProvince: MA
Postal Code: 02780
Country: United States of America
Voice Phone: 1 508 823 0707
Fax Phone: 1 508 823 4434
Contact #1 : Glenn 1. Walters
Business Description: Security label business.

****

A. H. Prismatic, Inc.
1175 Chess Drive, Suite D
City: Foster City
State/Province: CA
Postal Code: 94404-1 108
Country: United States of America
Voice Phone: 1 415 345 4855
Fax Phone: 1 415 345 4854
Contact #1: Sheila Bagley
Business Description: Manufacturers of exclusive ranges of holographic gifts, toys, jewelry, photopolymers, and film holograms. Licensed products

available: Star Wars. Call for U.S.A. wholesale catalog. SEE OURADVERTISEMENT INSIDE BACK COVER

****

A.H. Prismatic, Ltd.
New England House
New England Street
City: Brighton
State/Province: England
Postal Code: BNI 4GH
Country: United Kingdom
Voice Phone: 44 1273686966
Fax Phone: 44 1273 676 692.
Contact # 1 : Ian Dayus
Business Description: Manufacturers of exclusive ranges of holographic gifts, toys, jewelry, photopolymers, and film holograms. Licensed products available: Star Trek; StarTrek: The Next Generation; Star Trek: Deep Space Nine; Star Wars. SEE OURADVERTISEMENT IN-SIDE BACK COVER

****

AB Rueck Holoart
Roepersweide 26
City: Hamburg
Country: Germany
Voice Phone: +49 (0)40 8807151
Contact # 1 : A. B. Rueck
Business Description: Holograms for conventional presentations, refection and transmission.

****

Academy of Media Arts Cologne
Peter-Welter Platz 2
City: Cologne
Postal Code: 50676
Country: Germany
Voice Phone: +49 (0)221 201 89 1 15
Fax Phone: +49 (0)22120189124
Contact # 1: Dieter Jung
Business Description: International Academy for Media Arts. Extensive holography lab and teaching.

****

Acme Holography
12 Sunset Road
City: West Somerville
State/Province: MA
Postal Code: 02144
Country: United States of America
Voice Phone: 1 617623 0578
email: bconn@media.mit.edu
Contact # 1: Betsy Connors
Business Description: Acme Holography is Boston's first private holography lab. We offer full service in reflection, transmission and computer generated holography, including design consultation and large-scale environmental holography.

****

Action Tapes
Unit 5
Broundry Road
City: Brackley
State/Province: England
Postal Code: NN13 7ES
Country: United Kingdom
Voice Phone: 44 1280700591
Fax Phone: 44 1208 700 590
email: action.tapes@dial.pipex.com
Contact #1: Alan 1. Phillips
Business Description: Adhesive tapes for industry: close tolerance custom sl it rolls; varying length stroke diameter; high speed die cut; large and small volume runs. Fast and experienced service. Experienced staff to provide professional and technical advise.

****

AD 2000, Inc.
948 State Street
City: New Haven
State/Province: CT
Postal Code: 06511
Country: United States of America
Voice Phone: 1 203 624 6405
Voice Phone: 1 800 334 4633
Fax Phone: 1 203 624 1780
email: pscheir@ad2000.com
Web Address: ad2000.com/ad2000/
Contact # 1: Peter Scheir
Business Description: Fully custom & customized stock inlage embossed and photopolymer holograms. Our HOLOBANK contains the world's largest selection of stock image embossed holograms - available plain, as labels, foil, magnets, pins, roll stock, etc. Low run security applications are our specialty!
SEE OURADVERTISEMENT

****

Ad-Holograms Oy
Niinisaari
City: Puumala
Postal Code: FIN-52200
Country: Finland
Voice Phone: 358 5 488161
Fax Phone: 358 5 488164
Contact # 1: lkka Hyttinen

****

Adlas G.M.B.H. & Co Kg.
Seeland Strasse 9
City: Lubeck
Postal Code: 23569
Country: Germany
Voice Phone: 49451 3909300
Fax Phone: 49 451 390 9399
Business Description: Established 1986. Manufacturer of diode laser-pumped solid state lasers which operate in CW and pulsed mode with wavelengths in JR, visible and UV Branch office: 636 Great Road, Stow, MA 01775, USA. Owned by Coherent, Santa Clara, CA, USA

****

Advance Photonics
A-147 Ghatkopar Industrial Est
Ghatkopar
City: Bombay
Postal Code: 400 086
Country: India
Voice Phone: 91 22 582 204
Fax Phone: 91 22 202 4202
Business Description: Branch office of Newport Corporation.

****

Advanced Holographic Laboratories
(a division of Astor Universal Ltd.)
Astor Road / Eccles New Road
City: Salford
Postal Code: M5 2DA
Country: United Kingdom
Voice Phone: 44 (0) 1617898131
Fax Phone: 44 (0)161 787 8348
Contact # 1: Francis Tuffy
Business Description: Worldwide designers, originators and manufacturers of holographic pattern and image foi ls and film. See Astor Universal advertisement for further details.
SEE OURADVERTISEMENT

****

Advanced Holographic Laboratories
(a division of Astor Universal Corp.)
3841 Greenway Circle
City: Lawrence

State/Province: KS
Postal Code: 66046-5444
Country: United States of America
Voice Phone: 1 800 255 4605
Voice Phone: 1 9138427674
Fax Phone: 1 913 842 9748
Contact # 1: John Thoma
Business Description: Worldwide designers, originators and manufacturers of holographic pattern and image foils and films.
SEE OURADVERTISEMENT
****

Advanced Optics, Inc.
5 East Old Shakopee Road
City: Minneapolis
State/Province: MN
Postal Code: 55420
Country: United States of America
Voice Phone: 1 515 964 5050
Contact # 1: Wendy Heil
Business Description: Manufactures of custom and precision optics, first surface mirrors for use in holographic equipment.
****

Advanced Technology Program
Bldg. 101 - Room A430
City: Gaithersburg
State/Province: MD
Postal Code: 20899
Country: United States of America
Voice Phone: 1 800 287 3863
Web Address: atp.nist.gov
Business Description: United States Government program providing funding for R&D projects.
****

Aerospatiale
Ets D'Aquitaine
Saint-Medard-En-Jalles
City: Bordeaux
Postal Code: F-33165
Country: France
Voice Phone: 33 56 57 34 80
Fax Phone: 33 56 57 30 70
Contact # 1: C. Lefloc'H
Business Description: Scientific and industrial research, NDT testing
****

Aerotech Inc.
Electro-Optical Division
101 Zeta Drive
City: Pittsburgh
State/Province: PA
Postal Code: 15238
Country: United States of America
Voice Phone: 1 412 963 7470
Fax Phone: 1 412963 7459
Contact # 1: Steve A. Botos
Business Description: Manufacturers of helium neon tubes, power supplies and complete systems for OEM and end users. Other product lines include optical table positioners and precision rotary and linear positioning systems. Subsidiary Companies:Aerotech Ltd.- England, Aerotech GmbH-Germany.
****

Ag Electro-Optics Ltd.
Tarporley Business Centre
City: TarporJey
State/Province: England
Postal Code: CW6 9UY
Country: United Kingdom
Voice Phone: 44 1829 733 305
Fax Phone: 44 1829 733 679
email: 100064.3242@compuserve.com
Contact # 1: lA. Gibson

Business Description: Distributor of lasers, optics, lab equipment and fiber optics.
****

Agfa - a division of Bayer Corp.
US Headquarters
100 Challenger Road
City: Ridgefield Park
StatelProvince: NJ
Postal Code: 07660
Country: United States of America
Voice Phone: 1 201 4402500
Fax Phone: 1 201 4408401
Contact # I: Mark Redzikowski
Business Description: Manufacturer of recording materials suitable for holography. The major supplier of red sensitive and green sensitive silver halide emulsions on glass plates and film. Call to locate nearest distributor.
****

Agfa - Gevaert N. V
(a division of Bayer)
Septestraat 27
City: Mortsel
Postal Code: B 2640
Country: Belgium
Voice Phone: 32 3 444 8251
Fax Phone: 32 3 444 8243
email: bebay2uk@ibmmai!.com
Contact # 1: Frank Mortier
Business Description: Headquarters. Manufacturer of silver halide recording materials.
****

AHT 3D-Medien
Association for Hologram Techniques & 3D Media(AH
Niederesch 28
City: Bad Rothenfelde
Postal Code: 49214
Country: Germany
Voice Phone: +49 (0)542 5365
Fax Phone: +49 (0)542 5359
Contact # 1: Gunter Deutschmann
Business Description: Association of German hologram manufacturers. Association for hologram techniques and 3D media, engaged in the commercialization of holographic products. Realization of projects and campaigns in co-operaSon between the member companies. From consultation, design, mastering, hologram production to overprinting, converting, integration into the finished product, etc.
****

AK Rehman Traders
M 10013 Block C, Galino 3
City: Karachi
Postal Code: 75730
Country: Pakistan
Voice Phone: 924920 54142
Fax Phone: 92 21 4920719
Contact # 1: M. Zafar
****

AKS Holographie-Galerie GmbH
Potsdamer Strasse 10
City: Essen
Postal Code: 45145
Country: Germany
Voice Phone: +49 (0)201 756455
Fax Phone: +49 (0)201 753582
email: akshol@aol.com
Web Address: http ://members.aol.com/akshol/ hhome_d.htm
Contact #1: Detlev Abendroth
Business Description: Producing all kinds of holograms incl. computer-generated up to a size of 80 x 100 cm. One of the world's greatest film-

hologram-collections. Internationally represented by distributing partners.
****

AKS Holographie-Gallerie GmbH
Posts darner Strabe 10
4145 Eben
City: Ruhr
Country: Germany
Voice Phone: 49 201 756455
Fax Phone: 49 20 1 753582
Business Description: Holography Gallerie - various types of holograms.
****

Alabama A&M University
Center for Applied Optical Sciences
PO. Box 1268
City: Normal
State/Province: AL
Postal Code: 35762
Country: United States of America
Voice Phone: 1 205 851 5870
Fax Phone: 1 205 851 5622
email: caufield@caos.aamu.edu
Contact # 1: John Caulfield
Business Description: R&D in all applications of holography as well as real time holography.
****

Amagic Technologies Inc.
17915 Sky Park Circle #G
City: Irvine
State/Province: CA
Postal Code: 92714
Country: United States of America
Voice Phone: 1 71 44743978
Voice Phone: 1 800 262 4421
Fax Phone: 1 7144743979
Web Address: www.thomasregister.comlamagic
Contact #1: Marilyn Huff
Business Description: Fully integrated manufacturer of holographic custom images, stock images, diffraction/digitized patterns as polyester, PVC/ vinyl and polypropylene. Produces hot stamp foil, pressure sensitive labels, and film to 39.37 inches.
****

Amazing World Of Holograms.
Corrigan's Arcade Foreshore Road
South Bay
City: Scarborough
State/Province: England
Postal Code: YOII IPB
Country: United Kingdom
Voice Phone: 44 1723 500 696
Fax Phone: 44 1482492 286
Business Description: Exhibitors and retailers of film, glass, embossed, dichromate and related products. Permanent display of 200 holograms including mutliplexes which are updated and changed regularly. Main season May-October. Distributors of film & glass.
****

American Bank Note Holographics
399 Executive Blvd.
City: Elmsford
StatelProvince: NY
Postal Code: 10523
Country: United States of America
Voice Phone: 1 914592-2355
Fax Phone: 1 914592-3248
Contact # 1: Keith Woodward
Business Description: World leader in development of embossed holography for security and commercial applications. Produces embossed holograms in a range of fOlmats: foil, pressure-sensitive & tamper-evident labels, clear & wide-web laminates for packaging.

CORPORATE LISTINGS

****

American Holographic Inc.
601 River Street
City: Fitchburg
StatelProvince: MA
Postal Code: 01420
Country: United States of America
Voice Phone: 1 508 343 0096
Fax Phone: 1 508348 1864
Contact #1: Thomas Mikes
Business Description: Design, develop and manufacture optical components and instnunents for use in industrial and medical measurements. We are using holographic diffraction grating design and manufacture capability to produce components for unique measurement instruments.

****

American Laser Corporation
1832 South 3850 West
City: Salt Lake City
State/Province: UT
Postal Code: 84 104
Country: United States of America
Voice Phone: 1 801 972 1311
Fax Phone: 1 801 972 5251
email: 72731.1465@compuserve
Contact # 1: Dan Hoefer
Business Description: Established 1970. Manufacturer of Argon, Krypton and Mixed gas laser systems from 3 mw to 10W in air or water configuration.

****

American Paper Optics Inc.
2005 Nonconnah Blvd.
Suite 27
City: Memphis
StatelProvince: TN
Postal Code: 38 132
Country: United States of America
Voice Phone: 1 901 398611 1
Voice Phone: 1 800 767 8427
Fax Phone: 1 901 3986119
email: optics3D@aol.com
Contact # 1: John Jerit
Business Description: American Paper Optics is the leading manufacturer of paper 3D glasses in the world. The products include Di ffract ion glasses, Polarized and Chroma Depth glasses. The latest holographic 3D glass is called HoloSpex and creates an image of a logo when viewing a direct point of light.

****

American Propylaea Corporation
555 South Woodward, Suite 1109
City: Birmingham
State/Provi nce: MI
Postal Code: 48009-6626
Country: United States of America
Voice Phone: 1 810 642 7000
Fax Phone: 1 810 642 9886
email: dddimage@aol.com
Contact # 1: Hans Bjelkhagen
Business Description: Auto-stereoscopic realtime HOE based display devices for CAD, medicine, education, entertainment and defense applications. Capable of displaying full motion, projected imagery. Also color holograms and HOEs.

****

Ana MacArthur
PO. Box 15234
City: Santa Fe
State/Province: NM
Postal Code: 87506
Country: United States of America
Voice Phone: 1 505 438 8739

Fax Phone: 1 505 438 8224
email: ana.lit@zumacafe.com
Contact #1: Ana MacArthur
Business Description: Holographic in stallation artist - unique installations. Also produces limited edition dichromate holograms and holographic sculptures. Experimental use of realtime interferometry.

****

AndreasWappelt - Photonics Direct
Weisestrasse 11
City: Berlin
Postal Code: 12049
Country: Germany
Voice Phone: +49 (0)30 62709372
Voice Phone: +49 (0)3 62709370
Fax Phone: +49 (0)30 627093 71
email: wappelt@t-online.de
Web Address: http ://home.t-online.de/home/wappelt/
Contact #1: Andreas Wappelt
Business Description: Distribution of Holograms (fine art & entertainment), he-ne la sers, laser pointer, laser diodes, diode pumped blue and green solid-state lasers, holographic equipment
SEE OURADVERTISEMENT

****

Another Dimension
637 NW 12th Ave.
City: Deerfield Beach,
StatelProvince: FL
Postal Code: 33442
Country: United States of America
Voice Phone: 1 954 429 1166
Fax Phone: 1 954 421 2391
Contact # 1: Mark Anoff
Business Description: Worldwide distributor of 3-D products including holograms and new stateof-the-art stereo optic lenticular. Exclusive distributor for 3DIFX line. Numerous popular licensed products available.

****

Ap Holografika Studio
(Galvanart Bt)
POBox 113
City: Budapest.
Postal Code: H-1 70 1
Country: Hungary
Voice Phone: 36 1 282 492 1
Fax Phone: 36 1 282 4921
Contact #1: Tibor Balogh
Business Description: The AP Holographic Studio provides mastering, whole process for custom embossed holograms (mainly in security applications), optical design using HOEs. Wholesaler of holographic nove lties and diffraction foils.

****

APA Optics Inc.
2950 Northeast 84th Lane
City: Blaine
State/Province: MN
Postal Code: 55449
Country: United States of America
Voice Phone: 1 612 784 4995
Fax Phone: 1 612 784 2038
Contact # 1: Sumant Dumra
Business Description: Design and manufacture of custom computer generated holograms and binary optical elements. Manufacture IAT, an interferometer using CGHs to test aspherie optics.

****

Applied Holographics, Pic.
40 Phoenix Road
City: Crowther, Washington

State/Provi nce: England
Postal Code: NE38 OAD
Country: United Kingdom
Voice Phone: 44 191 417 5434
Fax Phone: 44 191 416 3053
email: 100407.1630@compuserve
Contact #1: David Tidmarsh
Business Description: Holography, stereograms, narrow and wide web embossing, hot-stamp foil, pressure sensitive, polyester, opp, acetate, large format holograms, compact discs, paper.

****

Applied Optics
2662 Valley Drive
City: Ann Arbor
State/Province: MI
Postal Code: 48103-2748
Country: United States of America
Voice Phone: 1 313 998 0425
Fax Phone: 1 313 998 0425
email: upatnks@applopt.com
Contact # 1: Juris Upatnieks
Business Description: Applied Optics provides consulting services and laboratory breadboard testing of optical systems in coherent optics, holography, dim'active optical elements, and light control. System analysis using the ZEMAX optical design program.

****

Arbeitskreis Holografie B.V
Boeckelter Weg 47
City: Geldern
Postal Code: 47608
Country: Germany
Voice Phone: +49 (0)283 3034
Contact #1: Herman-Josef Bianchi
Business Description: Artistic holography

****

Armin K1ix Holographie
Postfach 260218
City: Duesseldorf
Postal Code: 40095
Country: Germany
Voice Phone: +49 (0)211 317 775
Fax Phone: +49 (0)21 1 317 749
Contact #1: Armin Klix
Business Description: Anfertigung von Displayhologrammen im Auftrag von Werbung und Industrie und Einze lund Grosshandel. Katalog von Lieferbaren Hologrammen vorhanden. Einzelstucke und Grosserien.

****

Art Institute Of Chicago (The School of the ...)
Holography Department
112 South Michigan Ave.
City: Chicago
State/Province: IL
Postal Code: 60603
Country: United States of America
Voice Phone: 1 312 345 3998
Fax Phone: 1 312 345 3565
email: ewesly@artic.edu
Contact #1: Ed Wesly
Business Description: The School of the Art Institute of Chicago offers an MFA degree with a concentration in holography, and is equipped with three tables, with one containing a stereogram printer for recording computer generated imagery or live subjects. 1 instructor, 3 grad assistants.

****

Art Lab
1000 Richmond Terrace
City: Staten Island
State/Province: NY
Postal Code: 1030 I

Country: United States of America
Voice Phone: 1 7184478667
Contact #1: John Iovine
Business Description:TheArt Lab is an art school which offers classes and workshops in the art of holography. Free brochure available.
****

Art, Science & Teclmology Institute (ASTI)
2018 R Street, N.W
City: Washington
StatelProvince: DC
Postal Code: 20009
Country: United States of America
Voice Phone: 1 202 667-6322
Fax Phone: 1 202 265-8563
Contact #1: Laurent Bussaut
Business Description: Research and educational corporation dedicated to the advancement of art, science and technology in holography, laser, optics and photonics industries. The museum features a pennanent collection of masterpiece holograms including one of the largest holograms in the world. The museum is visited by guided tour only, and reservations are required.
****

Artbridge Light Studios
Madam Weg 77
City: Braunschweig
Postal Code: 38120
Country: Gennany
Voice Phone: +49 (0)531 352 816
Fax Phone: +49 (0)531 352816
Contact # 1: Odile Meulien
Business Description: Design & Production of Holograms and Light Show - Organization of special happenings in coordination with multi-disciplinary artists, engineers and scientists.
****

Asahi Glass Co.
R&D General Division,
2- 1-2 Marunouchi Chiyoda-Ku
City: Tokyo
Postal Code: 100
Country: Japan
Voice Phone: 81332185825
Fax Phone: 81 332145060
Contact #1: Fumihiko Koizumi
Business Description: R&D on holography and single copy holograms.
****

Astor Universal Ltd.
Astor Road
Eccles New Road
City: Salford
State/Province: England
Postal Code: M5 2DA
Country: United Kingdom
Voice Phone: 44 161 7898131
Fax Phone: 44 161 787-8348
Contact #1: Laurence Holden
Business Description: See listing for Advanced Holographic Laboratories and advertisement for AHLIAstor Universal.
****

Atelier Holographique De Paris
13, Passage Courtois
City: Paris
Postal Code: F-75011
Country: France
Voice Phone: 33 1 43 79 69 18
Fax Phone: 33 1 40 09 05 20
Contact # 1: Pascal Gauchet
Business Description: Artistic holography; Buying & Selling; Consulting
****

Australian Holographies
PO. Box 160
City: Kangarilla
Postal Code: S.A. 5157
Country: Australia
Voice Phone: 61 8 8383 7255
Fax Phone: 61 8 8383 7244
email: austholo@camtech.net.au
Web Address: camtech.com.aul-austholo
Contact # 1: Simon Edhouse
Business Description: Specialists in Large Format holography. Official commercial parmers to the South Australian Museum. Rainbow, lasertransmission, and Denisyuk holograms to 1.1 - 2.2 m. Reflections to 1.1 - 1.1 m. CW and Pulsed. Optical and laser sales. Studio rental. Stock image bank.
SEE OVRADVERTISEMENT
****

Autoadesivi Sri
Via Verdi 57
City: Soliera Modena
Postal Code: 41100
Country: Italy
Voice Phone: 39 59 56 1006
Fax Phone: 39 59 565 146
Contact # 1: Giorgio Korradai
****

Automated Holographic Systems
11 Stephanie Lane
City: Westfield
State/Province: MA
Postal Code: 0 1 085
Country: United States of America
Voice Phone: 1 413 568 2897
email: ahologram@earthlink.net
Contact #1: Jim Gibb
Business Description: Holographic consultant with 20 years of industry experience. Designs and manufacturers state of the art production equipment. Specializing in film DCG, embossed image ganging, and HOEs.
****

Avant-Garde Studio
34 North Rochdale Ave.
PO. Box 296
City: Roosevelt
StatelProvince: NJ ~
Postal Code: 08555-0296
Country: United States of America
Voice Phone: 1 609 448 6433
Fax Phone: 1 609 448 6433
email: avant3d@ao1.com
Contact # 1: Amy E. Medford
Business Description: Modelmakers. Sculptors, excelling in relief (forced perspective, texture, form), for holographic models. We will design and/or work from, photographs, drawings or other material.
****

Baier Praegepressen
Maschinenfabrik Gebr. Baier KG
Lindenthaler Str. 78
City: Rudersberg
Country: Gennany
Voice Phone: +49 (0)718 532
Fax Phone: +49 (0)718 3481
Business Description: Manufacturer of machines for production of embossed holograms.
****

Barilleaux, Rene Paul
Mississippi Musewn of Art
20 1 East Pascagoula St.
City: Jackson
State/Province: MS
Postal Code: 39201

Country: United States of America
Voice Phone: 1 60 1 960 1515
Contact #1 : Rene Paul Barilleaux
Business Description: Director of Exhibitions and Chief Curator. Organized New Directions in Holography: The Landscape Reinvented - a popular traveling holography exhibition. Currently working with M.I.T. Museum's holography exhibit.
****

Barr & Stroud, Ltd.
1 Linthouse Rd
City: Glasgow
State/Province: Scotland
Postal Code: G51 4BZ
Country: United Kingdom
Voice Phone: 44 141 4404000
Fax Phone: 44 141 4404001
Contact # 1: George Brown
Business Description: Several divisions to this business. We manufacturer Heads Up Display, Holographic gratings, and made to order HOE items.
****

Batelle Pacific Northwest National Laboratory
PO. Box 999
K8-09
City: Richland
State/Province: WA
Postal Code: 99352
Country: United States of America
Voice Phone: 1 5093754505
Fax Phone: 1 509 372 4373
email: ma_lind@pnl.gov
Contact #1: Michael Lind
Business Description: Theoretical and experimental research and development in applying holographic techniques to active and passive optical, radar, acollstic, ultrasound, EEG and EKG imaging.
****

Bbt Instnlmenter Aps.
Dronning Olgasvej 6
City: Fredetiksberg
Postal Code: DK-2000
Country: Denmark
Voice Phone: 45 0 1 198 208
Fax Phone: 45 01 198747
Business Description: Branch office of Newport Corporation.
****

Beddis Kenley (Machinery) Ltd.
Unit 3, Eiland Terrace
City: Holbeck, Leeds
StatelProvince: England
Postal code: LSI] 9NW
Country: United Kingdom
Voice Phone: 44 1132 465 979
Fax Phone: 44]132 425 400
Contact # 1: S.D. Smith
Business Description: Large sheet hot foil stamping machines for graphic enhancement and hologram application. MaximUl11 sheet size 29 x 41 (73cm x 104cm).
****

Beijing Fantastic Hologram Product Corp
PO. Box 2542
5 Xueyuan Road
City: Beijing
Postal Code: 100083
Country: People's Rep. of China
Voice Phone: 86 1 208 3810
Voice Phone: 86 1 205 2338
Fax Phone: 86 1 208 3810
Contact #1: Liu Zhiwei
****

CORPORATE LISTINGS

Beijing Hologram Printing Technology Co
PO, Box 9622 ext 2
82 Zhichuan Road
City: Beijing
Postal Code: 100086
Country: People's Rep, of China
Voice Phone: 86 1 837 9223
Fax Phone: 86 1 837 8302
Contact #1: Wang Quanhong
****

Beijing Inst. of Posts
PO, Box 163
1000088 Beijing Shi
City: Beijing
Postal Code: 100086
Country: People's Rep. of China
Voice Phone: 86 1 202 8561
Fax Phone: 86 1 202 8643
Contact # 1: Hsu Dahsiung
****

Beijing Institute Of Posts
And Teleconununications
Applied Physics (Holography)
City: Beijing
Postal Code: 10080
Country: People's Rep, of China
Voice Phone: 86 1 668 1255,
Contact # 1: Hsu Da-Hsiung
Business Description: College courses in holography
****

Beijing Normal University.
Analysis And Testing Centre
City: Beijing
Postal Code: 100875
Country: People's Rep, of China
Contact #1: Huang Wanyun
Business Description: Industrial and Scientific
research, Non-Destructive testing
****

Beijing Sanyou Laser Images Co
Room B0712 Huibin Building
8 Beichon Dong Road
City : Beijing
Postal Code: 100101
Country: People's Rep. of China
Voice Phone: 86 1 499 3987
Voice Phone: 86 1 499 3843
Fax Phone: 86 1 499 3945
Contact # 1: Pei Wen
****

Bellini, Victor
850 Howard Ave, # 5J
City: Staten Island
StatelProvince: NY
Postal Code: 1030 I
Country: United States of America
Voice Phone: 1 718 442 4726
Contact # I: Victor Bellini
Business Description: Extensive collection and
archive of holographic, lenticular, and other 3D
collectibles, Appraisals for trading cards,
****

Belmose Production Products
Wayz Goose Dr
City: Derby
State/Province: England
Postal Code: DE2 6XG
Country: United Kingdom
Voice Phone: 44 1332294242
Fax Phone: 44 133 229 0366
****

Berkhout, Rudie
223 West 21 st Street
City: New York

State/Province: NY
Postal Code: 100 II
Country: United States of America
Voice Phone: 1 212 255 7569
Voice Phone: 1 2127270532
Contact # 1: Rudie Berkhout
Business Description: Holographic Fine Artist
who has had work exhibited worldwide, includ-
ing at the Whitney Museum of American Art
(New York), Also teaches holography at the
School ofVisualArts (NY),
****

Better Labels Mnfg Co
No 1 100 Feet Taramani Rd
Ramagiri Nagir
City: Velacheri Madras
Postal Code: 60042
Country: India
Voice Phone: 91 442352777
Fax Phone: 91 44235 1777
Contact # 1: OK, Sutaria
****

BIAS
(Bremer Institute Applied Beam)
Klagenfurter Str 2
City: Bremen
Postal Code: 28359
Country: Germany
Voice Phone: +49 (0)421 218 5002
Fax Phone: +49 (0)421 21~ 5059
Contact # 1: Werner Prof. Juptner
Business Description: Industrial research; holo-
graphic non-destmctive testing,
****

Blue Ridge Holographics, Inc.
511 Stewart St.
City: Charlottesville
State/Province: VA
Postal Code: 22902
Country: United States of America
Voice Phone: 1 804 296 1110
Fax Phone: 1 804 296 1182
Contact # 1: Steve Provence
Business Description: Mastering facility for the
production of embossed holograms, Consulting
and design services offered, Extensive client list.
****

Bobst Group
146 Harrison Avenue
City: Roseland
State/Province: NJ
Postal Code: 07068
Country: United States of America
Voice Phone: 1 201 2268000
Fax Phone: 1 20 1 226 8625
Contact # 1: Doug Herr
Business Description: One of the world's largest
manufacturers of hologram hot stamp machinery,
****

Booth, Roberta
5326 Sunset Blvd,
City: Los Angeles
State/Province: CA
Postal Code: 90027
Country: United States of America
Voice Phone: 1 2134665767
Fax Phone: 1 213 465 5767
Contact # 1: Roberta Booth
Business Description: 1 am a holographic artist
working in transmission and reflection hologra-
phy. 1 also work as a consultant for holographic
projects and curate holography shows,
****

Boyd, Patrick

18 Whiteley Rd
City: London
State/Province: England
Postal Code: SEI9 IJT
Country: United Kingdom
Voice Phone: 44 976 298 578
Fax Phone: 44 181 670 7810
email: patrick@geijubu.tjokuba.ac.jp
Contact # 1: Patrick Boyd
Business Description: Holographic Fine Artist.
Extensive portfolio,
****

Brainet Corporation
4F Asset Bldg
3-3 1-5 Honkomagome Bunkyo-Ku
City: Tokyo
Postal Code: 113
Country: Japan
Voice Phone: 81 3 5395 7030
Fax Phone: 81 3 5395 7029
Contact # 1: Yutaka Inoue
Business Description: Distributor and producer
of all types offilm and glass holograms and other
holographic gifts and stationery goods,
****

Brandtjen & Kluge, Inc"
539 Blanding Woods Road
City: St. Croix Falls
StatelProvince: WI
Postal Code: 54025
Country: United States of America
Voice Phone: 1 715 483 3265
Voice Phone: 1 800 826 7320
Fax Phone: 1 715 483 1640
Contact #1: John Edgar
Business Description: With over 75 years experi-
ence in manufacturing what are acknowledged as
the industry's most reliable presses, Brandtjen &
Kluge is uniquely qualified to deliver presses for
efficient, trouble free application of hologram foils
****

Bridgestone Graphic Technologies, Inc.
375 Howard Ave.
City: Bridgeport
StatelProvince: CT
Postal Code: 06605
Country: United States of America
Voice Phone: 1 203 366 1595
Fax Phone: 1 203 366 1667
Contact # 1: Richard Zucker
Business Description: Product authentication
systems, Anti-counterfeiting technology, Fully
integrated provider of security printed prod-
ucts and services including holography, micro-
tracers, biocoding and field investigation, SEE
OUR ADVERTISEMENT
****

British Aerospace Pic,
Sowerby Research Centre
Fpc: 267 PO Box 5
City: Filton
State/Province: England
Postal Code: BSI2 7QW
Country: United Kingdom
Voice Phone: 44 1179 366 842
Fax Phone: 44 1179 363 733
Contact # 1: Dr. Steve Parker
Business Description: Sheer Holography and
NDT interferometry as it applies to material
stress. Will work on projects that are of mutual
benefit to British Aerospace and client.
****

Broadbent Consulting Services
1075 Linda Vista Dr,
Suite H

City: San Marcos
State/Province: CA
Postal Code: 92069
Country: United States of America
Voice Phone: 1 619752 1039
email: hologram@fia.net
Contact #1: Donald C. Broadbent
Business Description: An independent, privately owned holographic facility producing HOEs and display holograms in various recording materials. Donald Broadbent has 31 years experience in holography.
****

BTG pic
101 Newington Causeway
City: London
State/Province: England
Postal Code: SEI 6BU
Country: United Kingdom
Voice Phone: 44 171 403 6666
Fax Phone: 44 171 403 7586
email: btguk@btgplc.com
Contact # 1: Dr. Eugene Sweeney
Business Description: Leading international technology transfer service company which provides companies with new technologies and processes to remain competitive in the marketplace and to help research organizations and companies license out technologies.
****

Burleigh Instrwnents, Inc.
Burleigh Park
City: Fishers
State/Province: NY
Postal Code: 14453
Country: United States of America
Voice Phone: 1 7169249355
Fax Phone: 1 716 924 9072
Contact # 1: Patty Payne
Business Description: Burleigh Instruments, Inc. is a manufacturer of electro-optical equipment including wavelength meters, laser spectrum analyzers, interferometers, and nanopositioning devices.
****

Cambridge Laser Labs
853 Brown Road
City: Fremont
State/Province: CA
Postal Code: 94539
Country: United States of America
Voice Phone: 1 5106510110
Fax Phone: 1510651 1690
Contact # 1: Brian Bohan
Business Description: World renowned specialist in ion laser repair. Rental systems and used laser system sales. Price guide furnished on request.
****

Canon Inc. R&D Headquarters
890, Kawasaki-Shi
Saiwai-Ku Kawasaki
City: Kanagawa
Postal Code: 211
Country: Japan
Voice Phone: 81 44 549 5424
Contact #1: Tetsuro Kuwayama
Business Description: Research. Courses in holography.
****

Capitano College
Physics Department - Holography Lab
2055 Purcell Way
City: N. Vancouver
Postal Code: V7J 3H5
Country: Canada

Voice Phone: 1 604 986 191 I
Fax Phone: 1 604 984 4985
Contact # 1: Milessa Crenshaw
Business Description: Digital hologram R&D
****

Carl M. Rodia And Associates
13 Locust St.
City: Tnunbull
State/Province: CT
Postal Code: 06611
Country: United States of America
Voice Phone: 1 203 261 1365
Fax Phone: 1 203 268 1619
email: carlrodia@aol.com
Contact # 1: Carl M. Rodia
Business Description: Comprehensive engineering consultation services in precision hologram manufacturing. Plant design and engineering, process engineering, troubleshooting and seminar training of manufacturing personnel.
****

Casdin-Silver Holography
99 Pond Avenue Suite D403
City: Brookline
StatelProvince: MA
Postal Code: 02146
Country: United States of America
Voice Phone: 1 617 739 6869
Voice Phone: 1 617 423 4717
Fax Phone: 1 617 739 6869
Contact # 1: Harriet Casdin-Sitver
Business Description: 1 have been creating holographic art and interactive holographic installations since 1968. Our company specializes in original holograms for advertising, architectural and theater settings, and expositions. We are also consultants and exhibition organizers/designers.
****

Cavomit
22 Pipinou Steet
City: Athens
Postal Code: 11257
Country: Greece
Voice Phone: 30 1 823 2355
Fax Phone: 30 1 231 4499
Contact # 1: Alkis Lembessus
Business Description: Hot-stamping equipment (cylinders - platen), hologram registration systems, foils and consumables. Local distributor for Astor Universal, Xluge, Light Impressions, Applied Holographics, Revere Graphic Products.
****

Central Glass Co., Ltd.
Kowa-Hitosubashi Bldg
7-1 Kanda-Nishikicho 3-Chom
City: Tokyo (Chiyoda-Ku)
Postal Code: 101
Country: Japan
Voice Phone: 81 3 3259 7354
Contact # 1: Chikara Hashimoto
Business Description: Heads Up Display
****

Centre d' Art Holographique et Photonique
College de Maisonneuve (c-2200)
3800 Sherbrooke est
City: Montreal
Postal Code: HIX 2A2
Country: Canada
Voice Phone: 1 514 254 7131 ex4509
Fax Phone: 1 514 253 8909
Contact # 1: Eric Bosco
Business Description: We are situated in a CEGEP (post-secondary school). We have 2 fully equipped tables with HeNe's and an argon (with etalons). We do all types of holograms up to 2 feet by 3 feet. Courses,

workshops, production work and rent lab space.
****

Centro de Arte Holografico
Juan Bautista Alberdi 2784
City: Buenos Aires
Postal Code: 1406
Country: Argentina
Voice Phone: 54 54 612 9163
Contact # 1: Mrs. Cintia Aranguren
****

Centro De Investifaciones Opticas
Casilla de Correro 124
1900 La Plata
Postal Code: 1900
Country: Argentina
Voice Phone: 54 54 218 40280
Fax Phone: 5454217 12771
Contact # 1: Nestor Bolognini
****

Centro de Investigaciones en Optica, A.C.
Lorna del Bosque liS
Col. Lomas del Campestre
City: Leon
Country: Mexico
Voice Phone: 52 47 731017
Fax Phone: 52 47 175000
email: dfa@riscl.cio.mx
Contact #1: Fernando Mendoza, Ph.D.
Business Description: Our Optical Research Center offers: optical design and construction of a wide variety of optical components (lenses, mirrors, prisms, etc.); high standards in R&D in optical NDT for industrial applications, optical fiber sensors, rare earth doped fibers and optical shop testing.
****

CFCApplied Holographies
500 State St.
City: Chicago Heights
StatelProvince: IL
Postal Code: 60411
Country: United States of America
Voice Phone: 1 800 438 4656
Voice Phone: 1 708 891 3456
Fax Phone: 1 708 758 5989
email: 104347.1654@compuserve.com
Contact # 1: Dave Beeching
Business Description: For security, decoration and packaging applications; we offer a full range of tamper evident and authentication labels, hot stamping foils, and release and size coat combinations. Print treatments and custom metallizing available.
SEE OURADVERTISEMENT
****

Checkpoint
Chatham Street
City; Reading
State/Province: England
Postal Code: RG 1 7 JX
Country: United Kingdom
Voice Phone: 44 173 425 825 1
Fax Phone: 44 173 456 9988
email: sales@checkpoint.co.uk
Contact # 1: Keith Hatton
Business Description: Manufacturers Security Hologram Applicator which is used to apply hologram seals to checks for security at point of issue.
****

Cherry Optical Holography
2047 Blucher Valley Road
City: Sebastopol
StatelProvince: CA
Postal Code: 95472
Country: United States of America
Voice Phone: 1 707 823 7171

CORPORATE LISTINGS

Fax Phone: 1 707 823 8073
Contact # 1: Greg Cherry
Business Description: Highest quality display holography avai lable. Stock and custom reflection/ transmission holograms on glass plates or film up to 40 x 72 in size. Open and Limited edition fine art holograms. Custom mastering services offered for silver halide and photopolymer replication.
SEE OURADVERTISEMENT
****

Chiba University
Faculty Of Engineering
1-33 Yayoi-Cho
City: Chiba
Postal Code: 260
Country: Japan
Voice Phone: 81 472 511 III
Contact # 1: Jumpei Tsujili.:hi.
Business Description: Scientific research; holographic
****

China Ann Arbor Holographical Institute
Rm. 403 Bldg. 22
Zhong Lou Xing Chen
City: Jiangsu
Postal Code: 215006
Country: People's Rep. of China
Voice Phone: 86 512 227 461
Contact # 1: Prof. Yaguang Jiang
Jiang
****

CHIRONTechnoias GmbH
Max-Planck Strasse 6
City: Dornach
Postal Code: 85609
Country: Gennany
Voice Phone: +49 (0)89 945514 0
Fax Phone: +49 (0)89 945514 70
Contact # 1: Mr. Junger
Business Description: Ophthalmologic Systems.
****

Chromagem Inc.
573 South Schenley
City: Youngstown
StatefProvince: OH
Postal Code: 44509
Country: United States of America
Voice Phone: 1 330 793 35 15
Fax Phone: 1 330 793 35 15
Contact #1: Thomas 1. Cvetkovich
Business Description: Established in 1981. Hologram mastering facility specializing in shooting photoresist masters for use in commercial mass-production of embossed holograms: 2D, 3D, stereogram and dot matrix. Experience working with major corporate accounts. Also provide design and consultation services.
****

Chronomotion
424 Ninth St.
City: Santa Monica
State/Province: CA
Postal Code: 90402
Country : United States of America
Voice Phone: 1 310 393 9859
Fax Phone: 1 31 0 458 6269
email: mburney@ix.netcom.com
Contact # 1: Michael Burney
Business Description: Developed and patented a general process for producing electronic holograms with a real image projected into the room with the viewer.
****

Cifelli, Dan
712 Bancroft Road # 332
City: Walnut Creek

State/Province: CA
Postal Code: 94598
Country: United States of America
Voice Phone: 1 510 930 8033
Fax Phone: 1 5 10 9308033
email: 74544.1630@compuserve.com
Contact # 1: Dan Cifelli
Business Description: Holography consulting and brokering since 1979 for stock and custom products. Evaluate, match, and develop you product for holography mastering and production techniques (photopolymer, embossed, and dichromate); sales potential and market positioning (ASI, premiums, specialty, secwity/anti-coWlterfeiting etc.); manufacturing and converting processes/materials; cost analysis; and patent/licensing potential.
****

Cise Spa Teclmologie Innovative
P.O. Box 1208 1
City: Milano
Postal Code: 1-20134
Country: Italy
Voice Phone: 39 2 2167 2634
Fax Phone: 39 2 2167 2620
Contact # 1: M. Luciana Rizzi
Business Description: Various R&D on HOEs and NDT holography.
****

CISE Tecnologie Inn"vative SpA
Holographic Lab.
P.O. Box 12081
City: Milano
Postal Code: 1-20134
Country: Italy
Voice Phone: 39 2 2 167 2634
Fax Phone: 39 2 2167 2620
Contact # 1: Pierino Delvo
****

City Chemical
100 Hoboken Ave.
City: Jersey City
StatefP rovince: NJ
Postal Code: 07310
Country: United Slates of America
Voice Phone: 1 201 653 6900
Voice Phone: 1 800 2482436
Fax Phone: 1 201 653 4468
Business Descrip tion: Photo-chemicals and chemical supplies.
****

Claudette Abrams
22 Bayview Avenue
City: Toronto
Postal Code: M5J lZI
Country: Canada
Voice Phone: 1 416 203 7243
Fax Phone: 1 416 203 7243
Contact # 1 : Claudette Abrams
Business Description: Holographic artist and technician.
****

Clemenger Perth Fry Ltd
128 Hay Street
City: Subiaco
Postal Code: WA 6008
Country: Australia
Voice Phone: 61 93807777
Fax Phone: 61 9 380 77 11
Contact # 1 : Rick Schaefer
****

Coated Specialties
330 Kaliandas Udyog Bhavan
Century Bazar Lane
City: Worli Bombay

Postal Code: 400205
Country: India
Voice Phone: 91 22 430 9583
Voice Phone: 91 43 04033
Fax Phone: 91 22 430 4033
Contact # 1: Hemant Jain
****

Coburn Corporation
1650 Corporate Road West
City: Lakewood
State/Province: NJ
Postal Code: 0870 I
Country: United States of America
Voice Phone: 1 908 367 55 11
Fax Phone: 1 908 367 2908
email: cobumcorp@aol.com
Web Address: coburn.com
Contact # 1: John White
Business Description: Offers a wide range of stock repeating geometric holographic designs in pressure sensitive film; conventionally printable substrate; various traditional and designer oriented colors avai lable.
SEE OURADVERTISEMENT
****

Coherent Luebeck GmbH
Seelandstrasse 9
City: Luebeck
Postal Code: 23569
Country: Germany
Voice Phone: +49 (0)45 1 39 09300
Fax Phone: +49 (0)451 39 5725
Business Description: Manufacturer of diode laserpumped solid state lasers which operate in CW and pulsed mode with wavelengths in IR, visible and UV Branch office: 636 Great Road, Stow, MA 01775, USA. Owned by Coherent, Santa Clara, CA, USA
****

Coherent, Inc. - Laser Group
5100 Patrick Henry Drive
City: Santa Clara
State/Provi nce: CA
Postal Code: 95054
Country: United States of America
Voice Phone: 1 408 7644000
Voice Phone: 1 800 527 3786
Fax Phone: 1 408 764 4800
email: tech_sales@cohr.com
Web Address: cohr.com
Contact # 1: Paul Ginouves
Business Description: Coherent is the world leader in high-power ion and diode-pumped solid-state (DPSS) lasers. Our products for professional holography include argon ion lasers (up to 8 W at 488.0 nm, krypton lasers (up to 3.5 W at 647.1 nm), and DPSS lasers (up to 5 W at 532 nm). SEE OURADVERTISEMENT ON THE BACK COVER
****

Concordia University
Communications Studies
7 141 Sherbrooke St. W
City: Montreal
Postal Code: PQ H4B IR6
Country: Canada
Voice Phone: 1 5148482539
Voice Phone: 1 514 848 2424
Fax Phone: 1 514 848 3492
email: hal@vax2.concordia.ca
Contact # 1: Hal M. Thwaites
Business Description: 3 Dimension Research Center
****

Conte M. Vulcano
Via San Crescenziano 19
City: Roma

Postal Code: 1-0 0199
Country: Italy
Voice Phone: 39 6 686 1159
Fax Phone: 39 6 686 21159
Contact #1: Marino Vulcano
****

Continental Optical
15 Power Drive
City: Hauppauge
StatelProvince: NY
Postal Code: 11788
Country: United States of America
Voice Phone: 1 516 582 3388
Fax Phone: 1 516 582 1054
Business Description: Optics and custom orders.
****

Control Module Inc.
380 Enfield Street
City: Enfield
State/Province: CT
Postal Code: 06082
Country: United States of America
Voice Phone: 1 860 745 2433
Fax Phone: 1 860741 6064
email: rbilleri@controlmod.com
Web Address: controlmod.com
Contact # 1: Ralph Billeri
Business Description: Experience in the design and manufacture of automatic data collection equipment and systems, CMI offers exciting innovations in 1996, including Holonetics (TM), a machine-readable hologram, offering the highest security protection available for Access Control and Labor Management.
****

Control Optics Corp.
13111 Brooks Drive, Unit J
City: Baldwin Park
StatelProvince: CA
Postal Code: 91706
Country: United States of America
Voice Phone: 1 818813 1991
Fax Phone: 1 818813 1993
email: liucoc@interserv.com
Web Address: http://ourworld.compuserv!home page/control optics
Contact #1: Wai-Min Liu
Business Description: Provides full-service optical engineering supporting industry and education. Offers full range of holographic table top optics, positioning devices and mounts. New products include holography and fiberoptic experimenter's kits.
SEE OURTWOADVERTISEMENTS
****

Corion Corp.
8 East Forge Parkway
City: Franklin
State/Province: MA
Postal Code: 02038
Country: United States of America
Voice Phone: 1 508 528 44 11
Fax Phone: 1 508 520 7583
Contact #1: Don McLeod
Business Description: Corion Corp. manufactures volume and one-of-a-kind custom and stock optical components including coatings, filters, optics and optical assemblies for use in the UV-Visible-IR spectrum. Mostly biomedical applications.
****

Courtauld Acetate
PO. Box 5, Spondon
City: Derby
Postal Code: DE2 7BP
Country: United Kingdom
Voice Phone: 44 133 266 1422
Fax Phone: 44 133 266 0178
Contact # 1: Richard Ford

Business Description: die acetate film
****

Creative Holography Index, The
The International Catalog for Holography
46 Crosby Road
City: West Bridgford
Postal Code: NG2 5GH
Country: United Kingdom
Voice Phone: 44 7050 133 624
Fax Phone: 44 7050 133 625
email: pepper@monand.demon.co.uk
Contact # 1: Andrew Pepper
Business Description: The Creative Holography Index is an international catalogue, in colour, and includes an artist produced hologram. Available as the complete collection. It features altists working with holography as a creative medium and includes critical essays, biogr phies, statements and a hologram. Cost 89.95 Sterling.
****

Creative Label
2450 Estes Dri ve
City: Elk Grove Village
State/Province: IL
Postal Code: 60007
Country: United States of America
Voice Phone: 1 847 956 6960
Fax Phone: 1 847956 8755
Contact # 1: Jerry Koril
Business Description: Full range decorative graphic finishers. Large volume capability. Bindery application of holograms to paper, cardboard, and plastics. Kluge (2 stream) and Bobst (4 stream) machines. Call for more information.
****

Crown Roll Leaf, Inc.
Holo-Grafx Division
12 Columbia Ave.
City: Paterson
State/Province: NJ
Postal Code: 07503
Country: United States of America
Voice Phone: 1 201 7424000
Voice Phone: 1 800 631 3831
Fax Phone: 1 20 1 742 0219
Contact #1: Kathy Kas~over
Contact #2: Stewal1 Glazer
Business Description: Crown Roll Leaf is a major manufacturer of hot stamp foils suited for holographic applications. In addition, our in house production facilities are capable of full origination through finishing and converting.
SEE OURADVERTISEMENT
****

Curt Abramzik
Goethestr. 67
City: Offenbach
Country: Gennany
Voice Phone: +49 (0)69 8849 11
Contact #1: Curt Abramzik
Business Description: Manufacturer, importer and exporter of magician equipment and novelties. Holograms up to 100 x 100 cm, reflection and transmission holograms on film and glass, multiplex holograms, dichromate-and embossed holograms, diffraction foils.
****

Customer Service Instrumentation
Meadowfield Park South
City: Stocksfield
State/Province: England
Postal Code: NE43 7QA
Country: United Kingdom
Voice Phone: 441661 842741
Fax Phone: 44 1661 842288
email: ghscott@netcom.co.uk
Business Description: Manufacture front sur-

face mirrors and optics for holography.
****

CVI Laser Corporation
200 Dorado Place
City: Albuquerque
StatelProvince: NM
Postal Code: 87192
Country: United States of America
Voice Phone: 1 505 296 9541
Fax Phone: 1 505 298 9908
email: cvi@cvilaser.com
Contact # 1: Bob Soales
Business Description: Manufactures holographic quality single and multiple element lenses, mirrors, windows, and beam splitters for all standard holographic laser sources. Free 104-page catalog available.
****

Czechoslovak Academy Of Science
Institute Of Physics
Na Siovance 2
City: Prague
Postal Code: 180 40
Country: Czech Republic
Voice Phone: 42 84 22 419
Contact # 1: Josef Horvath
Business Description: Holography research
****

Dai Nippon Printing Co., Ltd.
Central Research Institute
250-1 Aza-Kahasawa
City: Kashiwa-City
State/Province: Chiba
Postal Code: 277
Country: Japan
Voice Phone: 81 471340512
Fax Phone: 81 4 7133 2540
Contact # 1: Takashi Wada
Business Description: Central research center. Embossed holography research.
****

Daimler Benz Aerospace
Dornier Medizinteclmik GmbH
Industriestrasse 15
City: Germering
Postal Code: 82110
Country: Gemlany
Voice Phone: 49 89 84108 0
Fax Phone: 49 89 84108 575
Contact #1: Ms. Thiemon
Business Descript ion: Industrial Research; holographic non-destmctive testing. HOE research.
****

Dan Han Optics
188-261 An Nyeong-Ri Tean-Eup
City: Hwasong-Gun
Country: South Korea
Voice Phone: 82 0331 351 030
Fax Phone: 82 0331 351 031
Contact # 1: Chung Song
Business Description: General optical supplies.
****

Datacard Corporation
11111 Bren Road West
City: Minneapolis
State/Province: MN
Postal Code: 55440
Country: United States of America
Voice Phone: 1 612933 1223
Fax Phone: 1 612 931 0418
Contact # 1: Mark Iverson
Business Description: Security and authentication applications utilizing holographic technologies. Capable of high volume runs for government and commercial users.
****

CORPORATE LISTINGS

Datasights Ltd.
Alma Road
Ponders End
City: Enfield
State|Province: England
Postal Code: EN3 7BB
Country: United Kingdom
Voice Phone: 44 181 8054151
Fax Phone: 44 181 805 8084
Contact # 1: Frank Sharpe
Business Description: Manufacture mirrors for use in holography. Beamsplitters and gratings also available.
****

Db Electronic Instruments S.R.L.
Via Teano 2
City: Milano
Postal Code: 1-20161
Country: Italy
Voice Phone: 39 02 646 934
Fax Phone: 39 02 645 6632.
Business Description: Newport Co. branch office. Laser supplier.
****

De La Rue Holographics Ltd.
Stroudley Road
Daneshill Industrial Estate
City: Basingstoke
State/Province: England
Postal Code: RG24 8FW
Country: United Kingdom
Voice Phone: 44 1256463000
Fax Phone: 44 1256460800
Contact # 1: Phillip M.G. Hudson
Business Description: De La Rue Holographics is the brand protection company of De La Rue pic, offering customers high technology protection against product counterfeiting and tampering at the point-of-sale, through the production of optical microstructures (holograms) and KINEGRAMs.
****

Deem, Rebecca
709 1/2 West Glen Oaks Blvd
City: Glendale
State/Province: CA
Postal Code: 91202
Country: United States of America
Voice Phone: 1 818 549 0534
Fax Phone: 1 818 549 0534
Contact # 1 : Rebecca Deem
Business Description: Holographic artist. Originates masters for mass production holograms in embossed, DCG and photopolymer materials . Both pulsed and CW lasers available.
****

Deep Space Holographics
1070 Moss Street # [05
City: Victoria
Postal Code: V8V 4P3
Country: Canada
Voice Phone: 1 6043843927
email: marcus@freenet.victoria.bc .ca
Contact #1: Marcus de Roos
Business Description: Exotic fine art/commercial sculpture/animation, conceptual/industrial design, display merchandising, exhibits and special effects. Since 1980 secured worldwide distribution of our DCG designs via Holocrafts, including Star Trek holograms design.
****

DeFreitas, Frank
PO. Box 9035
City: Allentown
State/Province: PA
Postal Code: 18105-9035

122 Holography MarketPlace - 6th edition
Country: United States of America
Voice Phone: 1 6107700341
email: director@holoworld.com
Web Address: holoworld.com
Contact #1: Frank DeFreitas
Business Description: Designer of the Internet Webseum of Holography - a multi-award winning web site dedicated to amateur and hobbyist holography.
****

Dell Optics Company, Inc.
25 Bergen Blvd.
City: Fairview
State|Province: NJ
Postal Code: 07022
Country: United States of America
Voice Phone: 1 201 941 [010
Fax Phone: 1 20 1 941 9524
Contact # 1: Belle Steinfeld
Business Description: Custom working of precision optical components. Established 1950. 15 Employees at this address.
****

Denisyuk, Yuri N.
A.F.Ioffe Physicotechnical Institute
Politechnicheskaya 26
City: St. Petersburg
Postal Code: 194021
Country: Russia
Voice Phone: 7 812 247 9384
Contact # 1: Yuri N. Denisyuk
Business Description: Holography teacher. One of the founders of holography.
****

Deutsche Gesellschaft fur Holografie
Geschaeftsstelle
Lutterdamm 82
City: Bramsche
Postal Code: 49565
Country: Germany
Voice Phone: +49 (0)546 91124
Fax Phone: +49 (0)546 91122
Contact # 1: Brigitte Burgmer
Business Description: The society was founded to promote awareness of holography, and its members are mainly holographers and artists. To this end, the group intends to organize exhibitions. Interferenzen is a periodical published by this organization.
****

Dialectica Ab
Skanegatan 87
6Tr
City: Stockholm
Postal Code: S-116 37
Country: Sweden
Contact # 1: Ambjorn Naeve
Business Description: Artistic holography.
****

Diamond Images, Inc.
PO. Box 1701
City: Miami
State/Province: FL
Postal Code: 33133
Country: United States of America
Voice Phone: 1 305 323 8406
Fax Phone: 1 305 770 1977
email: mdiamond@concentric.net
Web Address: DiamondImages.com
Contact #1: Mark Diamond
Business Description: Holographer Mark Diamond brings 23 years experience to full color stereograms. Work is featured in museums and collections in 15 countries. Founding member of Museum of Holography New York. Specializing

in portraiture and animated digital compositing.
****

Diaures S.A. Holography Division
Via 1 Maggio 262/ A
1-41019 Soliera
City: (Modena)
Country: Italy
Voice Phone: 39 059 567 274.
Business Description: Artistic holography; embossed holography; equipment & supplies.
****

Diavy sri
Via Vivaldi 108
City: soliera (Modena)
Postal Code: 41019
Country: Italy
Voice Phone: 39 59 565758
Fax Phone: 39 59 566074
Contact # 1: Alesandro Dondi
Business Description: Subsidiary ofDiaures; producer of holographic metallic paper as part of a venture with Scharr Industries, USA.
****

Die Dritte Dimension
Frankfurter Strasse 132-134
City: Neu Isenburg
Postal Code: 63263
Country: Germany
Voice Phone: +49 (0)610 33367
Fax Phone: +49 (0)610 326709
Contact #1: Elke Hein
Business Description: Greatest specialized shop for holography in Germany. Always over 1,000 different holograms in stock. Very comprehensive fine art section. Branch office: Nordwest-Zentrum, Tituscorso, 60439 Frankfurtt. Germany.
****

Dietmar Oehlmann
Bortfelder Stieg 4
City: Braunschweig
Postal Code: 38116
Country: Germany
Voice Phone: +49 (0)531 352 816
Fax Phone: +49 (0)531 352 816
Contact # 1: Dietmar Oehlmann
Business Description: Master of Arts in Holography from the Royal College of Arts, with own light creation lab to design and produce special effects in holography for artworks, performances and stage decoration.
****

Diffraction Ltd.
PO. Box 909
Route 100
City: Waitsfield
State|Province: VT
Postal Code: 05673
Country: United States of America
Voice Phone: 1 802 496 6642
Fax Phone: 1 802 496 6644
email: hologram@madriver.com
Contact # 1: Bill Parker
Business Description: Products and services relating to diffractive optics and holographic optical elements (HOEs) including micro fabrication and photomask production.
****

Dimension 3
3380 Francis-Hughes St.
City: Laval
Postal Code: H7L 5A 7
Country: Canada
Voice Phone: 1 514 662 0610
Fax Phone: 1 514 662 0047
email: dimension3@ctnet.net

Contact #1: Piene Gougeon
Business Description: We offer creative solutions to holograp hic projects. We are a full holographic production house (DCG, foil , transmission, photopolymer), large fonnat and micro embossed with animation and colour control Photograms (TM) (lenticular photography/ printing).
****

DimensionalArts
40 1 Carver Road
City: Las Cmces
State/Province: NM
Postal Code: 88005
Country: United States of America
Voice Phone: 1 505 527 9183
Fax Phone: 1 505 527 9927
email: al1s@holo.com
Web Address: www.holo.com
Contact # 1: Ken Hanis
Business Description: Exclusive manufacturer of the Light Machine, a patent protected digital origination system. Custom stock Dotz(r) dot matrix patterns available. Capable of 2D, 3D and full color stereogram work. Can transfer technology worldwide.
****

Dimensional Cinematography Co.
PO. Box 9994
City: North Hollywood
State: CA
Postal Code: 91609
Country: United States of America
Voice Phone: 1 818367-8080
Contact # I: Glenn Gustafson
email: gustaf@primenet.com
Business Description: Cinema-Photographer with experience originating holographic stereograms.
****

Dimensional Foods Co.
8 Faneuil Hall Market Place
City: Boston
State/Province: MA
Postal Code: 02109
Country: United States of America
Voice Phone: 1 617 973 6465
Fax Phone: 1 617 973 6406
Contact # 1: Erich Begleiter
Business Description : Scientific and artistic research. Licensing a proprietary micro relief process to food manufacturers for producing chocolate and hard candy holograms.
****

Dimensions
Taj Pura
City: Sialkot
Country: Pakistan
Voice Phone: 92 432 85 197
Voice Phone: 92 432 66006
Fax Phone: 92 432 558336
Contact #1: Mr. Shahjahan
Business Description: International agents and importers of holograms, diffraction foils and other holographic products.
****

Dimuken
33 Stapledon Rd
Orton Southgate
City: Peterborough
State/Province: England
Postal Code: PE2 6TD
Country: United Kingdom
Voice Phone: 44 1733 230 044
Fax Phone: 44 1733 230 012
Contact # 1: John Bentley
Business Description: Manufactures holographic hot stamping machinery which can do

hot stamping or blind embossing by switching machinery components.
****

Direct Holographies
PO Box 295
City: Strasburg
StatelProvince: PA
Postal Code: 17579
Country: United States of America
Voice Phone: 1 717 687 9422
Voice Phone: 1 888 43Dspex
Fax Phone: 1 717 687 9423
email: directholo@cpcnet.com
Contact #1: Jacque Phillips
Business Description: Manufacturer of SHOCK SPEX (holographic sunglasses), SHOCK MUGS, SHOCK STEINS and SHOCKFOBS. Distributor of a large range of silver-halide, embossed and photopolymer holograms. Line includes embossed stickers and magnets. Custom inquires invited.
SEE OURADVERTISEMENT
****

Doris Vila Holographics
445 Grand Street
City: Brooklyn
State/Province: NY
Postal Code: 11211
Country: United States of America
Voice Phone: 1 718 338 6533
email: vi la@dorsai.org
Contact # I: Doris Vila
Business Description: Custom holography in state-of-the-art in-house lab, silver halide limited editions, architectural -scale & fine-art orginals, mastering and transfers, consultations and classes available by appointment.
****

Dornier Medizintechnik GmbH
Industriestrasse 15
City: Gennering
Postal Code: 821 10
Country: Gennany
Voice Phone: +49 (0)89 84 108 0
Fax Phone: +49 (0)89 84108 575
Contact # 1: W Langer
Business Description: Industrial Research; holographic non-destructive testing. HOE research. Medical systems in Lithotripsy, Surgery, Orthopaedics.
****

DuPont
(see E.I. DuPont De Nemours & Co.)
State/Province: DE
Country: United States of America
****

Dutch Holographic Laboratory BV
Kanaaldijk Noord 61
City: Eindhoven
Postal Code: 5642JA
Country: Netherlands
Voice Phone: 31 40 281 7250
Fax Phone: 31 40 281 4865
email: walter@IAEhv. nl
Web Address: euroweb.comlDHL
Contact # 1: Walter Spierings
Business Description: Manufacturer of Holoprinter and Holotrack equipment. Production of holograms on silver halide, photoresist and photopolymer. Computer-generated holograms and multiple photo-generated holograms (MPGH). Also traditional recording techniques.
SEE OURADVERTISEMENT
****

E.I. DuPont De Nemours & Co.
Holographic Materials Division

P O. Box 80352
City: Wilmington
State/Province: DE
Postal Code: 19880-0352
Country: United States of America
Voice Phone: 1 302 695 4893
Fax Phone: 1 302 6954635
Contact # 1: Paula Bobeck
Business Description: Manufacturer of photopolymer emulsions for sale to holography businesses.
SEE OURADVERTISEMENT
****

Ealing Electro-Optics Inc.
89 Doug Brown Way
City: Holliston
State/Province: MA
Postal Code: 01746
Country: United States of America
Voice Phone: 1 5084298370
Voice Phone: 1 8003434912
Fax Phone: 1 508 429 7893
Business Description: Ealing manufactures and distributes optical tables and benches, subminiature manual and controlled positioners, optical mounts, custom optics, pinholes, filters, HeNe Lasers, spatial filters, laser mounts, Tungsten sources, interferometers, textbooks, and off the shelf lenses and minors.
****

Eastman Kodak Company
343 State St.
City: Rochester
State/Province: NY
Postal Code: 14650-0811
Country: United States of America
Voice Phone: 1 800 242 2424
Voice Phone: 1 800 823 4474
Fax Phone: 1 800 4979
Business Description: Manufacturer ofsilver halide recording materials. Glass plates & film. See chapter Recording Materials in this book for further infonnation.
****

Edmund Scientific Company
10 1 East Gloucester Pike
Ci ty: Banington
State/Province: NJ
Postal Code: 08007
Country: United States of America
Voice Phone: 1 609 547 3488
Fax Phone: 1 609 573 6295
email: scientifics@edsci.com
Business Description: Mail-order catalogue, wholesale, and retail. We offer one of the largest selection, of precision optics and optical components and accessories for the optical lab. Holography products for schools, science fairs, etc.
****

EI Don Engineering
4629 Platt Rd.
City: Ann Arbor
StatelProvince: MI
Postal Code: 48408
Country: United States of America
Voice Phone: 1 313 973 0330
Contact # 1: Don Gillespie
Business Description: Surplus and refurbi shed lasers for the holographer. Full warranty. Techni cal request calls welcomed.
****

Electro Optical Industries, Inc.
859 Ward Drive
City: Santa Barbara
State/Province: CA
Postal Code: 93 111

CORPORATE LISTINGS

Country: United States of America
Voice Phone: 1 805 964 6701
Fax Phone: 1 805 967 8590
Contact # 1: Joseph Lansing
Business Description: Manufacturer of infrared test and calibration instmmentation including: collimators, choppers, blackbody sources, differential temperature sources, FUR test equipment, radiometers and LLL-TV target simulators.
****

Electro Optics Developments Ltd.
Howards Chase
Pipps Hill Industrial Estate
City: Basildon
State/Province: England
Postal Code: SS14 3BE
Country: United Kingdom
Voice Phone: 44 1268 531 344
Fax Phone: 44 1268 531 342
Contact # 1: Chris Varney
Business Description: Made to order optics such as HOE items, gratings, etc.
****

Electro-Optics Lab, NECTEC
King Mongkut's Institute of Technology
Chalongkrung Road, Ladkrabang
City: Bangkok
Postal Code: 10520
Country: Thailand
Voice Phone: 66 2 326 9045
Fax Phone: 66 2 326 9045
email: fkh@nwg.nectec.or.th
Business Description: A national lab that is also Thailand's first hologram manufacturer. Produces embossed holograms and photopolymer holograms. Provides service in training, consulting and hologram mastering. Also conducts academic research in holography, photonics and optoelectronics.
****

Elusive Image
603 Munger Street, # 213
City: Dallas
State/Province: TX
Postal Code: 75202
Country: United States of America
Voice Phone: 1 214 720 6060
Contact # 1: Argelia Lopez
Business Description: Holography gallely and store located in Dallas 's West End Marketplace. Extensive collection available for exhibition and sale.
****

Embossing Technology Ltd
Steepmarsh, Nr Petersfield
City: Hants
State/Province: England
Postal Code: GU32 2BN
Country: United Kingdom
Voice Phone: 44 1730 895 390
Fax Phone: 44 1730 894 383
Business Description: Wide web embossing by contract. Also stock images. Also for sale is complete system for originating embossed holograms including laser.
****

Engineering Animation, Inc.
2321 N. Loop Drive
City: Ames
State/Province: IA
Postal Code: 50010
Country: United States of America
Voice Phone: 1 515 296 9408
Fax Phone: 1 515 296 7025
Contact # 1: Brad Shafer
Business Description: EAI develops, produces

and sells 3D animation products that address visualization, animation and graphics needs of its customers. Products include: 3D interactive software titles on CD ROM; animation software (UNIX); and custom 3D computer animations.
****

Environmental Education and Information Ctr.
ap.16 st.Resnitskaya 9a
City: Kiev
Postal Code: 252011
Counny: Ukraine
Voice Phone: 380 044 265 9968
email: eeic@gluk.apc.org
Contact # 1: Alexander Gnatovskii
Business Description: Holography correlators, R&D HOE, interferometry, holographic correction of laser beams.
****

ETA-Optik Gmbh
Niethausener Strasse 15
City: Heinsberg
Postal Code: 52525
Country: Germany
Voice Phone: +49 (0)245 66654
Fax Phone: +49 (0)245 64433
Contact # 1: Wilbert Dr. Windeln
Business Description: DCG pendants, diffraction gratings and custom HOE
****

Evolution Design, Inc.',
570 SW 181st Way
City: Pembroke Pines
State/Province: FL
Postal Code: 33029
Country: United States of America
Voice Phone: 1 305 534 9808
Fax Phone: 1 305 534 9808
email: pliberato@aol.com
Contact #1: Pablo Liberato
Business Description: A design, production and management firm that is dedicated to the integration and conceptualization of product design packaging and holography (embossed or photopolymer) from the beginning up to completion of the process.
****

Excitek Inc.
277 Coit Street
City: Irvington
State/Province: NJ
Postal Code: 07111
Country: United States of America
Voice Phone: 1 201 372 1669
Fax Phone: 1 20 1 372 8551
Contact # 1: George Cubberly
Business Description: Supplier of re-manufactured argon and krypton ion laser tubes, and used laser systems. Established in 1984. 10 Employees at this address.
****

Expanded Optics Limited
Moon Lane
City: Barnet
State/Province: England
Postal Code: EN5 5ST
Country: United Kingdom
Voice Phone: 44 181 441 2283
Fax Phone: 44 181 449 6143
Contact # 1: T.R. Hollinsworth
Business Description: Manufacturer of medical and industrial endoscopes; HOEs used in microprecision optics for medical viewing.
****

Fantastic Holograms
PO. BOX 492026

8400 Pena Blvd. (OIA Terminal - Level 5)
City: Denver
StatelProvince: CO
Postal Code: 80249
Country: United States of America
Voice Phone: 1 303 342 3440
Fax Phone: 1 303 342 3440
Contact # 1: RB Osada
Business Description: Well stocked holography store selling a variety of unique giftware including holographic pictures, jewelry, executive gifts, books, and optical novelties.
****

Far East Holographics
12/F Hang Wai Commercial Bldg
231-233 Queen's Road East
Ciry: Wanchai
Country: Hong Kong
Voice Phone: 852 2 893 9773
Fax Phone: 852 2 893 0640
Contact # 1: Adrian J. Halkes
Business Description: Finisher and distributor of holograms and holographic products.
****

Fast Light lnc.
940 1 Santa Fe Dr.- Suite 103
City: Overland Park
State/Province: KS
Postal Code: 662 12
Country: United States of America
Voice Phone: 1 913 649 4666
Fax Phone: 1 913 649 6751
Contact # 1: Gene Davis
Business Description: Creates specialty products which integrate holograms. Custom manufacturing services offered.
****

Feofaniya Ltd.
PO. Box 164
Ciry: Kiev
Postal Code: 252191
Country: Ukraine
Voice Phone: 3800442665108
Voice Phone: 380 044 261 4343
Fax Phone: 380 044 261 4343
email: eeic@gluk.apc.org
Contact # 1: Sergey Kornienko
Business Description: Non destmctive testing, pulse holography, embossing & shim making, production of holography stickers. Holography portraits studio.
****

Feroe, James
1420 45th Street #33
City: Emeryville
StatelProvince: CA
Postal Code: 94608
Country: United States of America
Voice Phone: 1 510 658 9787
Fax Phone: 1 510 658 9788
Contact # I: James Feroe
Business Description: Consultant with 16 years hands-on experience in holography: silver-halide reflect ion and transmission, photoresist and embossed.
****

Fielmann-Verwaltung KG
Weidestrasse 1 18a
Ciry: Hamburg
Country: Gennany
Voice Phone: +49 (0)40 2707 60
Fax Phone: +49 (0)40 2707 6399
Contact # 1 : Uta Kerpen
Business Description: Optician, 159 stores, Collector of Holograms, Exhibitions.
****

Fisher Scientific
Educational Materials Division
485 South Frontage Road
City: Burr Ridge
State/Provi nce: IL
Postal Code: 60521
Country: United States of America
Voice Phone: 1 800 955 1177
Voice Phone: 1 6306554410
Fax Phone: 1 630 655 4335
Web Address: fisheredu.com
Business Description: Supplies science lab equipment including holography kits, lab manuals, lasers and laser related equipment.
****

FLEXcon
I FLEXcon Industrial Park
City: Spencer
State/Province: MA
Postal Code: 01562-2642
Country: United States of America
Voice Phone: 1 508 885 8440
Fax Phone: 1 508 885 8399
Business Description: Manufacturer of holographic and prismatic materials used for authentication and decoration. Holograms can be combined with overt and covert security features to provide unique solutions to graphic films, packaging and security applications. Wide web embossing in excess of 60 inch width.
****

Flight Dynamics
16600 SW 72nd Ave.
City: Portland
State/Province: OR
Postal Code: 97224
Country: United States of America
Voice Phone: 1 503 684 5384
Fax Phone: 1 503 684 0169
Business Description: Manufacturer of HOEs and Head-Up Displays.
****

Focal Image Ltd.
Number 20 Conduit Place
City: London
State/Province: England
Postal Code: W2 1HZ
Country: United Kingdom
Voice Phone: 44 171 706 2221
Fax Phone: 44 171 706 2223
email: kaveh@focal.demon.co.uk
Contact # 1: Kaveh Bazargan
Business Description: Consultancy in holography; display holograms; computer graphics; electronic publishing.
****

Foil Stamping and Embossing Association
PO. Box 12090
City: Portland
State/Province: OR
Postal Code: 97212
Country: United States of America
Voice Phone: 1 503 331 6221
Fax Phone: 1 503 33 1 6928
email: fsea@aol.com
Contact #1: Mary Fuller
Business Description: A non-profit international trade association of the foil stamping, embossing, die cutting and other graphic finishing industries. It 's purpose is to develop a cohesive alliance within the trade for the advancement of the entire finishing industry.
****

FoilMark Holographic Images
(a division of Foilmark, Inc.)

5 Malcolm Hoyt Drive
City: Newburypo1l
State/Province: MA
Postal Code: 01950
Country: United States of America
Voice Phone: 1 508 462 7300
Fax Phone: 1 508 462 0831
Web Address: foilmark.com
Contact # I: David Dion
Business Description: FoilMark Holographic Images, a division of FoilMark, Inc., is a manufacturer of diffraction embossed films. These films are printable and are available in many different mediums; for example, unsupported film, film laminated to paper or board, pressure sensitive material, static cling products, and hot stamping foils.
****

Fong Teng Technology
No 41 , Lane 63, Hwa Chen Road
City: Hsin Chuang, Taipei
Country: Taiwan
Voice Phone: 886229984760
Fax Phone: 886 2 2 992 1240
Contact # 1: Mark Chiang
Business Description: 60 inch hologram and dotmatrix pattern foil manufacturer, service from origination to finished product.
****

Foreign Dimension
Manley Conunercial Bldg Rm 190 I
367-375 Queen's Road Central
City: Hong Kong
Country: Hong Kong
Voice Phone: 852 2 542 0282
Fax Phone: 852 2 541 60 1 I
Contact # 1: Frederic Schwal1zman
Business Description: Specialists in manufacturing all kinds of holographic and illusion products (watches, keyrings, etc.). If you are a hologram manufacturer, we can also make top quality products at unbeatable prices using your holograms! SEE OURADVERTISEMENT
****

Fornari,Arthur David
813 Eighth Avenue
City: Brooklyn
State/Province: NY ~
Postal Code: 11215
Country: United States of America
Voice Phone: 1 718 965 3956
Contact # 1: Arthur David Fornari
Business Description: Artistic holographer; silver halide transmission & reflection holograms.
****

Forth Dimension Holographics
36 East Franklin Street
City: Nashville
StatelProvince: IN
Postal Code: 47448
Country: United States of America
Voice Phone: 1 812 988 82 12
Fax Phone: 1 812 921 I
Contact # 1: Rob Taylor
Business Description: Holographic gallery and retail shop. Hologram mastering/consulting. Small run silver halide transmission, reflection holograms.
****

Foundation Ideecentrum.
PO Box 222
5600 Mk
City: Eindhoven
Country: Netherlands
Business Description: Gallery.
****

Frank DeFreitas Holography Studio
815 Allen Street
City: Allentown
State/Province: PA
Postal Code: 18102
Country: United States of America
Voice Phone: 1 800 458 3525
Fax Phone: 1 800458 3525
email: director@holoworld.com
Web Address: holoworld.com
Contact # 1: Frank DeFreitas
Business Description: A full service holography studio family owned and operated since 1983.
****

Free University Of Brussels.
Faculty Of Applied Sciences
Alna-Tw Pleinlaan 2
City: Brussels
Postal Code: B-I050
Country: Belgium
Voice Phone: 32 2 629 3452
Fax Phone: 32 2 629 3450
Contact # 1: Erik Styns
Business Description: Academic and Scientific research on diffractive elements and HOE's.
****

Fresnel Technologies Inc.
101 West Morningside Drive
City: Fort Worth
StatelProvince: TX
Postal Code: 76110
Country: United States of America
Voice Phone: 1 8 17 926 7474
Fax Phone: 1 817926 7146
Contact # 1: Linda H. Claytor
Business Description: Manufactures plastic Fresnel lenses & lens arrays from its POLY JR plastics for use into the infrared; also other optical products for use into the ultraviolet from acrylic & other plastics.
****

Fringe Research Holographics
Interference Hologram Gallery
1179A King Street West, Suite 010
City: Toronto
Postal Code: M6K 3C5
Country: Canada
Voice Phone: 1 4165352323
Contact # 1: Michael Sowdon
Business Description: Artistic holography; silver halide holograms; pulse portraits; gallery; workshops; traveling exhibit.
****

Fuji Electric Co. Ltd
Mecatnll1ics Division
1-12-1 Yuraku-Cho Chiyoda-Ku
City: Tokyo
Postal Code: 190·
COWllry: Japan
Voice Phone: 81 3211 7111
Business Description: Manufactures CO_2 lasers and related equipment.
****

Fujitsu Laboratories Ltd.
Electronic Systems Division
10-1 Wakamiya Morinosato
City: Atsugi
Postal Code: 243-0
Country: Japan
Voice Phone: 81 462 48 311 I
Fax Phone: 81 462 48 3233
Contact # 1: Takehumi Inagaki
Business Description: Embossed Hologram Manufacturer.
****

CORPORATE LISTINGS

G.M. Vacuum Coating Lab, Inc.
882 Production Place
City: Newport Beach
State/Province: CA
Postal Code: 92663
Country: United States of America
Voice Phone: 1 7146425446
Fax Phone: 1 714642 7530
Contact # 1: Dan Coursen
Business Description: Custom manufacturing only, usually on your substrate. Will do coatings for front surface mirrors, beamsplitters, etc. for holographic use.
****

Galaxies Unlimited, Inc.
637 N. W. 12th Ave.
City: Deerfield Beach
StatelProvince: FL
Postal Code: 33442
Country: United States of America
Voice Phone: 1 305 429 1017
Fax Phone: 1 305 421 2391
Contact # 1: Larry DeBerry
Business Description: Parent company for Zero Gravity retail stores.
****

Galerie Musoria
Schwarztorstrasse 70
City: Bern
Postal Code: CH-3007
Country: Switzerland
Voice Phone: 41 31 381 773 1
Fax Phone: 41 31 381 7731
Contact # 1: Sandro del-Prete
Business Description: Gallery featuring holograms.
****
****

Galvoptics Ltd.
Harvey Road
Burnt Mills Industrial Estate
City: Basildon
State/Province: England
Postal Code: SSl3 IES
Country: United Kingdom
Voice Phone: 44 1268 728 077
Fax Phone: 44 1268 590 445
Contact # 1: R. D. Wale
Business Description: Optics; mirrors, lenses.
****

General Design
2005 - 18th Street
City: San Francisco
StatelProvince: CA
Postal Code: 94107
Country: United States of America
Voice Phone: 1 415 550 9193
email: bk@sfo.com
Web Address: sfo.com/-bk
Contact # 1: Brian Kane
Business Description: Creative services - Computer graphics for print, video and holography. 3D Computer Modeling and 2D Computer Composition. General image design and construction.
****

General Holographics, Inc.
PO. Box 82247
City: Burnaby
Postal Code: V5C 5P7
Country: Canada
Voice Phone: 1 604 685 6666
Voice Phone: 1 800 667 9669
Fax Phone: 1 604 685 6678
Contact # 1: Paula Simson
Business Description: Distributor of dichromate

& embossed gift and jewelry items, silver halide wall and desk decor, and photopolymer for the Canadian market. Custom and stock.
****

Geola
PO. Box 343
City: Vilnius
Postal Code: 2006
Country: Lithuania
Voice Phone: 370 2 232 73 7
Fax Phone: 3702232838
email: mike@lmc.elnet.lt
Web Address: camtech.com.aul-austholo
Contact # 1: Mike Grichine
Business Description: Original manufacturers of Pulsed Nd:YLF / glass holography lasers. Rental of ultra-modern pulsed holography studio. Large format - our specialty. Systems rental worldwide. Optics sales. Custom Design. Partners withAustralian Holographics.
SEE OURADVERTISEMENT
****

Glaser - Technical Consulting
24 Hashnayim Street
City: Givatayim
Postal Code: 53239
Country: Israel
Voice Phone: 971 3 6720895
Fax Phone: 972 3 6732734
email: feglaser@weizmann.weizmann.ac.il
Contact # 1: Shelly Glaser
Business Description: Technical consulting on holography (HOE, display, etc.), diffractive optics (DOE and systems containing DOEs), nonconventional optical systems (lenslet array based etc.), and image processing (specifically imaging optics for image processing). Services include feasibility studies, system design and evaluation, courses, etc.
****

Glass Mountain Optics
9517 Old McNeil Road
City: Austin
StatelProvince: TX
Postal Code: 78758-5225
Country: United States of America
Voice Phone: 1 5121397442
Fax Phone: 1 5123390589'·
Contact # 1: Don Conklin
Business Description: Specialize in manufacturing collimating mirrors.
****

Global Images
I Northumberland Ave
City: London
State/Province: England
Postal Code: WC2N 5BW
Country: United Kingdom
Voice Phone: 44 171 872 5452
Fax Phone: 44 171 872 5611
Contact # 1: Walter Clarke
Business Description: Specialists in high volume, low cost, quality embossing equipment. ISO 9002 Qualification.
****

Gorglione, Nancy
2047 Blucher Valley Road
City: Sabastopol
StatelProvi nce: CA
Postal Code: 95472
Country: United States of America
Voice Phone: 1 707 823 7171
Fax Phone: 1 707 823 8073
email: gorglione@aol.com
Contact #1: Nancy Gorglione
Business Description: Holographic artist spe-

cializing in architectural installations and public art environments. Extensive portfolio of one-of-a kind fine artworks.
****

Graham Saxby
3 Honor Ave.
Goldthorn Park
City: Wolverhampton
State/Province: England
Postal Code: WVI INJ
Country: United Kingdom
Voice Phone: 44 902 341 291
Fax Phone: 44 902 321 944
Contact # I: Graham Saxby
Business Description: Research scientist; author of Practical Holography
****

Gresser, E., KG
An Der Warth 10
City: Ochsenfurt
Postal Code: 97199
Country: Gennany
Voice Phone: +49 (0)933 22 77
Fax Phone: +49 (0)933 78 41
Contact # 1: Joachim Mueller
Business Description: Laser measurement techniques, Lasers, medical
****

Hologrammi
(a division of Laser Scene)
36 rue Emile DECORPS
City: Villeurbanne
Postal Code: 69100
Country: France
Voice Phone: 33 04 78 5445 34
Fax Phone: 33 04 72 36 95 65
Contact # 1: Francois Mazzero
Business Description: Holographic laboratory. Creation, mastering and production of large format transmission holograms up to 1 12 x 224 cm (44x88) and reflection holograms up to 1 12 x 164 cm (44x 64). Multicolor and 3 colours LCD stereograms fully animated and digitized from rendering or video.
****

Hallmark Capital Corp.
230 Park Avenue Suite 510
City: New York
State/Province: NY
Postal Code: 10169
Country: United States of America
Voice Phone: 1 212 249 9634
Fax Phone: 1 212 249 9537
Contact # 1: Patricia M. Hall
Business Description: New York based investment banking firm , specializing in raising debt and equity financing for privately-held companies. Mergers & acquisitions are a strong secondary activity. Hallmark has raised capital for both public and private holography companies.
****

Harvard Apparatus, Canada
6010 Vanden Abeele
City: St. Lau rent
Postal Code: H4S 1 R9
Country: Canada
Voice Phone: 1 514 335 0792
Fax Phone: 1 5143353482
email: 102263.2131@compuserve.com
Contact #1: Elle Massuda
Business Description: Mirrors, pri sms, optical items for holography.
****

Hellenic Institute Of Holography
28 Dionyssou Street
City: Chalandri

Postal Code: GR-15234
Country: Greece
Voice Phone: 30 1 684 6776
Fax Phone: 30 1 685 0807
Contact #1: Alkis Lembessis
Business Description: Established in 1987, the Institute aims at the overall introduction and promotion of holography in Greece. Exhibitions, courses, vocational training and mastering laboratory.
****

Heptagon Oy
Otaniemi Science and Technolgry Park
Tekniikantie 12
City: Espoo
Postal Code: FIN-02150
Country: Finland
Voice Phone: 358 943542041
Fax Phone: 358943542041
email: info@heptagon.fi
Web Address: heptagon.fi
Contact # 1: Jyrki Saarinen
Business Description: Heptagon provides complete design services for designing diffractive optical elements (DOEs) to customer requirements. Heptagon also offers consulting services and engineering assistance to the fabrication and exploitation of DOEs.
****

Hi-Glo Holo Images Pvt. Ltd.
32- Green View, Rajbaha Road
City: Patiala
Postal Code: 147001
Country: India
Voice Phone: 91 175 220 243
Voice Phone: 91 175 72807
Fax Phone: 91 175 220 243
****

Hiat Image Technology Group, Inc.
2F, No. 16 Lane 6, Sec. 1, Hang Chou S. Rd.
City: Taipei
Country: Taiwan
Voice Phone: 886 2 393 0306
Fax Phone: 886 2 395 8122
Contact # 1: Billy Chou
****

High Tech Network
Skeppsbron 2
City: Malmo
Postal Code: S-211 20
Country: Sweden
Voice Phone: 46 040 350 75
Fax Phone: 46 040 237 667
Contact #1: Christer Agehall.
Business Description: Art in holography; security applications.
****

HODIC Holographic Display Artists & Engineers Club
Engineering Department Chiba University
1-33, Yayoi-cho
City: Chiba
Postal Code: 263
Country: Japan
Voice Phone: 81 472 511 III
Fax Phone: 81 472 517 337
Contact # 1: Miss Tomoko Sakai
Business Description: Regular meetings 4 times a year with oral presentations on holography given. Publishes HODIC circular. Membership open to all.
****

HOL 3, Galerie fur Holographie GmbH
Europa Center
City: Berlin

Postal Code: 10789
Country: Germany
Voice Phone: 49 30 261 4490
Fax Phone: 49 30 344 6379
Contact #1: Valeska Cordner-Guled
Business Description: Exhibition of mainly stock holograms of various producers. Sometimes feature one man shows. Sales of holograms and related holographic items.
****

Holage
1881 Eighth Avenue
City: San Francisco
State/Province: CA
Postal Code: 94122
Country: United States of America
Voice Phone: 1 415 564 1840
Contact # 1: Brad Cantos
Business Description: Microlithography and photoresist consulting.
****

Holart Consultants
18 Bonview Street
City: San Francisco
StatelProvince: CA
Postal Code: 94110
Country: United States of America
Voice Phone: 1 4152823646
Fax Phone: 14152824013
email: gaz@ix.netcom.com
Web Address: holo.comlgazl
Contact # 1: Gary Zellerbach
Business Description: Expert appraisals ofholographic art works. Consulting in all aspects of creating, displaying, and marketing custom and stock holograms.
****

Holicon Corporation.
3312 Belle Plain Ave. #2
City: Chicago
State/Province: IL
Postal Code: 606 18
Country: United States of America
Voice Phone: 1 312 267 9288
Fax Phone: 1 312 267 9288
Contact #1: Richard Bruck
Business Description: Holograms produced with pulsed lasers. Holographic portraits. Large format reflection and rainbow holograms. Mass production of silver halide holograms.
****

Holo 3
7 Rue du Gal Cassagnou
City: Saint-Louis
Postal Code: 68300
Country: France
Voice Phone: 33 89 69 82 08
Fax Phone: 33 89 67 74 06
Business Description: Industrial applications of holography: shock and vibration, non-destructive testing, microholography flow visualization, contouring. R&D: study and development of new tools for industrial applications.
****

HoloGmbH
Lutterdamm 82
City: Bramsche
Postal Code: 49565
Country: Gemlany
Voice Phone: +49 (0)546 91123
Fax Phone: +49 (0)546 91122
Contact # 1: Thomas Lucy
Business Description: Holograms up to 1 x 1 m; embossed holography. Holo-design.
****

Holo Images Tech Co., Ltd.
17, Alley 20, Lane 7, Jong Hwa Road
City: Yung Kang City, Tainan County
Country: Taiwan
Voice Phone: 886 66 237 3896
Fax Phone: 886662384641
Contact #1: Craig Chiou
Business Description: Embossed holograms and products.
****

Holo Impressions
47-1 Wu Chuan Road
Wu-Ku Industrial Park
City: Taipei Shien
Country: Taiwan
Voice Phone: 886 2 299 7576
Fax Phone: 886 2 299 7050
Contact # 1: Billy Chiou
****

Holo Impressions Inc
47-1 Wu Chuan Rd
Wu-Ku Industrial Park
City: Taipei Shein
Country: Taiwan
Voice Phone: 886 2 299 7576
Fax Phone: 886 2 299 7050
Contact #1: Jonathan Hsu
Business Description: Embossed holography.
****

Hol0 Sciences, LLC
480 East Rudasill Road
City: Tucson
State/Province: AZ
Postal Code: 85704
Country: United States of America
Voice Phone: 1 520 696 0773
Fax Phone: 1 520 696 0773
email: deck_O@azstamet.com
Contact # 1: Chuck Hassen
Business Description: HI Mastering and stereogram creation from video or computer graphic source imagery. Special capabilities to produce full-parallax, animated stereograms. Custom 3-D computer modeling and animation services on request. SEE OURADVERTISEMENT
****

Holo-Laser
6, Rue De La Mission
Ecole
City: Miserey
Postal Code: 25480
Country: France
Voice Phone: 33 1 45 5246 52
Fax Phone: 33 1 45 52 46 81
Contact' # 1: DrJean Louis Tribillon
Business Description: Embossed holography and equipment; artistic holography; buying and selling; education.
****

Holo-OrLtd
PO Box 1051
Kiryat Weizmann
City: Rehovot
Country: Israel
Voice Phone: 972 8 469 687
Fax Phone: 972 8 466 378
Contact # 1: Uri Levy
Business Description: Manufactures computer-generated diffractive optical elements by VLSI techniques. Catalogue elements and custom designs. Substrates include AnSe, GaAs, various glasses. DOE work station--dedicated workstation for element design, mask generation.
****

CORPORATE LISTINGS

Holo-Service
Neuensteinerstrasse 19
City: Basel
Postal Code: CH-4153
CountlY: Switzerland
Voice Phone: 41 502 287
Contact # 1: Edgar Bar
Business Description: Artistic holography.
****

Holo-Service.Fries
Therwi ler Strasse 26
City: Basel
Postal Code: CH-4045
Country: Switzerland
Voice Phone: 41 61 281 0917
Fax Phone: 41612810917
Contact #1: Urs Fries
Business Description: Artistic. holography, ho-
logram projects.
****

Holo-Spectra
7742 Gloria Ave.
City: Van Nuys
State/Province: CA
Postal Code: 91406
Country: United States of America
Voice Phone: 1 800 275 4880
Voice Phone: 1 8189949577
Fax Phone: 1 818 994 4709
Contact # 1: Bill Arkin
Business Description: Embossed hologram pro-
duction. Embossed mastering equipment, laser
repairs. Optical table and holographic equip-
ment resold.
****

Holo/Source Corporation
1 1930 Farmington Rd
City: Livonia
StatelProvince: MI
Postal Code: 48150
Country: United States of America
Voice Phone: 1 313 427 1530
Fax Phone: 1 313 525 8520
email: hscorpl530@aol.com
Contact # 1: Lee Lacey
Business Description: Paperboard sheets of ho-
lographic film and paper. Holographic image
mastering of all types and finished fl exo printed
holographic labels.
****

HoloCom
401 Carver Road
City: Las Cruces
State/Province: NM
Postal Code: 88005
Country: United States of America
Voice Phone: 1 505 527 9183
Fax Phone: 1 505 527 9927
email: arts@holo.com
Web Address: holo.com
Contact #1: Ken Harris
Business Description: Holographic web site and
web provider. Holographic technology transfer
and holographic photoresist training. Custom
originations in photoresist.
****

Holocrafts
Canadian Holographic Developments Ltd.
Box 1035
City: Delta
Postal Code: V4M 3T2
CountlY: Canada
Voice Phone: 1 604 946 1926
Fax Phone: 1 604 946 1648

Contact # 1: Karoline Cullen
Business Description: Holocrafts manufacturers dichromate holograms. Offering stock and
custom production in a variety of fonnats such
as plain discs. watches, keychains, pendants and
3 x 3 plates. Providing a tradition of excellence
since 1979. SEE OURADVERTISEMENT
****

Holocrafts Europe Limited.
Barton Mill House
Barton Mill Road
City: Canterbury
State/Province: England
Postal Code: CTI IBY
Country: United Kingdom
Voice Phone: 44 1227463 223
Fax Phone: 44 1227 450 399
Contact # 1: Chris Luton
Business Description: Specialists in manufacture
of dichromate reflection holograms. Also pro-
duce holographic gift products as well as selling
photopolymer. SEE OURADVERTISEMENT
****

Holodesign Studies
Rebenstrasse 20
City: Riehen
Postal Code: CH-4125
Country: Switzerland
Business Description: Marketing consulting.
****

Holofar Lab (Sri)
Piazza Acilia 103 lnt 3
City: Rome
Postal Code: 00199
Country: Italy
Business Description: Artistic holography
****

Holografia Polska
ul.sw.Mikolaja 161I 7
City: Wroclaw
Postal Code: 50- 128
Country: Poland
Voice Phone: 48 71 343 46
Fax Phone: 48 71 339 48
Contact # 1: Boguslaw Stich
Business Description: Practical application of ho-
lography and importers of holographic products.
****

Holografica
30 1 South Light Street
City: Baltimore
State/Province: MD
Postal Code: 21202
Country: United States of America
Voice Phone: 1 410 685 3331
Contact # 1: Renee Fee
Business Description: Retail store with full
range of holographic products.
****

Hologram Company RAKO GmbH
Moellner Landstrasse 15
City: Witzhave
Postal Code: 22969
Country: Germany
Voice Phone: +49 (0)410 693 250
Fax Phone: +49 (0)410 693 249
Contact # 1: Wilfried Schipper
Business Description: Specializing in produc-
tion of embossed holograms and the sa le of
embossing equipment.
****

Hologram Development Corp.
37 Standish Ave.
City: Toronto

Postal Code: M4W 3B2
Country: Canada
Voice Phone: 1 416 925 5569
Contact # 1: Ed Burke
Business Description: General hologram services.
****

Hologram Fantastic
368 West Market
Mall of America
City: Bloomington
State/Province: MN
Postal Code: 55425
Country: United States of America
Voice Phone: 1 612 858 9416
Voice Phone: 1 612 722 4423
Contact #1: Nina Bourque
Business Description: Retail store in high traffic
mall selling a full range of holographic giftware
and related optical novelties.
****

Hologram Industries
42/44 Rue De Tntcy
City: Fontenay-Sous-Bois
Postal Code: 94 120
Country: France
Voice Phone: 33 1439419 19
Fax Phone: 33 143 94 00 32
Contact # 1: Hughes Sou paris
Business Description: Communication holograms.
****

Hologram Land
284 E. Broadway
Mall Of America
City: Bloomington
StatelProvince: MN
Postal Code: 55425
Country: United States of America
Voice Phone: 1 6128549344
Fax Phone: 1 612 854 7857
Contact # 1: Sue Rickert
Business Description: Retail store specializing
in holographic and scientif ic products. framing
and lighting accessories provided.
****

Hologr am Research, Inc.
25 East Loop Road
City: Stony Brook
State/Province: NY
Postal Code: 11790-3350
Country: United States of America
Voice Phone: 1 516 444 8839
Voice Phone: 1 516 674 3130
Fax Phone: 1 516 444 8825
email: hologram@lihti.org
Web Address: hologramres.com
Contact # 1: Joseph Bums
Business Description: Exclusive source for Ilford silver halide plates and fi lm. Custom holograms to 42 x 72, stereograms and limited editions in silver halide, photoresist, embossed and
injection-molded from our argon and he-ne laser
labs. Exhibition services available from our extensive hologram collect ion; brokering services
for the New York Metro area. Fully equipped
studio for rent. SEE OURADVERTISEMENT
****

Hologram Varga Miklos
Kiraly u. l 02 .
City: Budapest
Postal Code: 1068
Counlly: Hungary
Voice Phone: 36 1 351 4725
Voice Phone: 36 20 342 076
Fax Phone: 36 1 227 354

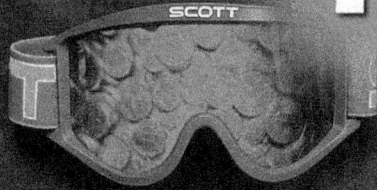

Contact # 1: Miklos Varga
Business Description: Stock and custom made holograms on best quality photopolymer.
****

Hologram World, Inc.
1860 Berkshire Lane North
City: Plymouth
State/Province: MN
Postal Code: 55441
Country: United States of America
Voice Phone: 1 612 559 5539
Voice Phone: 1 800 882 4656
Fax Phone: 1 612 559 2286
Contact #1: Jim Paletz
Business Description: One of the largest wholesale distributors of holographic novelties. We represent over 50 holographic manufacturers. Specialize in helping the new retail store owner in all stages of development from start to finish.
SEE OUR ADVERTISEMENT
****

Hologramas, S.A. de C.V.
PINO 343-3
COL. Atlampa
City: Mexico D.F.
State/Province: Mexico D.F.
Postal Code: 06450
Country: Mexico
Voice Phone: 52 5 541 1791
Fax Phone: 52 5 547 9046
email: holomex@holomex.com.mx
Web Address: www.holomex.com.mx
Contact #1: Dan Liebennan
Business Description: Holography can be used on: packaging, literature inserts, wrapping paper, security labels, bar codes, security paper, stickers, point-of-sale displays, etc. Call us today for quotes and consultation. We offer origination, mastering, mass production and finishing services. Extensive commercial portfolio.
SEE OUR ADVERTISEMENT
****

Hologranun Werkstatt & Galerie
Gallerie Fur Hologramme
Via Principale 30, Ch
City: Castesegna
Postal Code: 7649
Country: Switzerland
Voice Phone: 41 411 824 1718
Fax Phone: 414118241268
Contact #1: Horst Gutekunst
Business Description: Creative workshop, developments, looking for new and attractive ways for hologram making.
****

Holograms 3D
286 Earl's Court Road
City: London
State/Province: England
Postal Code: SW5 9AS
Country: United Kingdom
Voice Phone: 44 171 370 2239
Fax Phone: 44 171 3702239
Contact # I: Jonathan Ross
Business Description: Jonathan Ross has a personal holography collection available for touring shows. He also deals privately in holographic
****

Holograms and Lasers International
Retail Merchandise Division
1200 McKinney, Suite 433
City: Houston
State/Province: TX

Postal Code: 77010
Country: United States of America
Voice Phone: 1 713 650 9204
Fax Phone: 1 713 650 9204
email : felix@electrotex.com
Web Address: holoshop.com
Contact #1: Perry Felix
Business Description: Holograms and Lasers International operates the largest retail hologram shop in Houston, Texas with a complete library of over 750 hologram images and products to view on the Internet's World Wide Web located at www.holoshop.com
****

Holograms and Lasers International
Hologram Production Facilities
1200 McKinney, Suite 433
City: Houston
State/Province: TX
Postal Code: 77010
Country: United States of America
Voice Phone: 1 713 650 9204
Fax Phone: 1 713 650 9204
email: felix@electrotex.com
Web Address: holoshop.com
Contact #1: Perry Felix
Business Description: A full service Ruby Pulse and CW laser hologram production lab providing complete origination services for reflection, transmission and mass production, holograms, specializing in fine quality large format hologram portraits and trade show exhibits
****

Holograms Fantastic and Optical Illusions
P.O. Box 765
City: Bayswater
State/Province: Victoria
Postal Code: 3153
Country: Australia
Voice Phone: 61 18776226
Fax Phone: 61 3 9729 6020
Contact #1: Trevor McGaw
Business Description: Glass, film and foil (opp, PET & PVC) 2D, 2D/3D, 3D & multi images & patterns. Services to printers, hot-stampers, packaging, label, security marketing, sales promotion and advertising. Specialists in foil holography.
****

Holograms International
211 18th Street
City: Huntington Beach
State/Province: CA
Postal Code: 92648
Country: United States of America
Voice Phone: 1 714 536 0608
Fax Phone: 1 714 536 0608
Contact #1: Dave Krueger
Business Description: Distributor of all kinds of holograms to retail stores and wholesale accounts. We are known for our fast delivery, friendly consulting and factory-direct prices. Call or write for quote or catalogue.
****

Holographic Applications
21 Woodland Way
City: Greenbelt
State/Province: MD
Postal Code: 20770-1728
Country: United States of America
Voice Phone: 1 30 1 345 4652
Fax Phone: 1 301 345 4653
Contact # I: Suzanne SI. Cyr
Business Description: Design and product engineeri ng services for consumer products and

licensed promotions using 3-D imaging technologies. Product specifications and quality assurance. Vendor selection and product management. General contractor delivering finished, packaged product.
****

Holographic Design Systems
1134 We& Washington Blvd.
City: Chicago
State/Province: IL
Postal Code: 60607
Country: United States of America
Voice Phone: 1 312 829 2292
Fax Phone: 1 312 829 9636
email: museumh@concentric.net
Web Address: http://www.cris.coml-museumh
Contact #1: Robert Billings
Business Description: Unrivaled creativity, combining artistic imagination with complete technical mastery of all fonns of holography resulting in a worldwide reputation for excellence. The most complete labs in the industry with the most powerful and advanced lasers and computers. Our clients include the most innovative and sophisticated companies in the US and abroad.
SEE OUR ADVERTISEMENT
****

Holographic Dimensions
16115 S. W. 1 I 7th Avenue
Unit A-21
City: Miami
State/Province: FL
Postal Code: 33177-1615
Country: United States of America
Voice Phone: 1 305 255 4247
Fax Phone: 1 305 255 0339
email: holodi@shadow.net
Web Address: shadow.netl-holodi
Contact #1: Kevin Brown
Business Description: Holographic Dimensions, Inc. is a vertically integrated manufacturer of holographic imagery, with extensive experience in high volume security and authentication holograms. A recently, traded Public Company, it has complete in-house facilities from artwork origination to embossing.
****

Holographic Dimensions
Zone 5, Box 520 I
City: Panama City
Country: Panama
email: jimmywoo@sinfo.net
Contact # I: James Woolford
Business Description: See above
****

Holographic Dimensions, Poland S.A.
UI. Traugutta- '2S
City: Lodz .
Postal Code: 90-950
Country: Poland
Contact # I: Grace Golen, VP
Business Description: Holographic Dimensions, Inc. is a vertically integrated manufacturer of holographic imagery.
****

Holographic Images Inc .
521 Michigan Ave.
City: Miami Beach
State/Province: FL
Postal Code: 33139
Country: United States of America
Voice Phone: 1 305 531 5465
Fax Phone: 1 305 532 4090

CORPORATE LISTINGS

Contact # I: Matthew Schrieber
Business Description: Mastering and replication facility dedicated to the production of mult color/full color limited-edition holographic artworks produced on silver halide film.
****

Holographic Impressions
96 North Almaden Blvd.
City: San Jose
State/Province: CA
Postal Code: 951 10-2490
Country: United States of America
Voice Phone: 1 408 292 890 I
email: smithmckay@aol.com
Contact #1: Dave McKay
Business Description: Manufacturer and distributor of unique line of greeting cards, stationary and business cards. Stock and custom products available.
****

Holographic Industries, Inc.
PO. Box 1109
City: Libertyville
State/Province: IL
Postal Code: 60048
Country: United States of America
Voice Phone: 1 847680 1884
Fax Phone: 1 847 680 0505
Contact # I: Robert Pricone
Business Description: Parent company of Lightwave retail. Consulting and distribution for giftware industry.
****

Holographic Label Converting (HLC)
7669 Washington Avenue South
City: Edina
State/Province: MN
Postal Code: 55439
Country: United States of America
Voice Phone: 1 6129447408
Fax Phone: 1 612 944 7210
Contact # I: Scott Labelle
Business Description: Full service capabilities, 2D/3D holography, designing, embossing, hotstamping, precision die-cutting, wide variety of foils. Custom holographic labeling, magnetic holograms, packaging and more.You think of it, and we can put it together.
****

Holographic Optics Inc.
358 Saw Mill River Rd
City: Millwood
State/Province: NY
Postal Code: 10546
Country: United States of America
Voice Phone: 19147621774
Fax Phone: 1 9147622557
Contact # I: Jose R. Magarinos
Business Description: Manufacturer of holographic optical elements , particularly holographic filters, holographic mirror and beamsplitters. Design and manufacture of prototypes.
****

Holographic Products
1711 St. Clair Ave.
City: St. Paul
State/Province: MN
Postal Code: 55 105
Country: United States of America
Voice Phone: 1 612 698 6893
Fax Phone: 1 6126981619
email: sugar001@tc.umn.edu
Contact #1: Stephen Sugarman
Business Description: Holographic Products is actively pursuing new product development in educational toys, intermedia print design, ad specialties, promotions, and premiums. Specialists in security laminates (for ID card application) and tamper evident labeling. Also conducts hands-on elementary school workshops, and 1 structional presentations.
****

Holographic Service
10 Via Civerchio
City: Milan
Postal Code: 1-20159
Country: Italy
Business Description: Consultant, holograms on packaging material.
****

Holographic Studios
240 East 26th Street
City: New York
State/Province: NY
Postal Code: 100 1 0
Country: United States of America
Voice Phone: 1 212 686 9397
Fax Phone: 1 212 481 8645
email: drlaser@interport.net
Web Address: hmt.comlholographylholostudios
Contact # 1 : Jason Sapan
Business Description: New York's only gallery and commercial holographic lab. Custom and stock holograms. Single or mass-produced. Integral portrait cinematography, mastering, and scan copies from small to large format. Computer generated holograms. Classes.
****

Holographic Systems Muenchen GmbH
Melchior-Huber-Strasse 25
City: Otters berg
Postal Code: 85652
Country: Germany
Voice Phone: +49 (0)812 9925-0
Fax Phone: +49 (0)8129925-99
Contact # 1: Gunther Dausmann
Business Description: Holographic production, including machines.
****

Holographics (Uk).1td.
12 Whidborne St
City: London
, StatelProvince: England
Postal Code: WCIH 8EU
Country: United Kingdom
Voice Phone: 44 171 833 2236
Fax Phone: 44 171 833 2237
Contact #1: Jon Vogel
Business Description: Holographic & 3-D multimedia, design origination and production specialists (est. 1982) providing comprehensive service for the corporate, retail, & leisure sectors.
****

Holographics Inc.
44-0 1 Eleventh Street
City: Long Island City
StatelProvince: NY
Postal Code: 11101
Country: United States of America
Voice Phone: 1 7187843435
Fax Phone: 1 718 706 0813
email: 74503.2250@compuserve.com
Business Description: Company involved in 3 main areas of research and development primarily involving pulsed holography: multi/full color artistic portraiture (Ana Maria Nicholson); NDT for government and corporate applications (Dr. John Webster, Tim Schmidt); and laser development (Peter Nicholson).
****

Holographics North Inc.
444 South Union Street
City: Burlington
StatelProvince: VT
Postal Code: 0540 I
Country: United States of America
Voice Phone: 1 802 658 2275
Fax Phone: 1 8026585471
email: perryjf@vbi .champlain.edu
Contact # 1: John Perry
Business Description: Designers/manufacturers of large format holograms for commercial, museum and fine art applications. Multicolor, animated holograms up to 44x72 inches (1.1 m x 1.8 m). Design, model building, production, installation and consulting services.
****

Holographie Anubis
Oberer Kaulberg 37
City: Bamberg
Postal Code: D-96049
Country: Germany
Voice Phone: 49 951 5795 1
Fax Phone: 49 951 59529
email: holographie.anubis@t-online.de
Contact # 1: M.T. Frieb
Business Description: We are producer and distributor of all formats of holograms. Import and export. Consulting and education. Mass production. Full pulse laser facilities. Fully pictured wholesale catalog (122 pages).
****

Holographie Fachstudio Bad Rothenfelde
Postfach 1304
Niederesch 28
City: Bad Rothenfelde
Postal Code: 49214
Country: Germany
Voice Phone: +49 (0)542 5365
Fax Phone: +49 (0)542 5359
Contact #1: Gunter Deutschmann
Business Description: Expert consultancy for integration of holographic products into finished advertising media, including application, overprinting etc. Founder and office of the AHT (Arbeitskreis Hologramm-Techniken & 3D Medien).
****

Holographie Labor
BertelsmannAG
Auf dem Eickholt 47
City: Gutersloh
Postal Code: 33334
Country: Gelmany
Voice Phone: 49 5241 580 192
Fax Phone: 49 5241 580 549
Contact # I: Saurda Uwe
Business Description: Holographic projects.
****

Holography and Media Institute of Quebec
1139 Ave des Laurentides
City: Quebec City
Postal Code: GIS 3C2
Country: Canada
Voice Phone: 1 418 687 2985
Fax Phone: 1 4 18 687 2985
email: Marie-Andree.Cossette@arv.ulaval .ca
Contact #1: Marie-Andree Cossette
Business Description: Established in 1990 as an international centre for the study and production of holographic art. Workshops and private tutorials. Residency programs offered on an invitation only basis. Director Marie-Andree Cossette is Associate Professor of Visual Art at Laval University, Quebec.
****

Holography Center of Austria
Kahlenbergstrasse 6
City: Wurmla
Postal Code: A-3042
Country: Austria
Voice Phone: 43 2275 8210
Fax Phone: 43 2275 82105
Contact #1: Irmfried Wober
Business Description: Our holography laboratory, founded in 1985, is the first in Austria and the most comprehensive in the region. We offer high quality pulsed and CW laser origination, mastering and production (transportable pulsed laser systems for portrait, stereo grams up 1 x 1 meter and computer generated holograms available). In addition, we organize exhibitions in Austria and Germany and sell embossed holograms. SEE OURADVERTISEMENT
****

Holography Division
Pt. Pura Brautman
Jl. Agil Kusumadya 203
City: Central Java
Country: Indonesia
Voice Phone: 62 62 29121121
Fax Phone: 62 62 291 21822
Contact # I: Rendy Roedianto
****

Holography Institute of San Francisco
PO Box 24-1 53
City: San Francisco
State/Province: CA
Postal Code: 94124-0153
Country: United States of America
Voice Phone: 1 415 822 7123
email: hologram@well.com
Contact #1: Jeffrey Murray
Business Description: Specializing in one-on-one instruction in display holography (design, optics, recording, processing, etc.). Courses can be customized to the required balance of theory and practice for artistic, technical or production applications. Classes for artists, scientist, young and old. No prerequisites for beginners. Call for class schedule. Research facilities include a dedicated holography studio with 5 isolation tables, HeNe lasers, and optics for image plane holography up to 50x60 cm.
****

Holography Israel
21 Hakomemiut Str.
City: Herzlia
Postal Code: 46683
Countly: Israel
Voice Phone: 972 09 572 387
Voice Phone: 972 09 559 766
Fax Phone: 972 09 570 569
Contact #1: Hameiri Shimon
Business Description: Holography Israel specializes in exhibitions-lectures and demonstrations to pupils and students-advertising, commission, sales and production of art holograms.
****

Holography Marketplace
(c/o Ross Books)
PO.Box 4340
City: Berkeley
State/Province: CA
Postal Code: 94704
Country: United States of America
Voice Phone: 1 510841 2474
Voice Phone: 1 800 367 0930
Fax Phone: 1 510 841 2695
email: staff@rossbooks
Web Address: holoinfo.com
Contact #1: Alan Rhody
Contact #2: Franz Ross
Business Description: **The world's most comprehensive and informative book for, and about, the holography industry.**
Published annually, each edition of THE HOLOGRAPHY MARKETPLACE includes:
- a worldwide corporate directory;
- chapters of useful reference material;
- interviews with industry experts;
- and a sample kit of actual holograms produced by major manufacturers.
Edition 1 Collector's Item! , Call for price.
Editions 2-5 $19.95 each or $50 for set.
Edition 6 $19.95
VISIT OURWEBSITE!
****

Holography Presses On (HPO)
201 North Fruitport Road, Box 193
City: Spring Lake
State/Province: MI
Postal Code: 49456-0193
Country: United States of America
Voice Phone: 1 6168425626
Fax Phone: 1 616 842 5653
Contact #1: Jan Bussard
Business Description: Holographic stock or custom shapes and sizes applied with heat or pressure for adhesion to all substrates. Specialize in. stickers and textile applications. Sealed edges prevent delamination in all weather; washable. Worldwide distributors sought. SEE OUR ADVERTISEMENT
****

Hololand S.c.
Batumi 6 m 43
City: Warszawa
Postal Code: 02-760
Country: Poland
Voice Phone: 48 22 427 463
Fax Phone: 48 2 625 5567
email: stepien@if.pw.edu.pl
Contact # 1: Pawel Stepien
Business Description: Low volume holographic labels, holographic consultancy, security CGH research & development.
****

Hololaser Gallery
PO Box 23386
City: Dubai
Country: United Arab Emerates
Voice Phone: 971 4 518 989
Fax Phone: 971 4 528 015
Contact # 1: Abdul Wahab Baghdadi
Business Description: Holography Gallery and holographic items. We are the first and only gallery in the Gulf Countries and we produce laser shows.
****

HoloMedia Ab/Hologram Museum.
PO Box 45012
City: Stockholm
Postal Code: 10460
Country: Sweden
Voice Phone: 46 8 411 1108
Fax Phone: 46 8 107638
Contact # 1: Mona Forsberg
Business Description: Broker for embossed and custom made artistic holography; buying & selling holograms; holography education; gallery. Display unit available. Hologram center
****

Holomedia France
16 rue Maurice Fontvielle
City: Toulouse
Postal Code: 31000
Country: France
Voice Phone: 33 62 27 17 04
Fax Phone: 33 62 27 17 04
Contact # 1: Luigi Castagna
Business Description: Wholesale and distribution of silver halide, jewelry and fine art holograms. Two retail shops in Toulouse and Lyon, France.
****

Holomex Ltd.
4 Borrowdale Avenue
City: Harrow
State/Province: England
Postal Code: HA3 7PZ
Country: United Kingdom
Voice Phone: 44 181 427 9685
Contact # 1: Mike Anderson
Business Description: Supplier of film processing kits and safelights. Designs and manufacturers a holographic camera and viewer.
****

Holophile, Inc.
56 Abner Lane
City: Killingworth
StatelProvince: CT
Postal Code: 06419
Country: United States of America
Voice Phone: 1 203 663 3030
Fax Phone: 1 203 663 3067
email: info@holophile.com
Web Address: www.connix.coml-barefoot
Contact #1: Paul D. Barefoot
Business Description : Founded in 1975 , Holophile provides consulting services in holography and spectral imagery (3-D projection of moving images) to corporations, museums and display builders. Our company is a producer of holography exhibitions for museums, science centers and children's museums.
****

Holoptics
Heidjerweg 13
City: Oldenburg
Postal Code: 26133
Country: Germany
Voice Phone: +49 (0)441 45166
Fax Phone: +49 (0)441 4860928
Contact #1: Joerg Schweer
Business Description: Production of silver halide holograms, origination of SHH (F.E.: Eye in Pyramid), mass production of small size SHH, distribution of holograms and holographic articles, sophisticated display-systems. Mobile exhibitions, listed supplier of holograms for fastidious department stores.
****

Holopublic Unbehaun
Hirschstrasse 84 •
City: Wuppertal .
Postal Code: 42285
Country: Gennany
Voice Phone: +49 (0)202 84118
Contact # 1: Klaus Unbehaun
Business Description: Consulting, education, newsletters Holography 3D Software and AHT Reflexionen, fine arts (Holofotografik) book Holo Show International, founding memberAHTAssociation for Holography and New Media.
****

HolosArt Galerie
4 Place Grenus
City: Geneva
Postal Code: 1201
Country: Switzerland
Voice Phone: 41 22 325 191

CORPORATE LISTINGS

Contact # 1: Pascal Barre
Business Description: Gallery, retail sales.
****

Holosco, Ernest Barnes
Bajada de Viladecols, 2
City: Barcelona
State/Province: Spain
Postal Code: 08002
Country: Spain
Voice Phone: 34 3 3 1 0 71 13
Fax Phone: 343319 1676
Business Description: Holography Lab. Reflection and transmission, transfer to photoresist - embossing facilities. Consulting services.
****

Holostik India Pvt. Ltd.
50, Adhchini
Sri Aurobindo Marg.
City: New Delhi
Postal Code: 110017
Country: India
Voice Phone: 91 11 665 690
Voice Phone: 91 11 669 725
Fax Phone: 9 1 11 686 8828
Contact # 1: Govind Sharma
Business Description: We are among the first to have set up a fully automated plant for manufacture of security and promotional holograms and films in India. Soon setting up master lab and 40 inch wide web machine for holographic packaging.
****

Holotek
(a division ofECRM Inc.)
205 Summit Point Drive
City: Hemietta
StatelProvince: NY
Postal Code: 14467
Country: United States of America
Voice Phone: 1 7 1 6 321 6000
Voice Phone: 1 888 465-6832
Fax Phone: 1 7163216001
Contact # 1: Roger O'Brien
Business Description: Engineering and design of sub systems for laser optic scanning devices. Commercial and industrial applications.
****

HoloVision
Tumblingerstr. 32
City: Munich
Postal Code: 80337
Country: Germany
Voice Phone: +49 (0)89 746 9336
Fax Phone: +49 (0)89 746 9382
email: 101 625.3552@comp
Contact # 1: Julian Fischer
Business Description: Production of holograms on silver halide up to 100 x 100 cm. Holographic stereograms, computer generated holograms, multiple color holograms, pulse holograms, traditional recording techniques, design, modelmaking, installation and consulting services.
****

Holovision AB
Box 70002
City: Stockholm
Postal Code: 10044
Country: Sweden
Voice Phone: 46 8 33 1 186
Fax Phone: 46 8 331 186
Contact # 1: Jonny Gustafsson.
Business Description: Specializing in silver halide holography with pulsed lasers. Denisyuk and transferred-type reflection holograms up to 30 x 40 cm. Rainbow holograms with pulsed laser up to 2 x 1 m.
****

Holovision Systems Inc .
119 South Main St.
City: Findlay
State/Province: OH
Postal Code: 45840
Country: Un ited States of America
Voice Phone: 1 4194223604
Fax Phone: 1 4 1 9 422 4270
email: alt@bright.net
Contact # I: Ronald L. Kirk
Business Description: Holovision Systems, Inc. specializes in technological development for holographic and 3-dimensional image display. Holovision currently manufactures and markets Real Image (TM) displays which produce live full color 3-dimensional video projections into 3-D space for point of sale, trade show and exhibit applications. It is also developing a higher level of technology called Holoview(TM) or real time holographic displays of medical CAD CAM and cinemagraphic applications.
****

HoloWebs, Inc.
City: San Diego
State/Province: CA
Country: United States of America
Voice Phone: 52 5 541 1791
Fax Phone: 52 5 547 9~6
email: holomex@holomex.com.mx
Web Address: holomwx.com.mx
Contact # 1: Dan Lieberman
Business Description: State of the art (wide web embossing up to 66 inches in width), full service hologram facility to open in 1997. Call Holograms of Mexico for further details.
****

Honeywell Technology Center
MN 65 - 2500
3660 Technology Drive
City: Minneapolis
StatelProvince: MN
Postal Code: 55418
Country: United States of America
Voice Phone' 1 611'951 7938
Fax Phone: 1 612 95 1 7438
, email: cox@src.honeywell.com
Contact # 1: Dr. I. Allen Cox
Business Description: Diffractive optics. Micromachining.
****

HRT Holographic Recording Technologies GmbH
Am Steinaubach 19
City: Steinau
Postal Code: 36396
Country: Gelmany
Voice Phone: +49 (0)666 7668
Fax Phone: +49 (0)666 7463
Contact # I: Richard Dr. Birenheide
Business Description: Own production and distribution of silver halide emulsions with low noise and high diffraction efficiency.
****

Hughes Power Products, Inc.
1925 East Maple Ave.
Mail Station SC/SI3/G343
Ci ty: EI Segundo
StatelProvince: CA
Postal Code: 90245
Country: United States of America
Voice Phone: 1 3 JO 414 7086
Fax Phone: 1 3 1 0 726 0008
email: info@powerholos.com
Web Address: powerholos.com

Contact # 1: John E. Gunther
Business Description: Hughes Power Products department (an affiliated subsidiary of Hughes Aircraft Company and General Motors Corporation) has been a pioneer in holography since 1974. We manufacture and se ll a complete line of wall decor holograms for the wholesale giftware and art markets. We offer a full range of custom services including origination, mastering, mass replication and finishing on photopolymer film. We also have extensive experience with the manufacturing of HOEs. SEE OUR ADVERTISEMENT
****

Hyogo Prefectual Museum of Modern Art
Art Curator
Kobe-3-8-3 Harada-Dori
City: Nada-Ku Kobe 657
Country: Japan
Voice Phone: 8 1 78 80 1 159 I
Fax Phone: 81 78 86 14731
Contact # I: Hitoshi Yamazaki
Business Description: 20th century Art, History of Art and Holography, Art and Optics, curating a exhibition of holography into Art.
****

J.S.Gill
214 Kailash Hills
East of Kailash
City: New Delhi
Postal Code: 1 10065
Country: India
Voice Phone: 9111 16840377
Voice Phone: 91 11 68470377
Business Description: Bindry - application of holographic foil and stickers.
****

IBM Almaden Research Center
K03/G2
650 Harry Road
City: San lose
StatelProvince: CA
Postal Code: 95120
Country: United States of America
Voice Phone: 1 408927 1283
Fax Phone: 1 408 927 30 II
email: mikeross@almaden.ibm.com
Contact #1 : Michael Ross
Business Description: Scientific holography research; holographic storage.
****

Ibsen Micro Structures N S
CAT, Frederiksborgvej 399
PO. Box 30
City: Roskilde
Postal Code: DK-4000
Country: Denmark
Voice Phone: 45 4 677 4551
Fax Phone: 45 4 675 4012
Contact # I: Hemik Madsen
Business Description: Producer of photoresist plates for holography and diffractive optic.
****

iC Holographics
8 Flitcroft St.
City: London
State/Province: England
Postal Code: WC2H 8Dl
Country: United Kingdom
Voice Phone: 44 171 240 6767
Fax Phone: 44 17 1 240 6768
email: 10041 3.3406@compuserve.com
Contact # 1: Chris Levine
Business Description: Holographic design and digital mastering.
****

ICl Polyester
(a division ofICl America)
Concord Plaza - 3411 Silverside Road
City: Wilmington
State/Province: DE
Postal Code: 19850
Country: United States of America
Voice Phone: 1 800 635 4639
Fax Phone: 1 302 887 5365
email: icipet.com
Web Address: icipolyseter.com
Business Description: Supplies polyester film,
polyethylene naphtha late and polyester resins
for metalizing and/or direct embossing.
****

Illinois Institute Of Technology
Mechanical/Materials & Aerospace Engineering
Engineering Building #1 Rm 252-B
City: Chicago
State/Province: IL
Postal Code: 60616
Country: United States of America
Voice Phone: 1 312 567 3220
Fax Phone: 1 312 567 7230
email: mesciammarella@mimna.iit.ezu
Contact # I: Cesar Sciammarella
Business Description: Holographic interferometry; in-
dustrial holographic research; non-destructive testing.
****

Illuminations
1252 7th Avenue
City: San Francisco
State/Province: CA
Postal Code: 94122
Country: United States of America
Voice Phone: 1 4156640694
Contact #1: Louis Brill
Business Description: Involved in developing & ex-
panding market & sales efforts for holographic re-
tail/wholesale product lines. Assist in preparation of
promotions and collateral sa les materials, identify
potential sales markets & implementation of sales.
****

Imagen Holography, Inc.
10 Park Ave. Suite 500
City: Basalt
StatelProvince: CO
Postal Code: 81621
Country: United States of America
Voice Phone: 1 970 927 0360
Fax Phone: 1 970 927 0359
Contact # 1: Alan P Morterud
Business Description: Specialized holographic
products for mainstream marketing applica-
tions, including HOLOTEX and Advanced Ho-
lographic Textiles - a soft, supple, fully wash-
able rendering of reflection holograms in both
2D & 3D, to a variety offabrics.
****

Imagenes Holograficas De Columbia
Avenida Quinta Norte No 17-23
AA 3076 PBX 685450
City: Cali
Country: Columbia
Fax Phone: 5790 15108412695
****

Images Company
PO. Box 140742
City: Staten Island
StatelProvince: NY
Postal Code: 10314
Country: United States of America
Voice Phone: 1 718 698 8305
Fax Phone: 1 7189826145

Web Address: he.net/-imagesco
Business Description: Sells holographic equipment and materials to educational institutions, students and private holographers. Equipment And
materials available: lasers, film, development kits,
mounting kits for lenses, beamsplitters, mirrors,
spatial filters, safelights, filters and display lights.
****

Imagination Plantation
2650 18th street
City: San Francisco
StatelProvince: CA
Postal Code: 94110
Country: United States of America
Voice Phone: 1 415487 0841
Fax Phone: 14154872103
email: ipd@iplant.com
Web Address: www.iplant.com
Contact #1: Noah Hurwitz
Business Description: 3D imaging and content
creation for all media. Experienced in modeling
for holographic applications, including direct
output to master.
****

JmEdge Technology
Eastview Technology Center
350 Main St.
City: White Plains
State/Province: NY
Postal Code: 10601..
Country: United States of America
Voice Phone: 1 9149465536
Fax Phone: 1 9149465460
email: mmetz@imedge.com
Contact # I: Michael Metz
Business Description: Research, development
and manufacturing of edge-lit holograms; cre-
ative and innovative holography and optics
problem solving; custom display volume holo-
grams; edge-lit holographic optical elements;
consulting; edge-lit fingerprint imaging device
and other industrial holographic products.
****

Imperial College Of Science
Optics Section
Blackett Laboratory \,
Clty: London .
State/Province: England
Postal Code: SW7 2BZ
Country: United Kingdom
Voice Phone: 44 171 589 5111
Business Description: Courses in holography; sci-
entific holography research; particle measurement.
****

Industrial Technology Institute
290 1 Hubbard Road
City: Ann Arbor
State/Province: MI
Postal Code: 48105
Country: United States of America
Voice Phone: 1 313 769-4156
Voice Phone: 1 313 769-4000
Fax Phone: 1 313 769-4064
email: kh@iti.org
Web Address: iti.org
Contact # 1: Kevin Harding
Business Description: IT! is a not-for-profit con-
tract R&D organization. We help manufacturers
identify and apply technology to solve produc-
tion problems. ITI provides technical consult-
ing, R&D, testing (including holographic NDT)
and training programs.
****

Industrial Technology Research Inst.

Holography Department
Bldg 44, 195 Chung Hsing Road, Section 4
City: Chutung
Postal Code: 310 15
Country: Taiwan
Voice Phone: 886 35 917 482
Fax Phone: 88635 917479
Contact # 1: Dr. J.J. S u
Business Description: Research in HUD (Head
Up Display) and Dot Matrix Hologram systems
****

!NETI - Institute of Information Technologies
LAER - Aerospace Laboratory
Estrada do Paco do Lumiar
City: Lisboa Codex
Postal Code: 1699
Country: Portugal
email: xana@laer.ineti.pt
Contact # I: Ana Alexandra Andrade
Business Description: R&D in Holography and
OVDs. Special interests in security features.
Consultants on technology and customized
production projects of embossed holograms. In-
house design and origination.
****

Infinity Laser Laboratories
68 11 Flanders Station
City: Polk City
State/Province: FL
Postal Code: 33868
Country: United States of America
Voice Phone: 1 941 984 3108
Fax Phone: 1 941 984 4244
email: infinity@digital.net
Contact # 1: Thad Cason
Business Description: Full service company from
concept and design to finished product. Laborato-
ries include photo polymer capability with pulsed,
continuous wave and solid state lasers including
Argon, Krypton, Nd:YAG:KTP and HeCd.
****

Infox Corporation
3rd floor No 283
Sec2 Fu-hsing South Road
City: Taipei
Country: Taiwan
Voice Phone: 886 2 7056699
Fax Phone: 886 2 7551800
Contact # 1: Alex C. T. Chen
Business Description: Maker of injected-mold-
ed holograms.
****

Infrared Optical Products, Inc.
PO Box 292
City: Farmingdale
StatelProvince: NY
Postal Code: 11735-0664
Country: Unife·d ~tates of America
Voice Phone: 1 516 694 6035
Fax Phone: 1 516 694 6049
Contact # I: Barry Bassin
Business Description: Manufacturer of infrared
lenses, windows, reflectors, beamsplitters, com-
puter-designed IR lens systems. Front surface
optical coatings for mirrors.
****

Innovative Technology Associates
3639 East Harbor Blvd. # 203 E
City: Ventura
StatelProvince: CA
Postal Code: 93001
Country: United States of America
Voice Phone: 1 805 650 9353
Fax Phone: 1 805 984 2979
Contact # I: Joseph Gaynor

CORPORATE LISTINGS

Business Description: Technical and business consultants specializing in materials and processes relating to deformable films, photoplastics and photorefractive polymers. Special knowledge of holographic memory storage technologies.
****

Inrad, Inc.
181 Legrand Avenue
City: Northvale
State/Province: NJ
Postal Code: 07647
Country: United States of America
Voice Phone: 1 201 767 1910
Fax Phone: 1 20 1 767 9644
Contact # 1: Maria Murray
Business Description: Manufacturer of nonlinear materials, harmonic generation systems, electro-optic and acousto-opdc devices and drivers. Also provides optical components, assemblies and optical coatings for the Uv, visible and IR.
****

Inside Finishing Magazine
PO. Box 12090
City: Portland
State/Province: OR
Postal Code: 97212
Country: United States of America
Voice Phone: 1 503 331 622 I
Fax Phone: 1 503 331 6928
email: fsea@aol.com
Contact # I: Jeff Peterson
Business Description: Only trade publication specifically targeting the graphic finishing industry. Editorial focus is on foil stamping, embossing, holograms, die cutting, folding/gluing, and coatings. Published quarterly by the Foil Stamping & Embossing Association.
****

Institut fur Angewandte Physik
Schlossgartenstr. 7
TH Darmstadt
City: Dannstadt
Postal Code: 64289
Country: Gernlany
Voice Phone: 49 615 1 162786
Fax Phone: 49 6151 164534
email: AndreasBillo@Physik.TH-Darmstadt.de/andreas
Web Address: http: //www.physik.thdannstadt.de/andreas
Contact # I: Andreas Billo
Business Description: Investigation of cavitation bubbles, sprays and droplets with high speed holographic particle image ve locimetry (HPIV).
****

Institute Of Optical Science
Central University
City: Chung-Li
Postal Code: 32054
Country: Taiwan
Voice Phone: 886 3 425 7681
Fax Phone: 886 3 425 8816
Contact # 1: Tang Yaw Tzong
Business Description: HOEs, academic research.
****

Institute Of Plasma Physics
And Laser Microfusion
PO Box 49
City: Wroclaw
Postal Code: 00-908
Country: Poland
Contact # 1: Zbigniew Sikorsky
Business Description: Academic research
****

Integraf
PO. Box 586
745 N. Waukegan Rd.
City: Lake Forest
StatelProvince: IL
Postal Code: 60045
Country: United States of America
Voice Phone: 1 8472343756
Fax Phone: 1 847 615 0835
email: jeong@lfc.edu
Contact # 1: T.H. Jeong
Business Description: Distributor of holographic films and plates, including Russian emulsions. Expert consultation on HOEs, NDT, system designs and other holographic projects. Also, educational materials and stock holograms. SEE OUR ADVERTISEMENT
****

Interactive Industries Inc.
696 Plank Road
City: Waterbury
StatelProvince: CT
Postal Code: 06705
Country: United States of America
Voice Phone: 1 203 755 21 1 I
Fax Phone: 1 203 755 3999
Contact # I: Ronald Phillips
Business Description: Holographic and lenticular animated promotional products including mousepads, bookmarks, magnets, buttons, keyrings, cards, rulers;. calendars, point-of-purchase displays, and other custom items.
****

Interferens Holografi D.A.
Museum, Gallery, Studio
Halvor Hoels Gt 6
City: Hamar
Postal Code: N-2300
Country: Norway
Voice Phone: 47 62 25050
Voice Phone: 47 62 30659
Fax Phone: 47 62 30659
Contact #1: Olav Skipnes
Business Description: Makes glass (mainly reflection) holograms of museum exhibits. Continuous wave laser. Norway's largest collection of holograms. OUI'Specialty: museum exhibits.
****

International Data Ltd.
Units 5 & 6
Station Industrial Estate, Oxford Rd
City: Wokingham
State/Province: England
Postal Code: RG41 2YQ
Country: United Kingdom
Voice Phone: 44 11 89772255
Fax Phone: 44 11 89 772296
email: 101 763. 1273@compuserve.com
Contact # 1 : Dawn Dreelaw
Business Description: Manufacturer of plastic cards. Capable of hot stamping holographic materials.
****

International Hologram Manufacturers Association
Runnymede Malthouse
Runnymede Road
City: Egham
State/Province: England
Postal Code: TW20 9BD
Country: United Kingdom
Voice Phone: 44 1784 497 008
Fax Phone: 44 1784 497 00 I
email: 100142.1164@compuserve.com
Business Description: IMHA was founded in 1993 to promote the interests of hologram man-ufacturers and the holography industry worldwide. It is a non profit membership organization open to all producers of holograms, suppliers of equipment and material for the manufacture of holograms and hologram converters and finishers. Annual meeting in November.
****

Intrepid World Communications
(a subsidiary of American Propylaea Corp.)
555 South Woodward, Suite 1109
City: Binningham
State/Province: MI
Postal Code: 48009
Country: United States of America
Voice Phone: 1 810642 9885
Fax Phone: 1 810 642 9886
Contact # 1 : Ann Marie Harrison
Business Description: Distributor of display holograms, in particular true-color holograms. Currently working on archiving Vatican treasures. Subsidiary of American Propylaea Corp.
****

Ion Laser Technology, Inc.
PO. Box 5128
City: Salt Lake City
State/Province: UT
Postal Code: 84 115
Country: United States of America
Voice Phone: 1 801 2880555
Fax Phone: 1 80 1 262 5770
Contact # 1: Don Gibb
Business Description: Manufacturer of industrial, scientific and medical lasers, including air and water cooled Argon lasers suitable for holography.
****

Ishii, Ms. Setsuko
#404,
1-23 26 Kohinata,Bunkyo-Ku
City: Tokyo
Postal Code: 102
Country: Japan
Voice Phone: 81 0339459017
Fax Phone: 81 03 3945 9068
Contact # I: Setsuko Ishii
Business Description: Fine Artist.
****

James River Products
800 Research Road
City: Richmond
State/Province: VA
Postal Code: 23236
Country: United States of America
Voice Phone: 1 804378 1800
Fax Phone: 1 804 378 5400
email: jrp@richmond.infi.net
Contact # 1: Mike Florence
Business Description: World leader in embossed hologram machinery. Products include: origination lab equipment, photoresist plate spinning, electrofonn facilities, embossing machines, die cutting equipment, supporting technology and training ~ plus custom embossed holograms.
SEE OURADVERTISEMENT
****

Japan Communication Arts Co.
Yonezawa Bldg2F
2-37 Suehirocho, Kita-Ku
City: Osaka
Postal Code: 530
Country: Japan
Voice Phone: 81063 141919
Fax Phone: 81 06315 1900
Contact #1: Mineko Fukuma
Business Description: Sales of cards with hologram.
****

Jayco Holographics
29-43 Sydney Road
City: Watford
State/Province: England
Postal Code: WD 1 7PY
Country: United Kingdom
Voice Phone: 44 1923 246 760
Fax Phone: 441923247769
Contact # 1: Rohit Mistry
Business Description: Complete production service for embossed holograms. Embossing masters through to fully finished product. Our years of experience enables Jayco to offer outstanding quality of product and service at competitive prices.

Jeffery Murray Custom Holography
PO. Box 24 - 153
City: San Francisco
State/Province: CA
Postal Code: 92124
Country: United States of America
Voice Phone: 141 5822-7123
email: hologram@well.com
Contact # I: Jeffery Murray
Business Description: Museum displays, artists collaborations, commercial advertising, linage research, HOE custom optics for visual display. Research specialties: high quality display holography, holographic optics, holographic recording systems. One-offs, limited editions and unusual projects. Research facilities: holography darkrooms, silver halide processing, 5 isolation tables with HeNe lasers. Able to produce HI masters, reflection and transmission image plane transfers, and Denisyuk holograms up to 50x60 cm.

Jimenez-Ceniceros, Antonio
Cerrada De Moroleon #12 Colonia Roma
Delegacion Cauhtemoc c.p 06700
City: Mexico
Country: Mexico
Voice Phone: 52 (5) 264 1313
Fax Phone: 52 (5) 264 1313
email: bety@sysull.ifisicacu.unam.mx
Contact #1: Antonio Jimenez-Ceniceros
Business Description: Physics Engineer, optics laboratory manager. Production of holographic molds: 2D-3D, 3D-2D, Dot-Matrix, Double-channel, laser readable, combination of these in same mold, holographic molds for security applications. Research projects in photoresist holography.

Jodon Inc.
62 Enterprise Drive
City: Ann Arbor
StatelProvince: MI
Postal Code: 48103
Country: United States of America
Voice Phone: 13137614044
Voice Phone: 1 800 989 jodon
Fax Phone: 1 313 761 3322
email: johng@wwn.com
Contact # 1: Mike Gillespie
Business Description: Manufacturer of HeNe lasers, laser systems, specialty laser tubes, optical and electro-optical instruments and systems. Holographic films, plates and chemicals. Engineering services.

JR Holographics
Suite 206
424 Kelton Ave.
City: Los Angeles
State/Province: CA

Postal Code: 90024
Country: United States of America
Voice Phone: 1 310 393 2388
Fax Phone: 1 310 393 86 11
Contact # 1: Judy Roberts
Business Description: Marketer and consultant for clients around the world who wish to display an image, logo, or product in three dimensions. Expertise in entertainment licensing.

K.C. Brown Holographics
17 Salisbury Road
New Malden
City: Surrey
State/Province: England
Postal Code: KT3 3HZ
Country: United Kingdom
Voice Phone: 44 181 942 8294
Fax Phone: 44 181 877 3400
email: 100306.2015@compuserve.com
Contact # 1: Kevin Brown
Business Description: Pulse portraits; artistic holography.

Kaiser Optical Systems, Inc.
PO Box 983
371 Parkland Plaza
City: Ann Arbor
State/Province: MI
Postal Code: 48106
Country: United States ofi\merica
Voice Phone: 1 313 665 8'083
Fax Phone: 1 313 665 8199
Contact # 1: Harry Owen
Business Description: Compact fiber coupled Raman spectrometers for routine quality control and remote, real time, in line process monitoring applications with microscope accessory for line applications. Fast f/ 1.8 volume transmission grating based Holographic Imaging Spectrographs for visible, fluorescence and Raman applications. Holographic Notch and Laser Bandpass filters for Raman laser induced fluorescence spectroscopy.

Kan, Mike
1786 Quesada Ave. "
City: San Francisco
StatelProvince: CA
Postal Code: 94124
Country: United States of America
email: c/o:cdemon@sirus.com
Contact #1: Mike Kan
Business Description: Very high quality, very affordable dielectric coated beamsplitters suitable for professionals and hobbyists. Back coating optimized for HeNe (0.1 % R - to 99% P @ 45 degrees. Loss < .01 %). Custom coating services up to 4 diameter also offered. Write for details and prices. Quantity discounts.

Karas Studios S.L.
Hospital, 12
City: Madrid
Postal Code: 28012
Country: Spain
Voice Phone: 34 1 530 89 88
Fax Phone: 34 1 530 89 88
Contact #1: Ramon Benito
Business Description: Established in 1988; art exhibitions, art gallery, private collection.

Karolinska Institutet
School Of Dentistry

Box 4064
City: Huddinge
Postal Code: S-14104
Country: Sweden
Voice Phone: 46 8 774 0080
Contact # 1: Hans Ryden
Business Description: Holography research applied to dentistry.

Kauffman, John
Box 477
City: Point Reyes Station
State/Province: CA
Postal Code: 94956
Country: United States of America
Voice Phone: 1 415663 12 16
Fax Phone: 1415663 1216
Contact # 1: John Kauffman
Business Description: Holographic Fine Artist. Specializes in multi color reflection holograms. Extensive portfolio.

Keio University
Dept Of Electrical Engineering
3-14-1 Hiyoshi Kohoku-Ku
City: Yokohama
Postal Code: 223
Country: Japan
Voice Phone: 81 045 563 1141
Fax Phone: 81 045 563 3421
Contact # 1: Dr. Masato Nakajima
Business Description: Research using HNDT

Kendall Hyde Ltd.
Kingsland Industrial Park
Stroud ley Road
City: Basingstoke
State/Province: England
Postal Code: RG24 8UG
Cowltry: United Kingdom
Voice Phone: 44 125 684 0830
Fax Phone: 44 125 684 0443
Contact # 1: M. Kendall
Business Description: Optical coating specialists. Front surface mirrors, etc.

Keystone Scientific Co.
PO Box 22
City: Thorndale
StatelProvince: PA
Postal Code: 19372-0022
Country: United States of America
Voice Phone: 1 610 269 9065
Fax Phone: 1 610 269 4855
Contact # 1: Ed Kelly
Business Description: Distributor for Agfa and Kodak holographic films, plates and chemicals. Manufacturer ·or.holography kits and automatic processors.

Kimmon Electric Co., Ltd.
TM21 Building
1-53-2 Itabashi
City: ltabashi-Ku
Postal Code: 173
Country: Japan
Voice Phone: 81 03 5248-4811
Fax Phone: 81 03 5248-0021
email: lasers@kimmon.com
Web Address: kimmon.com
Contact # 1: Shinji Ninomiya
Business Description: Manufacturer of Helium Cadmium lasers which are used in holography. Kimmon manufacturers the highest powered

CORPORATE LISTINGS

HeCd laser in the world. The mode11K41711-G is the laser of choice for holographers because of its 175mW @ 442 TEMoo specified output power and long lifetime.
****

Kinetic Systems. Inc.
20 Arboretum Road
City: Boston
State/Province: MA
Postal Code: 02131
Country: United States of America
Voice Phone: 1 617 522 8700
Voice Phone: 1 800 992 2884
Fax Phone: 1 617 522 6323
Contact #1: Moss Blosvem
Business Description: Manufacturers of Vibra-plane standard and special Honeycomb optical tables in four grades LIP to 5' x 16' x 24'. Larger sizes available by butt Splicing. Also vibration isolation support systems.
****

Kolbe-Druck mit Tochtergesel lschaften
1m Industriegelaende 50
City: Versmold
Postal Code: 33775
Country: Germany
Voice Phone: +49 (0)542 9670
Fax Phone: +49 (0)542 41 230
Contact #1: Roland Pahnke
Business Description: Bindry: Hot stamping holographic foil , Lenticular-Printing, 3-0, motion, change.
****

Kreischer Optics, Ltd.
906 North Draper Rd.
City: McHenry
State/Province: IL
Postal Code: 60050
Country: United States of America
Voice Phone: 1 815 344 4220
Fax Phone: 1 8 15 344 4221
email: optics@k.reisher.com
WebAddress: www.k.reisher.com
Contact # I: Cody Kreischer
Business Description: Custom manufacturer of lenses, condensers, cylinders, windows, filters, prisms, mirrors, beamsplitters, substrates. Consulting services in optical design.
****

Krystal Holographics International Inc.
555 West 57th Street
City: New York
State/Province: NY
Postal Code: 10019
Country: United States of America
Voice Phone: 1 212 261 0400
Voice Phone: 1 801 753 5775
Fax Phone: 12122620414
email: krystal@sunrem.com
Web Address: www.khiinc.com
Contact # 1: Marion Baker
Business Descri ption: World headquarters of Krystal Holographics Intemationallnc., a manufacturer and marketer of photopolymer holograms produced by proprietary technology. Experienced with large volume orders of custom holograms for commercial applications including packaging, giftware, security/authentication, advertising premiums, etc.
SEE OURADVERTISEMENT
****

Krystal Holographies International Inc.
U.S. Holographics Division
365 North 600 West

City: Logan
Statel Province: UT
Postal Code: 8432 1
Country: United States of America
Voice Phone: 1 801 753 5775
Voice Phone: 1 800 998 5775
Fax Phone: 1 80 1 753 5876
email: krystal@sunrem.com
Web Address: www.khiinc.com
Contact # I: Dave Rayfield
Business Description: Marketing/manufachlring for mass-produced and custom photopolymer and dichromate holograms for retail and ad specialty needs. Stock and custom products available. KHI is also a manufacturer of mastering and mass replication equipment. SEE OUR ADVERTISEMENT
****

Krystal Holographics Vertriebs-GmbH
Bimenweg 15
City: Reutlingen
Postal Code: 72766
Country: Germany
Voice Phone: +49 (0)712 9461 -0
Fax Phone: +49 (0)712 9461 10
Contact # I: Richard Stooss
Business Description: Distribution of innovative KrystalGram products in Western Europe: Higb quality 3D-Holograms on DuPont Omnidex photopolymer film, aimed for high volume promotional & OEM applications. Custom and stock designs available. Krystalmark security systems.
****

KTH
Dept. of Materials Processing
Industrial Metrology
City: Stockholm
Postal Code: 100 44
Country: Sweden
Voice Phone: 46 8 790 7832
Voice Phone: 46 8 790 8169
email: nilsa@matpr.kth.se
Contact # I: Nils Abramson
Business Description: Industrial Metrology comprises conventional engineering metrology and laser-based metrology, especially industrial applications of display holography, holographic interferometry and Light-in-Flight recording by holography, which if; largely the result of the research and development at the Department. The principal objective of the group is research and education; to develop new measurement principals for app lying lasers in the industry and to disseminate knowledge of known laser-based methods of measurement.
****

L.A.S.E.R. News
(see LaserArts Society for Education & Research)
State/Province: CA
Country: United States of America
****

Laboratories of Image Information
Science and Technology
LC Bldg. 10F 1-4-2 Shin-senri Higashi-machi
City: Toyonaka, Osaka
Postal Code: 565
Country: Japan
Voice Phone: 81 06 873 2053
Fax Phone: 81 06 873 2056
Business Description: R&D on Holographic Display, Dynamic Holography, Holographic Optical Elements
****

Laboratory for Optical Data Processing
Carnegie Mellon University
Dept. of Electrical And Computer Engineering

City: Pittsburg
State/Province: PA
Postal Code: 15213
COIllltry: United States of America
Voice Phone: 1 412 268 2464
Fax Phone: 1 412 268 6345
emai l: marlene@ece.cmu.edu
Contact # 1: David Casasent
Business Description: Research in optical data processing, pattem recognition, product inspection, neural nets. Processors, filters and feature extractors using computer generated holograms.
****

Laboratory Vinckiner
Holography Workshop Univ Gent
41 St Pi etersnieuwstraat
City: Gent
Postal Code: B-9000
Country: Belgium
Voice Phone: 32 9 264 3242
Fax Phone: 32 9 223 7326
Contact # I: Pierre Boone
Business Description: Consultancy, education, problem-solving for display holography. Museum applications and (mainly!) non-destructive testing.
****

Lake Forest College
Center for Photonics Studies
555 N. Sheridan Road
City: Lake Forest
Statel Province: IL
Postal Code: 60045
Country: United States of America
Voice Phone: 1 847 735 5160
Fax Phone: 1 847 615 0835
email: JEONG@LFC.EDU
Contact # 1: TH. Jeong.
Business Description: Each summer during Ju ly, Lake Forest College offers a 5-day hands-on workshop for participants who have no prior experience in holography. An advanced 6-day workshop follows. Write for information. SEE OUR ADVERTISEMENT
****

Larry Liebennan Holography
The Hologram Gallery & The Living Portrait Studio
1642 Euclid Ave.
City: Miami Beach
Statel Province: FL
Postal Code: 33139
Country: United States of America
Voice Phone: 1 305 604 9986
Fax Phone: 1 305 604 9998
email: lieber741 @aol.com
Contact # I: Larry Liebennan
Business Description: Mastering facility for full color holograms and stereogram portraiture. A collection of full color limited edition artworks available for sale and exhibition. Call for details.
****

Lasan Ltd.
2911 San Isidro Ct.
City: Santa Fe
State/Province: NM
Postal Code: 8750 I
Country: United States of America
Voice Phone: 1 505 438 8224
Fax Phone: 1 505 438 8224
Contact # I: August Muth
Business Description: Lasart Ltd. is a manufachirer of production holographic jewelry and gifts as well as one-of-a-kind holographic glass sculpture. We also specialize in unique corporate gifts, taking the project from modeling through completion of the product in our in house fac ilities.
****

Laser Affiliates
2047 Blucher Valley Road
City: Sebastopol
State!Province: CA
Postal Code: 95472
Country: United States of America
Voice Phone: 1 707 823 7171
Fax Phone: 1 707 823 8073
Contact # 1: Nancy Gorglione
Business Description: LaserAffiliates is an award-winning non-profit organization that designs innovative holographic and laser theatrical productions, installations and exhibitions. Services include curatorial guidance, videotapes and media lectures.

Laser and Motion Development Company
Professional Equipment Exchange
3101 Whipple Road
City: Union City
State/Province: CA
Postal Code: 94587-1216
Country: United States of America
Voice Phone: 1 510 429 1060
Fax Phone: 1 510 429 1065
email: em@mediacity.com
Web Address: lasermotion.com
Contact # 1: Ed Monberg
Business Descliption: LMDC is a buyer and seller of lasers, motion and optical equipment. We also integrate laser processing systems at significant savings to our customers. We offer experience and engineering in galvanometer, X-Y table, and laser based optical systems.

Laser Arts Society For Education and Research
PO Box 24-153
City: San Francisco
StatefProvince: CA
Postal Code: 94124 - 0153
Country: United States of America
Voice Phone: 1 415 822 7123
email: hologram@well.com
Contact # I: Jeffrey Murray
Business Description: Volunteer staffed non-profit organization dedicated to holography and laser education and research. Members receive the quarterly L.A.S.E.R. News. One year membership USA 530, outside USA $40.

Laser Drive Inc.
5465 WM Flynn Hwy.
City: Gibsonia
State/Province: PA
Postal Code: 15044
Country: United States of America
Voice Phone: 14124437688
Fax Phone: 14124446430
Contact # I: Carol Smith
Business Description: Manufacturer high voltage power supplies for Helium Neon, Argon Ion and C02 lasers.

Laser Focus World
(a division of Penwell Publishing)
10 Tara Boulevard - 5th floor
City: Nashua
State/Province: NH
Postal Code: 03062
Country: United States of America
Voice Phone: 1 603 891 0123
Fax Phone: 1 603 891 0574
Business Description: Trade publication coveing the field of optics, lasers, electro-optics, and related imaging research, as well as commercial applications.

Laser Holography Workshop
320 South Willard Street
City: New Buffalo
State/Province: MI
Postal Code: 49117
Country: United States of America
Voice Phone: 1 6164694658
fax Phone: 1 616 469 4658
email: 102017. 1330@compuserve.com
Contact #1: Joseph A. Farina
Business Description: Specializing in modelmaking for holography. CI ients sending us their reference material by facsimile will receive a noobligation proposal within 24 how·s. Call for our free brochure.

Laser 1 mages
P.O. Box 6873
City: Leawood
StatefProvince: KS
Postal Code: 66206
Country: United States of America
Voice Phone: 1 913 648 2525
Fax Phone: 1 913 648 6898
Contact # I: Steve Larson
Business Description: R&D and production capabilities in dichromate, silver halide, photoresist and photopolynwr. Manufacturers of stock and custom holograms and holographic products. Stock products include jewelry, watches, calculators, and framed art.

Laser Innovations
668 Flinn Ave. #22
City: Moorpark
State!Province: CA
Postal Code: 9302 1
Country: Un ited States of America
Voice Phone: 1 805 529 5864
fax Phone: 1 805 529 6621
Contact # 1: R. Eric King
Business Desc ription: Laser Innovations offers sales, repair, and support of ion laser systems. Specializing in the repair and service of CO HERENT lasers ; Laser~ Innovations stocks remanufactured INNOVA plasma tubes for fast and reliable support of your ion laser system.

Laser Inspeck
360 rue franquet, Bur 20
City: St. foy
Postal Code: G 1 P 4N3
Country: Canada
Voice Phone: 1 4186502112
fax Phone: 1 418 6502141
email: inspeck@riq.qc.ca
Contact # I: Li Song
Business Description: Marketer of the RoTech holographic interferometry camera and them10-plastic films. Also photo resist plates. Digital interferometry. 3-D digitizer (scans into computer the coordinates of object and color). Data is imported as DXf or other system independent data format. Works on PC and can be ported to SGJ, SUN, etc. for lise with any rendering program.

Laser In stinlte Of America
12424 Research Parkway #125
City: Orlando
StatefProvince: fL
Postal Code: 32826
Country: United States of America
Voice Phone: 1 407 380 1553
Fax Phone: 1 407 380 5588
emai l: lia@mail.creol.ucf.edu
Web Address: creol.ucf.edul-lial

Contact # I: Jackie Thomas
L
Business Descri ption: Laser safety training courses. Publishes Journal of Laser Applications. Hosts annual International Congress on Applications of Lasers and Electro-Optics (ICALEO), including holographic applications and International Laser Safety Conference. Call for membership details and publication catalog.

Laser Intemational
19 NOimanton Rise
Holbeck Hill
City: Scarborough
State/Province: England
Postal Code: YO 1 1 2XE
Country: United Kingdom
Voice Phone: 44 172 336 4452
Contact # I: Keith Dutton
Business Description: Specializing in laser display systems for exhibitions. Auto & manual control available. Full range of prices and features.

Laser Las Vegas
5725 N. Fort Apache
City: Las Vegas
State/Province: NV
Postal Code: 89129
COllntry: United States of America
Voice Phone: 1 702 645 0477
Fax Phone: 1 702645 0477
email: laserlv@aol.com
Contact # I: Bill Aymar
Business Description: Laser sales, repairs and rentals. Specialists in high power laser systems.

Laser Light Designs
2412 Kennedy Way
City: Antioch
State/Province: CA
Postal Code: 94509
Country: United States of America
Voice Phone: 1 5107543144
Contact # I: Michael Malott
Business Description: Embossed product.

Laser Light Ltd.
28 Old Fulton St.
City: Brooklyn Heights
StatefProvince: NY
Postal Code: 1120 I
Country: United States of America
Voice Phone: 12122267747
fax Phone: 1 718 858 2062
email: ddt-Iaser@aol.com
Contact #1: Abe Rezny
Business Description: All formats of holography. Designers/products of 3-D imaging.

Laser Media, Inc.
6383 Arizona Circle
City: Los Angeles
State!Province: CA
Postal Code: 90045
Country: United States of America
Voice Phone: 1 310 338 9200
fax Phone: 1 310 338 9221
email: Imilasers@aol.col11
Contact # 1: Kevin McCarthy
Business Description: Custom design of entertainment lighting and multimedia installations including lasers, waterworks, holograms, etc. Specialize in large corporate presentations.

CORPORATE LISTINGS

Laser Movement
ZA 57, Ie Trou Grillon
City: Saint Pierre du Perray
Postal Code: 91280
Country: France
Voice Phone: 33 1 60 75 67 27
Fax Phone: 33 1 69 89 08 43
Contact # 1: Patrice Pecheux
Business Description: Laser show.
****

Laser Optics, Inc.
III Wooster SI.
City: Bethel
State/Province: CT
Postal Code: 0680 I
Country: United States of America
Voice Phone: 1 203 7444160
Fax Phone: 1 203 798 7941.
Contact #1: Jim Larim
Business Description: A complete line of laser and optical components for ultraviolet, visible and infrared applications from 250 nrn 10 16 microns, including focusing lenses, windows, cavity components, prisms, beam splitters, mirrors and coatings.
****

Laser Reflections
25 North Second Street
City: San Jose
StatelProvince: CA
Postal Code: 95113
Country: United States of America
Voice Phone: 1 408 292 7484
Fax Phone: 1 408 292 8115
email: holograph@aol.com
Contact #1: Ron Olson
Business Desc ription: Highest quality holographic portraiture. Advanced Nd:YAG laser recording technology producing visibly superior holograms ofliving subjects up to 14 x 24. Sel standing displays and custom installations. Unique commercial applications for signage and advertisements. Extensive portfolio of limited fine art editions. Stock images available on silver halide and photopolymer. SEE OUR ADVERTISEMENT
****

Laser Resale Inc.
54 Balcom Road
City: Sudbury
StatelProvince: MA
Postal Code: 01776
Country: United States of America
Voice Phone: 1 508 443 8484
Fax Phone: 1 508 443 7620
email: LaseResale@aol.com
Web Address: laserresale.com
Contact # 1: Jack Kilpatrick
Business Description: Laser Resale provides a marketplace for buying and selling pre-owned lasers, laser systems, optical tables and associated equipment for holographers.
****

Laser Technical Services
1396 River Road, Box 248
City: Upper Black Eddy
State/Province: PA
Postal Code: 18972
Country: United States of America
Voice Phone: 1 6109820226
Fax Phone: 1 610 982 0226
email: lasertek@earthlink.net
Contact # 1: Dan Morrison
Business Description: Technical consultant and field repair of lasers. Full customer service of laser equipment - Specifically Pulsed Ruby holographic lenses. Specialize in Lumonics lasers. SEE OURADVERTISEMENT
****

Laser Technology, Inc.
1055 West Germantown Pike
City: Norristown
State/Province: PA
Postal Code: 19403
Country: United States of America
Voice Phone: 1 610631 5043
Fax Phone: 1 610 631 0934
Contact # 1: John Newman
Business Description: Manufacture equipment for laser-based NDT; Holography and Shearography equipment and inspection services. Portable and production units available.
****

Laserfilm Eckard Knuth
Milchstr. 12
City: Munich
Postal Code: 81667
Country: Germany
Voice Phone: 49 89 480 77 14
Fax Phone: 49 89 48 56 66
Contact # 1: Eckard Knuth
Business Description: 120/360 degree Multiplex holograms (stereograms). Moving images. Diameter 45 cm or 65 cm. Most representative work in 1996 - a 360 degree hologram with an imperial crown in original size for the exhibition Austria 996 -1996.
****

Lasermetrics, Inc.
(a division of Fastpulse Technology, Inc.)
220 Midland Ave.
City: Saddlebrook
State/Province: NJ
Postal Code: 07663
Cowltry: United States of America
Voice Phone: 1 201 4785757
Voice Phone: 1 800 449 FAST
Fax Phone: 1 201 4786115
Contact # I: Robert Goldstein
Business Description: Laser components and electronic drivers for lasers.
****

Lasersmith, Inc.(,Fhe) ..
1000 West Monroe Street
City: Chicago
State/Province: IL
Postal Code: 60607
Country: United States of America
Voice Phone: 1 312 733 5462
Fax Phone: 1 312 733 5926
email: steven@lasersmith.com
Web Address: www.lasersmity.com
Contact #1: Steven L. Smith
Business Description: Specialists in holographic imaging . In-house service: art work origination, photo shoots of art/object, 2D/3D full color 2D & 3D mastering. Stereogram filming; separations & mastering; computer modeling & rendering for full color. Stock items. Sequential numbering and bar coding. Complete security system in house.
****

Laserworks
PO Box 2408
City: Orange
StatelProvince: CA
Postal Code: 92859
Country: United States of America
Voice Phone: 1 7148322686
Fax Phone: 1714832 1451
Web Address: ourworld.compuserve.com;80/ homepage/laserworks/
Contact #1: Selwyn Lissack
Business Description: Company manufactures programmable laser scanning equipment for signage and display applications. Selwyn Lissack has been a holographic artist and researcher since 1969. He has produced numerous holographic exhibitions in addition to the Salvador Dali holograms.
****

Lasing S.A.,
Marques De Pico Velasco 64
E-28027
City: Madrid
Country: Spain
Voice Phone: 34 0 1 268 3643
Business Description: Branch office of Newport Corporation
****

Lasiris Inc.
Main Office
3549 Ashby
City: Ville St Laurent
Postal Code: H4R 2K3
Country: Canada
Voice Phone: 1 514335 1005
Fax Phone: 1 5143354576
Contact #1: Alain Beauregard
Business Description: HOE optics in stock gratings. HOE special projects. Beamsplitters.
****

Lauk & Partner GmbH
(a division of Lauk Kommunikation)
Augustinusstr 9B
City: Frechen
Postal Code: 50226
Country: Germany
Voice Phone: 49 02234 51055-66
Fax Phone: 49 02234 65019
email: 101624331@compuserve.com
Contact #1: Mathias Lauk
Business Description: Holograms, holographic projects, hologram museum.
****

Lawrence Berkeley Laboratory
University Of California
1 Cyclotron Road
City: Berkeley
State/Province: CA
Postal Code: 94720
Countt-y: United States of America
Voice Phone: 1 510 486 4000
Business Description: Holography R&D
****

Laza Holograms Ltd. / Spatial Imaging
6 Marlborough Road
City: Richmond, Surrey
State/Province: England
Postal Code: TWIO 6JR
Country: United Kingdom
Voice Phone: 44 181 332 1080
Fax Phone: 44 181 332 2990
Contact # 1: Jonathan Cope
Business Description: Manufacturer of high quality mass produced reflection holograms. 120 stock images in various sizes. Production of custom holograms. Distributor of a large range of holographic products.
****

Lazart Holographics
22 Erina Valley Road
City: Erina
StatelProvince: NSW
Postal Code: 2250
Country: Australia
Voice Phone: 61 43 676245
Fax Phone: 61 043 652 306
email: lazart@ozemail.com.au
Contact # 1: Brett Wilson

Business Description: Artistic holography; buying & selling holograms. Wholesale distribution and retail sales of artist editions and stock images. Production of jewelry items and novelty products from embossed images.

****

Lazer Wizardry
(a division of Sum on, LLC)
5805 West 6th Ave., Suite B
City: Lakewood
StatelProvince: CO
Postal Code: 80214
Country: United States of America
Voice Phone: 1 303 274 0706
Voice Phone: 1 800 793 0565
Fax Phone: 1 303 274 0733
Contact # 1: Elmer DeFillipo
Business Description: Wholesale distributor ofa wide range of holographic products and related giftware. Special expertise regarding the operations and inventory needs of retail holography specialty shops. Call for catalog.
SEE OURADVERTISEMENT

****

Lenox Laser
12530 Manor Road
City: GlenArm
State/Province: MD
Postal Code: 21057
Country: United States of America
Voice Phone: 1 410 592 3106
Fax Phone: 1 410 592 3362
email: sales@lenoxlasercom
Contact # 1: Joseph P D'Entremont
Business Description: Laser-systems laboratory specializing in laser drilling and etching. Manufactures pre-fabricated aperture kits and pinholes suitable for precision industrial/optical applications. Also sells used laser systems and low cost optical kits for holography.
SEE OURADVERTISEMENT

****

Leonhard Kurz GmbH
Schwabacher Strasse 482
City: Fuerth
Postal Code: 90763
Country: Germany
Voice Phone: +49 (0)91171410
Fax Phone: +49 (0)911 7141507
Contact # I: Werner Reinhart
Business Description: Manufacturer of embossing equipment; broker for hologram embossing.

****

Les Productions Hololab!
3970 Boulevarde St Laurent
City: Montreal
Postal Code: H2W 1 Y3
Country: Canada
Voice Phone: 1 514 8494325
Contact #1: Marie-Christiane Math ieu
Business Description: Artistic holography

****

Leseberg, Dr. Detlef
Kamener Str 172
City: Lunen-Beckinghausen
Postal Code: W4670
Country: Gennany
Voice Phone: +49 (0)230 1794
Fax Phone: +49 (0)230 1793
Contact #1: Detlef Dr. Leseberg
Business Description: Scientific holography research, HOE, computer-generated holography.

****

Letterhead Press, Inc.

W226 N880 Eastmound Drive
City: Waukesha
State/Province: WI
Postal Code: 53186-1 689
Country: United States of America
Voice Phone: 1 414574 1717
Fax Phone: 14145741719
Contact #1: Mark Mulvaney
Business Description: Full service trade finisher with 24-hour, 7-days/week manufacturing. Featuring 19 x 25 inch and 40 inch fonnats for holographic stamping. Complete projects from print to final bindery assuring single-source responsibility.

****

Lexel Laser, Inc.
48503 Milmont Drive
City: Fremont
State/Province: CA
Postal Code: 94538
Country: United States of America
Voice Phone: 1 510 770 0800
Voice Phone: 1 8005273795
Fax Phone: 1 510 651 6598
email: lexel@aol.com
Contact #1: Ben Graham
Business Description: Lexel produces the highest quality Argon, Krypton and mixed gas laser systems. In particular, Lexel specializes in production of single frequency systems which are very stable over a variety of environmental situations.

****

LiCONiX
3281 Scott Boulevard
City: Santa Clara
State/Province: CA
Postal Code: 95054
Country: United States of America
Voice Phone: 1 408 496 0300
Fax Phone: 1 408 492 1303
Contact #1: Mark Dowley
Business Description: LiCONiX has long been recognized as the leader in Helium Cadmium laser technology. SEE OURADVERTISEMENT

****

Light Dimension, Inc.
Sunfamily Hongo #403
5-10, Hongo 4-Chome~Bunkyo-ku
City: Tokyo
Postal Code: 113
Country: Japan
Voice Phone: 81 3 38129201
Fax Phone: 81 338129422
Contact #1: Mariko Oishi
Business Description: Handling the whole range of holograms/holographic products from embossed holograms to fine art images; also focusing on exhibitions on holography.

****

Light Impressions International, Ltd.
430 West Diversey Parkway - Suite 501
City: Chicago
State/Province: IL
Postal Code: 60614
Country: United States of America
Voice Phone: 1 773 665 1579
Fax Phone: 1 773665 1679
Contact # 1: Pamela Jamison
Business Description: A full service manufacturer offering the highest quality custom mastering. Embossed product available in hot stamp foil or pressure sensitive labels for security or promotional application. Stock images and holographic equipment available.
SEE OUR ADVERTISEMENT

****

Light Impressions International, Ltd.
5 Mole Business Park 3
City: Leatherhead
State/Province: England
Postal Code: KT22 7BA
Country: United Kingdom
Voice Phone: 44 1372 386 677
Fax Phone: 44 1372 386 548
email: 100331.326@compuserve.com
Web Address: euroweb.com/nlcon/b2b/ehol/limpress.htm
Contact #1: John Brown
Business Description: A full service manufacturer offering the highest quality custom mastering. Embossed product available in hot stamp foil or pressure sensitive labels for security or promotional application. Stock images and holographic equipment available. SEE OUR ADVERTISEMENT

****

Light Wave Gallery
North Pier
435 East Illinois Street
City: Chicago
State/Province: IL
Postal Code: 60611
Country: United States of America
Voice Phone: 1 312321 1123
Fax Phone: 1 312 321 0892
Contact #1: Jim Harden
Business Description: Gallery, retail shop. Complete line of holograms and related holographic giftware. Extensive collection of limited edition artworks available for sale or rental.

****

Lightrix, Inc.
377 Oyster Point Blvd. #11
City: South San Francisco
State/Province: CA
Postal Code: 94080
Country: United States of America
Voice Phone: 1 4152449791
Fax Phone: 1 415 244 9795
email: lightrix@creative.net
Web Address: ionserve.com/lightrix.html
Contact #1 : Deborah Robinson
Business Description: Lightrix manufactures and designs high quality holographic toys, gifts and wall decor. Lightrix offers state of the art holographic product development and graphic design. Lightrix holograms are available to the wholesale trade. We offer a full custom program for embossed and photopolymerholograms. SEE OURADVERTISEMENT ON INSIDE FRONT COVER

****

Linda Law Holographics
425 New York Ave., Suite 202
City: Huntington'.
State/Province: NY
Postal Code: 11743
Country: United States of America
Voice Phone: 1 5166733138
Fax Phone: 1 5166739127
email: Ilholo@I-2000.com
Contact #1 : Linda Law
Business Description: State of the art computer graphics for holography. Using Mac and SGI computers, artwork can be created for mass production holograms or large scale display holograms. SEE HOLOGRAM ON THE FRONT COVER OFTHIS BOOK

****

Lone Star Illusions
2901 Capital Of Texas Highway, #191
City: Austin

CORPORATE LISTINGS

State/Province: TX
Postal Code: 78746
Country: United States of America
Voice Phone: 1 512 328 3599
Contact #1: Alan Lifshen
Business Description:Austin 's only hologram gallery and retail shop which features a full range of holographic giftware and related optical novelties.
****

Lopez's Gallery International
500 North Michigan Ave., Suite 1920
City: Chicago
StatelProvince: IL
Postal Code: 60611-3704
Country: United States of America
Voice Phone: 1 312 975 2052
Voice Phone: 1 800 4-D FI~E-,~RT
Fax Phone: 1 3122489527
email: fineart4D@aol.com
Contact # 1: Jesus Lopez
Business Description: Creating the world's most collectible holographic Fine Art. Specialists in 4 dimensional female figure studies and holographic sculptures. SEE OURADVERTISEMENT
****

LOT Oriel
1 Mole Business Park
City: Leatherhead
State/Province: England
Postal Code: KT22 7 AU
Country: United Kingdom
Voice Phone: 44 1372378822
Fax Phone: 44 137237 5353
email: 10toriel@lotoriel.co.uk
Contact # I: John Green
Business Description: Artistic holography and laser products.
****

Loughborough Univ. Of Tech.
Dept Of Phys ics - Dept of Mechanical Engi-neerlIIg
City: Loughborough
State/Province: England
Postal Code: LEI 1 3TU
Country: United Kingdom
Voice Phone: 44 1509 263 171
Fax Phone: 44 1509 219 702
Contact # 1: N. Halliwell
Business Description: Scientific and industrial research.
****

Louis Paul Jonas Studios, Inc.
304 Miller Road
City: Hudson
State/Province: NY
Postal Code: 12534
Country: United States of America
Voice Phone: 1 5188512211
Fax Phone: 1 51 8 8851 2284
email: dmerritt@epix.net
Contact # I: Dave Men-itt
Business Description: Jonas Studios specializes in making models and miniatures for the museum and film industries. Services include: sculpting, model making, dioramas, EDM machining, CAD for rapid prototyping, laser cutting, mold making and casting in a wide va riety of materials.
****

Lulea University Of Technology
Dept Of Mechanical Engineering
City: Lulea
Postal Code: S-951 87
Country: Sweden
Contact # 1: Nil s-Erik Molin

Business Description: Industrial research; holographic non-destructive testing.
****

Lumenx Technologies, Inc.
PO Box 219
City: New Durham
State/Province: NH
Postal Code: 03855
Country: United States of America
Voice Phone: 1 603 859 3800
Fax Phone: 1 603 859 250 I
Contact #1: Ed Neister
Business Description: We manufacture and repair a variety of laser equipment, mostly for scientific and medical applications. Call for more details.
****

Luminer Printing and Converting
1400 1 ndustrial Way
City: Tom's River
State/Province: NJ
Postal Code: 08755
Country: United States of America
Voice Phone: 1 908 341 5727
Fax Phone: 1 908341 6175
Contact # 1: Thomas Spina
Business Description: Innovative printer and converter; labels and promotional materials. Expertise and technology for adhesive coating imaged holographic materials, including zone and patterned areas. Overprint, laminate, fold multiple webs. Complete design and origination services.
****

Lumonics Ltd.
Cos ford Lane
Swift Valley
City: Rugby
State/Province: England
Postal Code: CV21 1 QN
Country: United Kingdom
Voice Phone: 44 1788 570 321
Fax Phone: 44 1788 579 824
Contact # I: George Synowiec
Business Desc ription: Lumonics manufactures a variety of lasers and laser based systems for industrial applications. These include a range of pulsed ruby lasers specifically designed for holography with output energies spanning from 30 mJ per pulse to greater than 10 Loules per pulse.
****

Lund Institute Of Tech.
Department Of Physics
Box 118
City: Lund
Postal Code: S-221
Countly: Sweden
Voice Phone: 46 046 222 7656
Fax Phone: 46 046 222 40 1 7
email: seven-goran.pattersson@fysik.lth.se
Contact # I: Sven-Goran Pattersson
Business Description: Color H-I; holography education; academic research.
****

M.I.T. (Massachusetts Institute of Technology)
Media Laboratory/Spatial Imaging Group
20 Ames Street # EI5-416
City: Cambridge
State/Province: MA
Postal Code: 02139 - 4307
Country: United States of America
Voice Phone: 1 617 253 0632
Fax Phone: 1 617 253 8823
email: sab@media.mit.edu
Web Address: media.mit.edu/groups/spi/
Contact #1: Melissa Yoon

Business Desc ription: College holography courses; Computer Generated Holography research. Holographic hard copy printer research.
****

M.I.T. Museum
265 Massachusetts Ave.
City: Cambridge
StatelProvince: MA
Postal Code: 02139-4307
Country: United States of America
Voice Phone: 1 617 253 4462
Fax Phone: 1 617 253 8994
Contact # 1: Diego Garcia
Business Description: Museum has approximately 50 holograms on display from a inventory of approximately 1,000 holograms. Some of the most historically-significant holograms ever made are on di splay here. Museum shop has holograms for sale.
****

M.O.M.Inc.
2436 Forest Green Rd.
City: Baltimore
State/Province: MD
Postal Code: 21209
Country: United States of America
Contact # 1: Alan Evan
Business Description: Maryland Optical Manufacturing. Highest quality. 38 years experience.
****

MacShane Holography
C/O Laser Arts Programs
512 West Braeside Drive
City: Arlington Heights
State/Province: IL
Postal Code: 60004-2060
Country: United States of America
Voice Phone: 1 8473984983
Contact #1: Jim MacShane
Business Description: Design and manufacturing of Sun bows; sculptural, architectural, and gift embossed holographic products; educational programs and artistic holography.
****

Magic Laser
Quartier De r..: Horloge
4 Rue Brantome
City: Paris
Postal Code: 75003
Country: France
Voice Phone: 33 1 42 74 35 78
Fax Phone: 33 1 42 74 33 57
Contact # I: Anne-Marie Clu'istakis
Business Description: Importer and wholesaler of all holographic products-traveling exhibition.
****

Magick signs Holografie
August-Bebel-Str. 40
City: Egelsbach
Postal Code: 63329
Country: Germany
Voice Phone: +49 (0)61045544
Fax Phone: +49 (0)6 10 45548
Contact # 1: Andreas Wollenweber
Business Description: PRODUCTION of embossed holograms: Artwork, models, origination, embossing, many stock images (Trade mark). RETAIL in our four holographic magic gift stores (Frankfurt) .
****

Magick Signs Holografie
Isenburg-Zentrum
Shopteil Wost
City: Neu- Isenburg
Postal Code: D-63263

Country: Gennany
Voice Phone: 49 (0)610 328404
Business Description: Sells holograms of all kinds to the public.
****

Man/Environment, Inc.
2251 Federal Avenue - offices
2240 Federal Avenue - laboratory
City: Los Angeles
StatelProvince: CA
Postal Code: 90064
Country: United States of America
Voice Phone: 1 310 477 7922
Voice Phone: 1 310 477 8960
Fax Phone: 13104774910
email: metaphor@ix.netcom.com
Web Address: armchair.com
Contact # 1: Gary Fisher
Business Description: Silver halide and photo-polymer R&D projects. Design and manufacture optical printers and holographic systems. Complete website development. Check our website for additional information.
****

Margaret Benyon Holography Studio
40 Springdale Avenue
City: Broadstone
State/Province: England
Postal Code: BH 18 9EU
Country: United Kingdom
Voice Phone: 44 1202698067
Fax Phone: 44 1202 698067
Contact # 1: Margaret Benyon
Business Description: Holographic fine artist since 1968, with works included in a large number of private and public collections world wide. Works are available for exhibiti on, hire or sale.
****

Marks, Gerald
29 West 26th Street
City: New York
StatelProvince: NY
Postal Code: 100 1 0
Country: United States of America
Voice Phone: 1 212 889 5994
Fax Phone: 1 212 889 5926
email: pulltime3d@aol.com
Web Address: nttad.com/asci/gmwork.html
Contact # I: Gerald Marks
Business Description: Artist specializing in stereoscopic 3D of every type for over twenty years. He is best known for the 3D music videos he created for the Rolling Stones and his 3D museum exhibits, anaglyph prints and books, lenticulars, random dot stereograms, computer multimedia and computer generated holography.
****

Martinsson Elektronik Ab.
lnstrumentvagen 16
Box 9060
City: Hagersten
Postal Code: S-126 09
Country: Sweden
Voice Phone: 46 08 744 0300
Fax Phone: 46 08 744 3403
Contact # 1: Mikael Hell
Business Description: Equipment & suppl ies.
****

Marubun Corporation
8-1 Nihombashi Odemmacho
Chuo-Ku
City: Tokyo
Postal Code: 103
Country: Japan

Voice Phone: 81 03 648 8115
Fax Phone: 81 03 648 9398
Business Description: Branch office of Newport Corporation, Fountain Valley, CA USA
****

MasterPrint Holography, Inc.
250-P Executive Drive
City: Edgewood
State/Province: NY
Postal Code: 1 1717
Cow1try: United States of America
Voice Phone: 1 516243 0170
Fax Phone: 1 5162430180
Contact # 1: Michael Liu
Business Description: Specialists in photoresist mastering for mass production. R&D HOEs.
****

Mazda Motor Corp.
Technical Research Center
POBox 18
City: Hiroshima
Postal Code: 730 91
Country: Japan
Voice Phone: 81 082 282 1111
Fax Phone: 81 082252 5343
Contact # 1: Ichiro Masamori
Business Description: Holographic Interferometry
****

McCain Marketing & Graphic Design
10962 North Wauwatosa Road
76W
City: Mequon
State/Province: WI
Postal Code: 53097
Country: United States of America
Voice Phone: 1 414 242 4023
email: mc2nite@aol.com
Contact # 1: Richard McCain
Business Description: Graphic designer and litho printing consultant, 20 years experience. Extensive understanding of holographic effects to educate or sell product using holography with print. Set up project, customer buys direct from manufacturer.
****

McMahan Electro-Optic
2160 Park Avenue
(Orlando Division) "
City: Winter Park '
StatelProvince: FL
Postal Code: 32789
Country: United States of America
Voice Phone: 1 407 645 1000
Fax Phone: 1 407 644 9000
email: bobmcmahn@aol.com
Contact # 1: Robert McMahan
Business Description: McMahan Electro-Optics manufactures a laser-based NDT system for testing composite aerospace components and assemblies. Mobile unit.
****

Media Interface, Ltd,
215 Berkeley Place
City: Brooklyn
State/Province: NY
Postal Code: 11217
Country: United States of America
Voice Phone: 1 718 398 1136
Fax Phone: 1 718 398 1136
email: ronholog@bway.net
Web Address: bway.netl-ronholog
Contact # 1 : Ronald R. Erickson
Business Description: Consulting in holographic applications and mass produced and commercial holographic image design and production. Medi-

cal holography. Custom holographic optical configurations. Computer assisted holographic image design and production - research or commercial.
****

Melissa Crenshaw Holography Studio International
Jl RRI
City: Bowen Island
Postal Code: VON 1 GO
Country: Canada
Voice Phone: 1 604 645 2019
Voice Phone: 1 604 645 2019
email: mcrensha@capcollege.bc.ca
Contact # 1: Melissa Crenshaw
Business Description: Holographic fine artist with experience integrating holographic elements into commercial projects, including architectural and lighting design . Color reflection hologram mastering and mass production (12x 16, 4x5 film) services offered. Extensive portfolio.
****

Melles Griot
1770 Kettering Street
City: Irvine
StatelProvince: CA
Postal Code: 92 714
Country: United States of America
Voice Phone: 1 714261 5600
fax Phone: 1 714261 7589
email: mgtech@irvine.mellesgriot.com
Web Address: www. mellesgriot.com
Business Description: Melles Griot is a major manufacturer of off-the-shelf and custom tables and isolation equipment, laser, lenses, mounting hardware, positioners, polarizers, coated optics, detectors, collimators and spatial filters.
****

Melles Griot GmbH
Lilienthalstrasse 30-32
City: Bensheim
Postal Code: 64625
Country: Germany
Voice Phone: +49 (0)625 8406-0
Fax Phone: +49 (0)625 8406-22
Contact # I: Daniel Hinz
Business Description: Condensers (optics), fiber optics constmction components, laser diodes, laser optics,optical filters ,optical lenses, optical mirrors, optical parts, bulk, optoelectronic components, planar optics, planar parallel optics, prisms.
****

Meredith Instruments
5035 North 55th Avenue
Suite 5
City: Glendale
State/Proyince: AZ
Postal Code: 8530 I
Country: United States of America
Voice Phone: 1 602.934 9387
fax Phone: 1 602 934 9482
email: sales@lasersl.com
Web Address: www.mi-Iasers.com
Contact # I: Lee Toland
Business Description: Specializing in surplus inventories of HeNe lasers as well as argon and diode lasers, Meredith Instmments is the USA's largest laser discount dealer. Free catalogue. Laser repair.
****

Mesmerized Holographic Marketing
PO. Box 984
City: White Plains
StatelProvince: NY
Postal Code: 10602-0984
Country: United States of America
Voice Phone: 1 914 948 6138

CORPORATE LISTINGS

Fax Phone: 1 9149489509
email: sales@mesmerized.com
Contact #1: Jeff
Jeffrey Levine
Business Description: Design and manufacturing of finished hologram products, including desk premiums, awards, direct mailers, and displays. Has developed proprietary design, production and application techniques for producing the highest quality finished products at the lowest possible price.
****

MetroLaser
18006 Skypark Circle # 1 08
City: Irvine
State/Province: CA
Postal Code: 92614
Country: United States of America
Voice Phone: 1 714553 0688 -,
Fax Phone: 1 7145530495
email: metro@deltanet.com
Web Address : www.pages.prodgy.com/metrolaser/home.htm
Contact # I: James D. Trolinger
Business Description: Holographic non destructive testing services. Measurements and instruments based on holography and holographic interferometry Holographic particle and flow diagnostics.
****

Metrologic Instruments GmbH
Dornierstrasse 2
City: Puchheim
Postal Code: 82178
Country: Germany
Voice Phone: +49 (0)89 89019 0
Fax Phone: +49 (0)89 89019 200
Contact # I: Benny Noems
Business Description: see Metrologic Instruments Inc., USA. Please fax us.
****

Metrologic Instruments, Inc.
Coles Road at Route 42
City: Blackwood
State/Province: NJ
Postal Code: 08012
Country: United States of America
Voice Phone: 1 6092288100
Voice Phone: 1 800 IDMETRO
Fax Phone: 1 609 228 6673
email: cstserv@metrologic.com
Web Address: www.metrologic.com
Contact #1: Christen Kendall
Business Description: Metrologic Instruments manufacturers holographic laser bar code scanners or HoloTrak (TM). The Holotrak is omnidirectional with a 40 inch depth of field and 26 inch scan width. Applications include pallet scanning, unattended scanning, truck unloading and conveyor belts.
****

Meulien Odile
Mergesst. 16
City: Braunschweig
Postal Code: 38108
Country: Germany
Voice Phone: +49 (0)531 352816
Fax Phone: +49 (0)531 352 816
Contact # I: Odile Meulien
Business Description: Collector and analyst of the holographic trend since 10 years in the US and Europe. Manage a private collection - Conduct studies on future uses of holography - Coordinate holographic and light happenings.
****

J 42 Holography MarketPlace - 6th edition
MGM Converters Inc.
16604 Edwards Road
City: CelTitos
State/Province: CA
Postal Code: 90703
Country: United States of America
Voice Phone: 1 310 404 3779
Fax Phone: 1 310404 7408
Contact # I: Steve Meyer
Business Description: Full service converting services for the holography market. Wide web hot stamping, including continuous application.
****

Midwest Laser Products
PO Box 262
City: Frankfort
State/Province: IL
Postal Code: 60423
Country: United States of America
Voice Phone: 1 8154640085
Fax Phone: 1 8154640767
email: mlp@midwest-Iaser.com
We b Address: www.midwest-Iaser.com/lasers
Contact # I: Steve GalTett
Business Description: New and used laser equipment for holographers including: HeNe, Argon, HeCd, Nd:YAG and visible diode lasers. We also sell complete holography kits, including low-cost HeNe lasers and related materials.
SEE OUR ADVERTISEMENT
****

Ministry OfInternational Trade
Electrotechnical Laboratory
Optical Information Section
City: Tsukuba Science City
Postal Code: 305
Country: Japan
Voice Phone: 81 0298 58 5625
Fax Phone: 81 0298 58 5627
Contact # I: Dr. Satoshi Ishihara
Business Description: Research using HOEs
****

Mitsubishi Heavy Industries Ltd.
Nagasaki Technical Institute
1-1 Akunoura-Machi
City: Nagasaki
Postal Code: 850-91
Country: Japan
Contact #1: M. Murata
Business Description: Holographic non-destructive testing; industrial research.
****

Mitutoyo Measuring Instruments (MTI Corp.)
965 Corporate Blvd.
City: Aurora
State/Province: IL
Postal Code: 60504
Country: United States of America
Voice Phone: 1 630 820 9666
Fax Phone: 1 630 820 1393
Web Address: industry.netlmitutoyo
Contact #1: Bill Naaman
Business Description: Manufacturers of precision measuring instruments including highly accurate holographic linear tracking systems suitable for precision industrial and research applications.
****

Moonbeamers
Jl5 Gibbons Street
City: Telopea
StatelProvince: NSW
Postal Code: 2117
Country: Australia

Voice Phone: 61 612 890 1233
Fax Phone: 61 612 890 1243
Contact # I: John Tobin
Business Description: Since 1984, we have been producing commercial holograms and diffractions for security and display applications. We offer a complete service from artwork creation through application & printing within Australia and Asia.
****

Morning Light Holograms
106 Xi Huan Middle Street
Cang Zhou
City: He Bei
Postal Code: 06100 I
Country: People 's Rep. of China
Voice Phone: 86 317 226 164
Fax Phone: 86 317 226 167
Contact #1: Chen Guo Tong
Business Description: Largest Hologram producer in China (HN 2.95)
****

Mu's Laser Works
1328 Dunsterville Avenue
City: Victoria
Postal Code: V8Z 2X I
Country: Canada
Voice Phone: 1 604 479 4357
Contact # 1: Ron Meuse
Business Description: Holographic and 3-D photographic services, Can provide lab rental and technical assistance. Laser light show production and rental.
****

Multifacet
Paul Gilsonlaan 450
City: Drogenbos
Postal Code: 1620
Country: Belgium
Voice Phone: 32 2 331 2455
Fax Phone: 32 2 331 3009
Contact # 1 : Desmet Guy
Business Description: A division of Finiprint which concentrates on holographic films .
****

Multiplex Moving Holograms
746 Treat Street
City: San Francisco
State/Province: CA
Postal Code: 94110
Country: United States of America
Voice Phone: 1 415 285 9035
Fax Phone: 1 415206 1622
Contact # 1: Peter Claudius
Business Description: We are the originators of the Multiplex Hologram. We produce white light viewable moving holograms for trade shows and exhibits. 120, 360 degree and flat format white light viewable holograms made to your specifications. Stock images also available. Ask for our catalogue! In business since 1973!
****

Museu D' Holografia
Jaume 1,1
City: Barcelona
Postal Code: 08002
Country: Spain
Voice Phone: 343 3 102 172
Fax Phone: 343 3 319 1676
Business Description: Holographic Gallery, Itinerant exhibitions, sale of holograms. Teaching. Holographic courses. Holographic laboratory.
****

Museum Of Holography/Chicago
1134 West Washington Blvd.

City: Chicago
State/Province: IL
Postal Code: 60607
Country: United States of America
Voice Phone: 1 312 226 1007
Fax Phone: 1 312 829 9636
Contact # I: Loren Billings
Business Description: Founded in 1978, the MOHC is now the world's oldest institution devoted to the display, acquisition and maintenance of holography as well as education and research in the field. Permanent collection is now the largest in the world. At least two major exhibitions a year featuring artists from around the world.
****

MWK Industries
1269 West Pomona Road # 1 12
City: Corona
State/Province: CA
Postal Code: 91720
Country: United States of America
Voice Phone: 1 909 278 0563
Voice Phone: 1 8003567714
Fax Phone: 1 909 278 4887
email: mkennyl989@aol.com
WebAddress: www.pweb.comlmwklmain.htm
Contact # 1: Mike Kenny
Business Description: Large selection of surplus and used lasers from major manufacturers. Save 30% to 60% on brand name laser purchases. We offer the beginning, intermediate and advanced holographer a large selection of lasers suitable for holography (including HeNe lasers ranging up to 25 mw), as well as other related materials.
SEE OURADVERTISEMENT
****

National Physical Laboratory
Queens Road
City: Teddington
State/Province: England
Postal Code: TW11 OLW
Country: United Kingdom
Voice Phone: 44 181 977 3222
Fax Phone: 44 181 943 2155
email: library@newton.npl.co.uk
Web Address: http://www.npl.co.uk
Contact # 1: David Robinson
Business Description: Scientific and industrial research; holographic non-destructive testing.
****

Navidec Inc . (formerly ACT Systems, Inc.)
14 Inverness Drive East
Suite F-116
City: Englewood
State/Province: CO
Postal Code: 80112
Country: United States of America
Voice Phone: 1 303 790 7565
Voice Phone: 1 800 797 7565
Fax Phone: 1 303 790 8845
email: patt@navidec.com
Contact # I: Patrick Townsend
Business Description: Exclusive agent (US, Canada, Mexico) for Kimmon Helium Cadmium laser systems used for holography.
SEE OURADVERTISEMENT
****

NeoVision Productions
PO Box 74277
City: Los Angeles
State/Province: CA
Postal Code: 90004
Country: United States of America
Voice Phone: 1 213 387 0461
Contact # 1: Bill Hillard
Business Description: Traveling show. Fine art

originals. Produce holograms for home and industry. Consulting.
****

New Dimension Holographics
27 Nurses Walk, The Rocks
City: Sydney
State/Province: NSW
Postal Code: 2000
Country: Australia
Voice Phone: 61 29 743 3767
Fax Phone: 61 297433241
Contact # I: Tony Butteriss
Business Description: Retail shop. Wholesale distribution. Origination consultant. Educational consultant. Gallery.
****

New Horizons (Thailand), Ltd.
460/33 Chiangmai Land Chang klan Road
City: Chiangmai
Postal Code: 50000
Country: Thailand
Voice Phone: 66 53 252 007
Fax Phone: 66 53 252 007
Contact # 1: Si lvio Aprile
Business Description: Mass produce dichromates for jewelry, etc.
****

New Light Industries
West 97 13 Sunset Hwy.
City: Spokane
State/Province: WA -
Postal Code: 99204
Country: United States of America
Voice Phone: 1 509 456 8321
Fax Phone: 1 509 456 8351
email: stevem@compch.iea.com
WebAddress: www.iea.coml..1lli
Contact # I: Steve McGrew
Business Description: Extensive experience with technology transfer, consulting and R&D for embossed holography. Complete origination and production system installations, worldwide.
****

New York Hall Of Science
47-01 III Th Street
City: Corona
State/Province: NY ~
Postal Code: 11368
Country: United States of America
Voice Phone: 1 7186990005
Contact # I: Beth Weinstein
Business Desc ription: The New York Hall of Science is New York's only hands-on science and technology museum. Lasers and optics demonstrated daily. Color hologram depicting quantum atom is on display.
****

New York Holographic Laboratories
P.O. Box 20391
Thomkins Square Station
City: New York
State/Province: NY
Postal Code: 10009
Country: United States of America
Voice Phone: 1 212674 1007
Fax Phone: 1 212677 6304
Contact #1: Dan Schweitzer
Business Description: Fine art editions. Tutorial courses, lectures and consultations.
****

Newport (Asian Office)
Kyokuto Boeki Kaisha
7Th Floor New Otemachi Bldg
City: 2-1, 2-Chome,
Postal Code: Tokyo 100-
Country: Japan

Business Description: Branch office of Newport Corp., Fountain Valley CA, USA.
****

Newport Corporation
1791 Deere Ave.
City: Irvine
State/Province: CA
Postal Code: 92606
Country: United States of America
Voice Phone: 1 800 222 9980
Voice Phone: 1714863-3144
Fax Phone: 1 7142531800
Web Address: newport.com
Contact #1: Gary Spiegel
Business Description: Designer and manufacturer of E/O components, optics, spatial filters, optical & beamsteering instruments, magnetic bases, fiber optic components, vibration isolation systems, and holographic recording materials.
****

Newport Gmbh
European Headquarters
Holzhofallee 19
City: Darmstadt
Postal Code: 64295
Country: Germany
Voice Phone: +49 (0)61536210
Fax Phone: +49 (0)615 362 152
Business Description: Designer and manufacturer of laserlholographic systems, E/O components, optics, spatial filters, optical & beamsteering instruments, magnetic bases, fiber optic components, vibration isolation systems, and holographic recording materials.
****

Nihon University
Dept Electronic Engineering
7-24-1 Narashinodai
City: Funabashi-Shi
State/Province: Chiba
Postal Code: 274
Country: Japan
Voice Phone: 81 0474695391
Fax Phone: 810474679683
Contact # I: Dr. Hiroshi Yoshikawa
Business Description: Research NDT
****

Nimbus Manufacturing, Inc.
(a division of Nimbus CD Intemational)
P.O. Box 7427
City: Charlottesville
State/Province: VA
Postal Code: 22906
Country: United States of America
Voice Phone: 1 800 231 0778 x457
Fax Phone: 1 804 985 4625
email: Ihaney@nimbuscd.com
Contact # 1: L()ffi. Haney
Business Description: Nimbus manufactures holographic CDs, CD ROMs, CDls, enhanced CDs, and DVD's, as well as providing packaging, prepress, print procurement, and spine labels.
****

Nippon Polaroid K.K.
Business Development Division -
Mori Bldg No. 30
3-2-2 Toranomon, Minato-ku
City: Tokyo
Postal Code: 105
Country: Japan
Voice Phone: 81 03 3438 8883
Fax Phone: 81 03 5473 8637
Contact #1: Makoto ide
Business Description: Subsidiary of Polaroid Corp., Cambridge, MA USA

optical items. Catalog available.
****

Ross Books
P.O. Box 4340
City: Berkeley
State/Province: CA
Postal Code: 94704
Country: United States of America
Voice Phone: 1 800 367 0930
Voice Phone: 1 510 841 2474
Fax Phone: 1 510 841 2695
email: staff@rossbooks.com
Web Address: www.holoinfo.com
Contact #1: Alan Rhody
Business Description: Publisher of the
HOLOGRAPHY MARKETPLACE
Editions 1-6 (a worldwide database and
sourcebook), the HOLOGRAPHY HAND-
BOOK - *Making Holograms the Easy Way*
(world's best selling laboratory manual),
and other related titles. Educational,
research and information services also
provided.
****

Rowland Institute For Science
1 vans   Land Blvd
         mbri

Since 1982
ROSS BOOKS

has been providing readers around the world with relevant information about holography.

In addition to our own publications, we also provide affordable consulting and research assistance to companies looking for general education or specific reference materials.

We have world's most comprehensive and extensive listing of companies and individuals involved with the holography industry. We also have thousands of pertinent publications in our database. We can provide reports, database searches and manuscripts custom-tailored to your needs.

Contact us for further details.

Consulting   Database Searches   Customized Reports   Manuscripts and Articles

Nippondenso Co., Ltd.
System Develop Engineering Dept
I-I Showa-Cho Kari ya-Slu
City: Aichi-Ken
Postal Code: 448
Country: Japan
Voice Phone: 81 0566 256 924
Contact #1: Toru Mizuno
Business Description: Hologram manufacturer.

Nissan Motor.
Central Research Lab
Natsushima Machi
City: Yokosuka
Postal Code: 237
Country: Japan
Voice Phone: 81 0468 625 182
Fax Phone: 81 046 654 183
Business Description: Hologram manufacturer head-up display

Norges Tekniske Hogskole.
Institute For Almen Fysikk
Sem Saelandsv 7 N-7034
City: Trondheim-Nth
Country: Norway
Voice Phone: 47 07 5
Fax Phone: 47 07 592 886
Business Description: Holographic Non-destructive testing

Norland Products, Inc.
PO Box 7145
City: North Brunswick
State/Province: NJ
Postal Code: 08902
Country: United States of America
Voice Phone: 1 908 545 7828
Fax Phone: 1 908 545 9542
Business Description : Optical adhesives (which cure with UV light). Used to adhere HOEs and fo r splicing fiber optic cables.

Northern Illinois University
Department Of Phys ics
City: Dekalb
State/Province: IL
Postal Code: 60 115
Country: United States of America
Voice Phone: 1 815753 1772
Contact # 1: Thomas Rossing
Business Description: Scientific holography research. Projects vary in nature. Holographic interferometry for studying vibration modes such as in musical instruments.

Numazu College Of Technology
Dept Of Mechanical Engineering
3600 Ooka
City: Numazu-City
State/Province: Shizuoka
Postal Code: 410
Country: Japan
Voice Phone: 81 0559 212 700
Contact # 1: Dr. Koji Lkegami
Business Description: Holographic research.

OIE Research
7227 Eastwood Street
City: Philadelphia
State/Province: PA
Postal Code: 19149

Country: United States of America
Voice Phone: 1 215 33 1 5067
Contact # I: Len Stockier
Business Description: Technical and educational consulting. Holography workshops, resource center. Conceptual design and production of display holograms. H-I and H-2 mastering. Custom optical table components. Touring programs, exhib its and workshops.

Odhner Holographies
PO Box 56-8574
City: Orlando
State/Province: FL
Postal Code: 32856-8574
Country: United States of America
Voice Phone: 1 407 856 7665
Fax Phone: 1 407 856 9003
Contact # 1: Jefferson E. Odhner
Business Description: Exclusive distributor of the Stabilock II inch fringe stab ilizer (used to make brighter holograms), manufacture of custom holograms (transirefl.). Specializing in HOE arrays (to 8 x 10) on silver halide.
SEE OUR ADVERTISEMENT

Ojasmit Holographics
409 Vardhman Market Sector 17, VASHI
City: New Bombay
Postal Code: 400703
Country: India
Voice Phone: 91 22 768 '3526
Voice Phone: 91 22 763 0373
Fax Phone: 91 22 763 2509
Contact # 1: Kailesh Shah
Business Description: Manufacturing and marketing of embossed holograms, photo polymers and dichromates for varied applications.

Om nichrome Corp.
Innovation, Reliability, Results
13580 Fifth Street
City: Ch ino
State/Province: CA
Postal Code: 91710
Country: United States of America
Vo ice Phone: 1 909 ~27 1594
Fax Phone: 1 909 59 1 8340
email: info@omnichrome.com
Contact # I: Kevin Rankin
Business Description: Manufachlrer of Argon, Krypton and HeCd lasers ranging in wavelength from 325nm to 752nm. Hands-off operation lasers find applications in semiconductors, optical disk mastering, medicine and holography.
SEE OUR ADVERTISEMENT

Ontario College Of Ali
100 McCaul Street
City: Toronto
Postal Code: M5T IWI
Country: Canada
Voice Phone: 1 416 977 5311
Fax Phone: 1 416 977 0235
email: page@astraJ.magic.ca
Contact # 1: Michael Page
Business Description: General holography courses .

Ontario Science Centre
770 Don Mills Road
City: Don Mills
Postal Code: M3C 1 T3

Country: Canada
Voice Phone: 1 416 429 4100 x 2820
Fax Phone: 1 416 696 3197
email: alena_kottova@fcgatel.osc.on.ea
Contact # 1: Alena Kottova
Business Description: We have gallery of 15 holograms on permanen t di splay and laser demonstration area. Holography workshops cover theory and practical uses of holography. Participants make their own reflection hologram.

Op-Graphics (Holography) Ltd.
Unit 4 Technorth
7 Harrogate Road
City: Leeds
State/Province: England
Postal Code: LS7 3NB
Country: Un ited Kingdom
Voice Phone : 44 113 262 8687
Fax Phone: 44 1132374 182
emai l: n.hard y@ukonline.co.uk
Contact #1: Valerie Love
Business Description: Manufacturer of display holograms. Large selection of stock images in variety of fonnats and sizes. Commissioned work undertaken. Copying work for holographers undertaken.

OpSec - Corporate Headquarters
4500 Cherry Creek Drive South, Suite 900
City: Denver
State/Province: CO
Postal Code: 80222
Country: United States of America
Voice Phone: 1 303 534 4500
Voice Phone: 1 3 1 2 665 8932
Fax Phone: 1 303 759 1046
Contact # 1: Yo ram Curiel
Business Description: Produce custom hol ographic labels as optical security devices for authentication applications. Full range of production services offered.

OpSec - England
4E/F Gelders Hall Road
City: Shepshed
State/ Province: England
Postal Code: LEl2 9NH
Country: United Kingdom
Voice Phone: 44 509 600 220
Fax Phone: 44 509 508 795
email: 100745.1342@compuserve.com
Contact # 1: Mark Turnage
Business Description: A total secure service from concept design artwork to finished product-specializing in customer service and delivering quality embossed holographic security and non-security work on time. (fo rmally Light Fantasfic ('Ic.)

OpSec - USA
38 Loveton Circle
City: Sparks
State/Province: MD
Postal Code: 21152
Country : United States of America
Voice Phone: 1 410 666 1144
Fax Phone: 1 410472 4911
Contact # 1: Dean Hill
Business Description: Embossed hologram manufacturing facility specializing in security and authentication applications. Custom work accepted. Stock items avai lab le, including 38 patterns of foil (16 colors) . Company

CORPORATE LISTINGS

pioneered mass replication of embossed holograms (formerly Diffraction Co.)
****

Optical Coating Laboratory GmbH
MMG Di vision
Alte Heerstrasse 14
City: Goslar
Postal Code: 38644
Country: Germany
Voice Phone: +49 (0)532 359 0
Fax Phone: +49 (0)532 359 103
Contact # 1: Mr. Koch
Business Description: Flat glass, refined front surface mirrors, glass components, mirrors, optical mirrors, surface-coated mirrors
****

Optical Corporation Of America
3-A Lyberty Way
City: Westford
State/Province: MA
Postal Code: 01886
Country: United States of America
Voice Phone: 1 508 692 3220
Fax Phone: 1 508 692 9416
Contact # I: Walt Lekki
Business Description: Products: Precision, large aperture (to 36 inch diameter) aspheric mirrors for holographic production systems.
****

Optical Research Services
3280 East Foothill Blvd. , Suite 300
City: Pasadena
State/Province: CA
Postal Code: 91 107-31 03
Country: United States of America
Voice Phone: 1 818 795 9 101
Fax Phone: 1 818 795 9102
email: service@opticalres.com
Web Address: www.opticalres.com
Contact # 1: Lia Titizian
Business Description: Optical Design Software. We sell the programs that allow you to create Holographic Optical Elements.
****

Optical Society of America (OSA)
2010 Mass Avenue NW
City: Washington
State/Province: DC
Postal Code: 20036-1023
Country: United States of America
Voice Phone: 1 202 223 8 130
Voice Phone: 1 800 762 6960
Fax Phone: 1 202 223 1096
email: osamem@osa.org
Web Address: www.osa.org
Business Description: Organization devoted to promoting optics and photonics research and applications. Publications include: Applied Optics, Optics Letters, Optics and Photonics News, Journal of Optical Society of America.
****

Optical Test Equipment
(a division of J.D. Moeller Optische Werke GmbH)
Rosengarten 10
City: Wedel
Postal Code: D-22880
Country: Germany
Voice Phone: 49 4103 709 345
Fax Phone: 49 4103 709 375
email: mail@moeller-wedel.com
Web Address: moeller-wedel.com
Contact # 1: Carsten Schlewitt
Business Description: Manufactures custom

optical components and optical test equipment including auto collimators, testing telescopes, focometers, goniometers, goniometer-spectrometers, and Fizeau-type interferometers.
****

Optical Works Ltd.
Ealing Science Centre
Treloggan Lane
City: Newquay
State/Province: England
Postal Code: TR7 IHX
Country: United Kingdom
Voice Phone: 44 1637 87 7222
Fax Phone: 44 1637 87 7211
Contact #1: E.O. Frisk
Business Description: Make optical components, lenses and scientific instruments.
****

Optics Plus Inc.
1369 East Edinger Avenue
City: Santa Ana
State/Province: CA
Postal Code: 92705
Country: United States of America
Voice Phone: 1 714 972 1948
Fax Phone: 1 714 835 6510
Contact # I: Allison Valdivia
Business Description: Manufacture optics; precision tool mounts (including lens and mechanical mounts),
****

Optilas B. V.
PO Box 222
2400 Ae Alphen
City: AID Rijn
Country: Netherlands
Voice Phone: 31 1720 31234
Fax Phone: 31 172 43414
Contact #1: A. Kooi
Business Description: Sales/service/engineering of electro-optical and vacuum related products.
****

Optimation
2235 East 10300 South
City: Sandy •
State/Province: UT
Postal Code: 84092
Country: United States of America
Voice Phone: 1 801 956 2990
Fax Phone: 1 80 1 956 2991
Contact # I: Dean Jorgensen
Business Description: Specialize in the manufacture of excellent quality, burr-free pinholes for holographic and re lated optical applications. Call for catalog.
SEE OUR ADVERTISEMENT
****

Optimation Holographics
3200 South Haskell, Suite 160
City: Lawrence
State/Province: KS
Postal Code: 66046
Country: United States of America
Voice Phone: 1 913 841 1642
Fax Phone: 1 9 13 841 0439
Contact # 1: Terry Faddis
Business Description: Large format holographic embossing facility. Complete in-house system including resist master and shim making.
****

Optineering
2247 E. La Mirada St.
City: Tucson
State/Province: AZ

Postal Code: 18719
Country: United States of America
Voice Phone: 1 520 882 2950
Fax Phone: 1 520 882 6976
email: kcreath@primenet.com
Contact # I: Kathy Creath
Business Description: Optical engineering consulting services specializing in optical testing, metrology, nondestructive evaluation, and optical design. Application areas include holography, NDT, speckle interferometry, microscopy, photography, and process control and monitoring.
****

Optische Fenomenen
Nederlandse Stichting Voor Waarn
Warenarburg 44
City: Capelle AID Ijssel
Postal Code: NL 2907 CL
Country: Netherlands
Contact # 1: Jan M. Broeders
Business Description: Monthly Newsletter - subsidiary of Dutch Foundation of Perception & Holography
****

Optitek
100 Ferguson Drive
Mailstop 5G61
City: Mountain View
State/Province: CA
Postal Code: 94039
Country: United States of America
Voice Phone: 1 415 966 3194
Fax Phone: 1 415 966 3200
Business Description: Holographic data storage research and development.
****

Optopol Panoramic Metrology Consulting
Csiksomlyo u. 4.
City: Budapest
Postal Code: H-I025
Country: Hungary
Voice Phone: 36 1 463 2518
Voice Phone: 36 1 335 5139
Fax Phone: 36 1 463 3178
email : gregyss@next-Ib.manuf.bme.hu
Contact # 1: Pal Greguss
Business Description: Nonmultiplexed single-shot 360 degree panoramic holograms, based on the Panoramic Annular Lens (Pa l-optic) invented by Dr. Greguss, are produced for metrological and other applications in science, technology and arts.
****

Oregon Institute of Technology
Laser Optical Engineering Technology
3201 Campus Drive
City: Klamath Falls
State/Province: OR
Postal Code: 97601-8 801
Country: United States of America
Voice Phone: 1 541 885 1698
Fax Phone: 1 541 885 1666
email: piecer@oit.osshe.edu
Web Address: oit.osshe.edu
Contact # I: Robert Pierce
Business Description: The LOET program provides state of the art education by combining an applied laboratory approach to optical engineering technology together with theoretical classroom discussion. For course information, contact Dr. Robert Pierce.
****

Oregon Laser Consultants
455 Hillside Ave.

City: Klamath Falls
State/Province: OR
Postal Code: 97601-2337
Country: United States of America
Voice Phone: 1 541 882 3295
email : olcbill@aip .org
Contact # 1: Bill Deutschman
Business Description: Specialists in laser safety consulting and laser safety training. ANSI services for laser users and CDRH services for laser manufacturers. Also laser safety audits, employee training and electronic consulting.
****

Oriel Instruments
250 Long Beach Boulevard
City: Stratford
State/Province: CT
Postal Code: 06497
Country: United States of America
Voice Phone: 1 203 377 8282
Fax Phone: 1 203 378 2457
email: res_sales@oriel.com
Web Address: www.oriel.com
Contact # I: Nancy Fernandez
Business Description: A full line of optical components for holographic and related laboratory applications. Call for our catalog.
****

OWlS Gmbh
1m Gaisgraben 7
City: Staufen
Postal Code: 0-7921 9
Country: Germany
Voice Phone: 49 7633 9504 0
Fax Phone: 49 7633 9504 44
Contact # 1: Hubert Munzer
Business Description: Optical benches in different sizes. Mirror mounts (gimbal and ki nematic), mirrors and lenses.
****

Oxford Holographics
71 High Street
City: Oxford
State/Provi nce: England
Postal Code: OXI 4BA
Country: United Kingdom
Voice Phone: 44 1865 250 505
Fax Phone: 44 1865 250 505
Contact # 1: Nick Cooper
Business Desc ription: Oxford Holographics has both a very well established retail and an expanding distribution operation , focusing on unusua l and unique giftware , including holograms.
****

P.S.A Peugeot Citroen
ISMEI CEI EVM
LaSam - Chemin de la Malmaison
City: Bievres
Postal Code: F-91578
Country: France
Voice Phone: 33 1 69 35 81 78
Fax Phone: 33 1 69 35 81 94
Contact # I: M. Feingold
Business Description: Industrial research; holographic non-destructive testing.
****

P.T. Pura Barutama
JL. Kresna, Jati Wetan
P.O. Box 29
City: Kudus
Postal Code: 59346
Country: Indonesia
Voice Phone: 62 291 32223
Voice Phone: 62 291 32483

Fax Phone: 62 291 31606
Contact # I: Albertus Busano
****

Pacific Holographics Inc .
503 Caledonia Street
City: Santa Cruz
State/Province: CA
Postal Code: 95062
Country: United States of America
Voice Phone: 1 408 425 4739
Fax Phone: 1 408 425 4739
Contact # I: Randy James
Business Description: Photo-resist mastering for embossed holography. Origination, design and consulting services offered. Extensive commercial portfolio.
****

Panatron Inc.
P.O. Box 2687
City: Pomona
State/Province: CA
Postal Code: 91 769-2687
Country: United States of America
Voice Phone: 1 909 629 0748
Fax Phone: 1 909 620 0378
email: panatron@aol.com
Business Description: Supplies complete support, parts and service on all lasers. Also manufactures mirrors, lenses, rods and other parts for lasers. Laser repair and used lasers.
****

Parallax Gallery
Shop R-I, Harbor Rocks Hotel
Nurses Walk, The Rocks
City: Sydney
Country: Australia
Voice Phone: 61 29 247 6382
Fax Phone: 61 29 247 6382
Contact # 1: Tony Butteriss
Business Description: Hologram Gallery. Large variety of holograms for sale to the public.
****

Pasco Scientific
10101 Foothills Blvd.
City: Roseville
State/Province: CA "
Postal Code: 95661 .
Country: United States of America
Voice Phone: 1 916 786 3800
Fax Phone: 1 916 786 8905
Business Description: Distributor for science supplies and educational materials. Catalog available.
****

Peacock Laboratories, In c.
1901 S. 54th St.
City: Philadelphia
State/Province : PA
Postal Code: 19143
Country: United States of America
Voice Phone: 1 215 729 4400
Fax Phone: 1 215 729 1380
Contact # 1: Sagar Venkateswaran
Business Description: Established 1930. Dedicated to developmental research in mirror manufacturing, si lver metalizing, and protective coatings . Innovators of silver spray processes and dual -nozzle spray guns. Consultants and suppliers of si lvering solutions and chemicals for electroconductive, decorative and reflective applications.
****

Pennsylvania Pulp & Paper Co.
2874 Lime Kiln Pike
City: Glenside

State/Province: PA
Postal Code: 19038
Country: United States of America
Voice Phone: 1 2 15 572 8600
Fax Phone: 1 215 572 8154
email: hologramer@ao l.com
Web Address: holoprism.com
Contact # 1 : Brian Monaghan
Business Description: Manufacturer of Pri smatic Illusions holographic paper line. Holographic paper and board suitable for high volume printing and packaging applications. 5 stock patterns as well as custom images avai lable (utilizing computer-generated dot matrix origination) Wide web up to 30 inch x 40 inch image size. Educational CD avai lable.
****

Pepper, Andrew
46 Crosby Road
City: West Bridgford
Postal Code: NG2 5GH
Country: United Kingdom
Voice Phone: 44 7050 133 624
Fax Phone: 44 7050 133 625
email: pepper@monand.demon.co.uk
Contact # 1: Andrew Pepper
Business Description: Fine art holography. Limited editions, unique pieces, collaborations.
****

Phantastica
Suchtener Strasse 4a
City: Arnsberg
Postal Code: 59757
Country: Germany
Voice Phone: +49 (0)293 8 191 7
Fax Phone: +49 (0)293 29441
Contact # 1 : Gerd M. Albrecht
Business Description: Makers and distributors of articles related to embossed holograms and diffraction foil , including earrings and other jewelry, badges, pens, mobiles. Main focus is street, crafts and Chri stmas markets.
****

Photon Cantina Ltd.
PO Box 1098
City: La Canada
State/Province: CA
Postal Code: 91012- 1098
Country: United States of America
Voice Phone: 1 818 790 6735
Fax Phone: 1 8 18 790 7081
email: cpax@ix.netcom.com
Contact # I: Roy Chiarot
Business Description : Full service producers of high quality artistic and commercial silver halide reflection holograms. Stock image catalog available. Custom origination, mastering and replication services offered.
****

Photon League Of Holographers Ontario
401 Richmond Street West Suite B03
City: Toronto
Postal Code: M5U3A8
Country: Canada
Voice Phone: 1 416 599 9332
Voice Phone: 1 416 203 7243
Contact # 1 : Claudette Abrams
Business Description: Artist run, non-profit holo grap hy studio. Workshop s. 2 tables .50mw HeNe. Copy Lab. Stereogram LCD HOP. Associate Membership Sl5/year Lab Users Program $60/year.
****

Photonics Spectra
Laurin Publish ing Co. Inc.

CORPORATE LISTINGS

PO Box 4949
City: Pittsfield
State/Province: MA
Postal Code: 01202
Country: United States of America
Voice Phone: 1 4134990514
Fax Phone: 1 4134423180
email: photonics@micmail.com
Business Description: Trade publication covering the field of optics, lasers, electro-optics, and related imaging research, as well as commercial applications.
****

Photonics Systems laboratory
7 Rue De l 'Universite
City: Strasbourg,
Postal Code: 67000
Country: France .
Voice Phone: 33 88 65 50 00 . Fax
Phone: 33 88 65 52 49
Contact # I: P. Meyrueis
Business Description: Education, NDT
****

Physical Optics Corporation.
20600 Gramercy Place Bid. 100
City: Torrance
State/Province: CA
Postal Code: 90501
Country: United States of America
Voice Phone: 1 3 10 320 3088
Fax Phone: 1 310 320 8067
Contact # 1: Rick Shie
Business Description: Photoni cs -based high tech company involved in research, development and manufacture of holographic optical elements, including reflection and transmission HOEs, diffraction gratings and display holograms.
****

Physics Inst itute. Latvian
Ssr Academy Of Sciences
City: Riga-Salaspils
Postal Code: 229021
Country: latvia
Voice Phone: 371 007 947 642
Contact # 1: Dr. Kurt Shvarts
Business Description: Scientific research on recording materials
****

Physik Instrumente (PI) GmbH & Co.
Polytecplatz 5-7
City: Waldbronn
Postal Code: 76337
Country: Germany
Voice Phone: +49 (0)724 604-100
Fax Phone: +49 (0)724 604-145
Contact # 1: Karl Dr. Spanner
Business Description: Holography, laser, optical components, mi sce llaneous, oscillation insulators, vibrating dampers, PZT actuators, sensors, PZT ceramics.
****

Pilkington Optronics
Glascoed Road
SI. Asaph
City: Clwyd
State/Provinc e: England
Postal Code: LLl7 Oll
Country: United Kingdom
Voice Phone: 44 745 588 344
Fax Phone: 44 745 584 258
Contact # I: Andrew Hurst
Business Desc ription: Manufacturer of DCG and photopolymer HOEs and related optical components for Heads Up Displays, etc.
****

Pink, Patty
PO Box 24-153
City: San Francisco
State/Province: CA
Postal Code: 94124
Country: United States of America
Voice Phone: 1 415 822 7 1 23
Contact # I: Patty Pink
Business Description: Fine arts holography. High quality glass plate transmission and reflection holograms for sale or exhibition.
****

Planet 3-D
201 Silver Fox lane
City: Downingtown
State/Province: PA
Postal Code: 19335
Country: United States of America
Voice Phone: 1 6108736192
Fax Phone: 1 6 10 873 6194
Contact # I: Rich Cossa
Business Descri ption: Marketing of holographic products.
****

Point Source Productions
14670 Highway 9
P.O. Box 55
City: Boulder Creek "
State/Province: CA
Postal Code: 95006
Country: United States of America
Voice Phone: 1 408 338 1304
Fax Phone: 1 408 338 3438
Contact #1: Bob Hess
Business Desc ription: We are an independent recording studio offering product design and technical imaging consultations, mastering and ganging services (specializing in si lver-halide and photopolymer), and short-run or limited edition transfer services.
****

Polaroid Corporation
2 Osborn Street - 2nd Floor
City: Cambridge •
State/Province: MA
Postal Code: 02139
'Country: United States of America
Voice Phone: 1 617 386 8676
Voice Phone: 1 800 237 5519
Fax Phone: 1 6173868671
email: neum@polaroid.com
Web Address: holoroid.com or www.poloroid.com
Contact #1: Martha Neu
Business Description: Fully integrated supplier of highest quality, mass produced photopolymer holograms. Se rvices include design, modeling, origination , manufacturing and converting. Our industrial division provides the highest efficiency, mass produced holographic optical elements available, including reflective and transmissive diffusers, projection screens and depixellators.
SEE OUR ADVERTISEMENT
****

Polymer Image
21411 North 11th Ave. , Suite 4
City: Phoenix
State/Province: AZ
Postal Code: 85027
Country: United States of America
Voice Phone: 1 602 780 4882
Fax Phone: 1 602 780 0360
email: dan@azlink.com

Contact # 1: Dan Norton
Business Description: Full-service holographic lab, including mastering and production of custom images.
****

Potomac Photonics, Inc.
4445 Nicole Drive
City: l anham
State/Province: MD
Postal Code: 20706
Country: United States of America
Voice Phone: 1 30 1 459-3033
Fax Phone: 1 301 459-3034
emai l: gbehrmann@potomac_laser.com
Web Address: potomac_laser. com
Contact # 1: Greg Behrmann
Business Description: Manufactures compact UV lasers and tabletop micro machining workstations. Potomac offers rapid prototyping of computer generated holograms and diffractive optical elements.
****

Print-M-Boss
5/24 Kirti Nagar Indl. Area
City: New Delhi
Postal Code: 11001 7
Country: India
Voice Phone: 91 11 530586
Fax Phone: 91 11 544 11 44
Contact # 1: Ravinder Singh
Business Description: Manufacturers of embossed holograms with in-house facility for shim making. Would be interested in buying copyrights for various images and patterns. Like to make contacts for mastering.
****

Process Technologies
436 West Rawson Ave.
City: Oak Creek
State/Province: WI
Postal Code: 53154
Country: United States of America
Voice Phone: 1 414 57 1 9200
Fax Phone: 1 414571 9202
emai l: pti@execpc.com
Web Address: www.exepc.coml-pti
Contact #1: Manfred Stelter
Business Description: Provides photoresist coated plates, ronchi rulings, reticles and masks.
****

PI. Pura Nusapersada
Unit Holografi
11. Dr. lukmonohadi No. 25
City: Kudus
Postal Code: 59318
Country: Indonesia
Voice Phone: 62 291 37 1 21
Voice Phone: 62 291 39253
Fax Phone: 62 291 31452
Contact # I: Edy Soewandi
****

Pull Time 3-D laboratories
29 West 26 Street
City: New York
State/Province: NY
Postal Code: 10010
Country: United States of America
Voice Phone: 1 212 889 5994
Fax Phone: 1 212 889 5926
email: pulltime3d@aol.com
Web Address: nttad.com/asci/gmwork.html
Contact # I: Gerald Marks
Business Description: PullTime 3-D Laboratories is responsible for 3D broadcast televi

sion and home video for clients including CBS Records, Fox Television, The Rolling Stones, AT &T, Howard Stern and Atlantic Records. Over 25 mi llion PullTime 3-D glasses produced to date.
****

Quantel
17, av de I' Atlantique
ZA de Couliaboeuf, BP 23
City: Les Ulis Orsay Cedex
Postal Code: 91941
Country: France
Voice Phone: 33 1 69 29 17 00
Fax Phone: 33 1 69 29 17 29
Contact # I: Alain Orszag
Business Description: Manufacturer of lasers.
****

Rainbow Symphony Inc.
6860 Canby Ave. # 120
City: Reseda
State/Province: CA
Postal Code: 91335
Country: United States of America
Voice Phone: 1 818 708 8400
Fax Phone: 1 818 708 8470
email: sjdc6la@prodigy.com
Web Address: www.rainbowsymphony.com
Contact # I: Mark Margolis
Business Description: Manufacturers of uniquely designed holographic and diffraction products for the gift, novelty, advertising, specialty, premium incentive, souvenir and museum markets.
SEE OUR ADVERTISEMENT
****

Ralcon
Box 142
8501 South 400 West
City: Parad ise
State/Province: UT
Postal Code: 84328
Country: United States of America
Voice Phone: 1 801 245 4623
Fax Phone: 1 801 245 6672
email: ralcon@cache.net
Contact # I: Richard Rallison
Business Description: Design, development and fabrication of volume holographic optical elements, (HOEs) including gratings, scanners, multi focus devices, heads up and down displays and notch fi lters formed in dichromated ge latin or photopolymer.
SEE OUR ADVERTISEMENT
****

Ralph Cullen Holographics
C/O Uk Optical Supply
84 Wimborne Road West
City: Wimborne
State/Province: England
Postal Code: BH21 2DP
Country: United Kingdom
Voice Phone: 44 202 886 831
Contact # I: Ralph Cullen
Business Description: A consultancy-design service which in associa tion with UK Optical Supplies (manufacturing) provide customized holographic opt ical components. Advice on component selection and laboratory/studios de signed to any budget is available. Specialists in embossed holography systems and production.
****

Real Image
PO Box 566
City: Pacifica

State/Province: CA
Postal Code: 94044
Country: United States of America
Voice Phone: 1 415 355 8897
Fax Phone: 1 415 355 5427
Contact # 1: Roy Bradshaw
Business Description: Incorporation of patented holographic designs into fishing tackle and fishing lures.
****

Reconnai ssance International Ltd .
3003 Arapahoe St.
Suite 213
City: Denver
State/Province: CO
Postal Code: 80205
Country: United States of America
Voice Phone: 1 303 293 3000
Fax Phone: 1 303 293 8661
Contact # 1: Lewis Kontnik
Business Description: North American office. We are an international consultancy for market and industry information and analysis Publisher of Holography News, Halo-Pack! Halo-print Guidebook and Authentica tion News. All clients studies are fully confidential.
****

Reconnaissance International Ltd.
Runnymede Ma lthouse
Runnymede RoacL
City: Egham .
State/ Province: England
Postal Code: TW20 9BD
Country: United Kingdom
Voice Phone: 44 1784 497008
Fax Phone: 44 1784 49700 I
email: 100142.1164@compuserve.com
Web Address: hmt.com/holography/hnews/hnhome.htm
Contact # I: Ian Lancaster
Business Description: We are the leading in ternational consultancy for market and industry information and analys is Publisher of Holography News, Holo-P ack/Holo-print Guidebook and Authentication News. All clients studies are fully confidential.
****

Red Beam, Inc.
90 II Skyline Blvd.
City: Oakland
State/Province: CA
Postal Code: 94611
Country: United States of America
Voice Phone: 1 510 482 3309
Fax Phone: 1 510482 1214
Contact # 1: Lon Moore
Busi ness Description: Mastering faci lity specializing in the design and production of masters suitable for high volume corporate applications, especially on photopolymer films and embossed materials. Clients include Activision, AT &T, NFL (Superbowl) and Polaroid. Also produces a line of trademarked giftware holograms distributed by Lightrix - (See ad on inside front cover and listing in this book).
****

Reva 's Holographic Illusions
446 South Main Street
City: Frankenmuth
State/Province: MI
Postal Code: 48734
Country: United States of America
Voice Phone: 1 517 652 3922
Fax Phone: 1 517 652 6503
Contact # I: Reva Krick

Business Description: Gallery/Retail store, with over 250 holograms on display. We feature a full line of holographic jewelry, gifts apparel, toys, etc. Established in 1992.
****

Reynolds Metals Co.
Flexib le Packaging Division
6603 West Broad St.
City: Richmond
State/Province: VA
Postal Code: 23230
Country: United States of America
Voice Phone: 1 804 281 2000
Voice Phone: 1 804 281 3969
Fax Phone: 1 804 281 2238
Web Address: nnC .com
Contact # 1: Rich Patterson
Business Description: Holographic specialty cartons and printed paper materials for distilled spirits and wine, pharmaceuticals, confections, personal care, and other consumer goods. Holographic flexible light web materials for pouches, lidding, and overwrap s. Full service from design to finishing.
****

Rice Systems
1150 Main street suite C
City: Irvi ne
State/Province: CA
Postal Code: 92614
Country: United States of America
Voice Phone: 1 714 553 8768
Fax Phone: 1 714 553 0307
email: RiceSys@prodigy.com
Contact # I: Colleen Fitzpatrick
Business Description: Laser metrology and diagnostic measurements, HNDT fluid measurements. Combustion diagnostics. Integrated optics and non linear optical material (R&D and product development). Very successful SBIR company.
****

Richard Bruck Holography
3312 West Belle Plaine #2
City: Chicago
State/ Province: IL
Postal Code: 60618-2316
Country: United States of America
Voice Phone: 1 3 12 267 9288
Fax Phone: 1 312 267 9288
Contact # I: Richard Bruck
Business Description: Specialists in large format holography. Extensive experience with live models and commercial work. We are accustomed to the advertising world , and know the importance of quality and service.
****

Richardson ,G4 1ting Laboratory
820 Linden Ave.
City: Rochester
State/Province: NY
Postal Code: 14625
CountlY: United States of America
Voice Phone: 1 716 262 1331
Voice Phone: 1 800 654 9955
Fax Phone: 1 716 454 1569
email: grating s@spectronic.com
Web Address: gratinglab.com
Contact # 1: Christopher Palmer
Business Description: The Richardson Grating Laboratory, formally part of Milton Roy Company, has been a world leader in the design and manufacture of ruled and holographic diffraction grating for fifty years.
****

CORPORATE LISTINGS

Richmond Development Group
(formerly Gray Scale Studios)
63 South 500 West
City: Richmond
State/Province: UT
Postal Code: 84333
Country: United States of America
Voice Phone: 1 801 258 0709
Fax Phone: 1 801 258 0709
Contact #1: George Sivy
Business Description: Specialists in the design
and creation of models and sculptures for ho-
lographic imaging, including digital origination
services for stereograms. Consultant services
offered, II years experience. Samples available
upon request.
****

Richmond Holographic Studios
6 Yorkton Street
City: London
State/Province: England
Postal Code: E2 8NH
Country: United Kingdom
Voice Phone: 44 171 739 9700
Fax Phone: 44 171 739 9707
email: rhs@augustin.demon.co.uk
Contact #1: Edwina Orr
Business Description: We make stock and cus-
tom large format silver halide holograms. We
also sell pulsed laser holographic systems for
mastering and transferring. Involved in the re-
search and invention of auto stereoscopic dis-
plays using holographic optical elements.
****

Robert Sherwood Holographic Design
1636 North Bell Ave.
City: Chicago
State/Province: IL
Postal Code: 60647
Country: United States of America
Voice Phone: 1 312 944 3200
email: coherence@aol.com
Contact # 1: Robert Sherwood
Business Description: Our designers, hologra-
phers, and account service personnel provide
you with the highest quality standards this new
and exciting technology can offer. Now offering
holography for apparel.
****

Rochester Inst. Of Technology
Center for Imaging Science
City: Rochester
State/Province: NY
Postal Code: 14623-5604
Country: United States of America
Voice Phone: 1 716475 6678
Fax Phone: 1 716475 5988
email: pzmpph@rit.edu
Contact # I: P. Mouroulis
Business Description: Research on HOEs, holo-
graphic materials, CGHs. Instruction in holog-
raphy and related topics in the Department of
Imaging and Photographic Technology, and the
Center for Imaging Science.
****

Rochester Photonics Corporation
330 Clay Road
City: Rochester
State/Province: NY
Postal Code: 14623
Country: United States of America
Voice Phone: 1 716 272 30 1 0
Fax Phone: 1 716272 9374
email: rpc@eznet.net

Web Address: http ://www.rphotonics.com
Contact # I: Dan McGarry
Business Description: RPC specializes in the
design and manufacturing of diffractive optical
components and subsystems. Precision diffractive mastering, molding, replication, and testing
services are provided. Products include: hybrid
refractive/diffractive lenses and subassemblies,
microlens arrays, diffractive phase plates, engineered diffusers, and holographic gratings.
****

Rofin-Sinar Laser GmbH
Berzeliusstrasse 85
City: Hamburg
Postal Code: 22113
Country: Germany
Voice Phone: +49 (0)40 733 630
Fax Phone: +49 (0)40 733 63 100
Business Description: C02 and Nd:YAG lasers
for materials processing, Laser components, La-
ser processing devices and machines.
****

Rolls-Royce Plc
Advanced Research Laboratory
POBox 31
City: Derby
State/Province: England
Postal Code: DE2 48BJ
Country: United Kingdom
Voice Phone: 44 133 224 2424
Fax Phone: 44 133 ~24 9936
Contact # 1: Ric Parker
Business Description: NDT for aircraft engines.
****

Rolyn Optics
706 Arrow Grand Circle
City: Covina
State/Province: CA
Postal Code: 91722-2199
Country: United States of America
Voice Phone: 1 818 915 5707
Fax Phone: 1 818915 1379
Business Description: General selection of opti-
cal items. Catalogue available.
****

Ross Books
P.O. Box 4340
City: Berkeley
State/Province: CA
Postal Code: 94704
Country: United States of America
Voice Phone: 1 800 367 0930
Voice Phone: 1 510841 2474
Fax Phone: 1 510 841 2695
email: staff@rossbooks.com
Web Address: www.holoinfo .com
Contact #1: Alan Rhody
Business Description: Publisher of the
HOLOGRAPHY MARKETPLACE Editions
1-6 (a worldwide database and sourcebook),
the HOLOGRAPHY HANDBOOK - Making
Holograms the Easy Way (world's best sell-
ing laboratory manual), and other related titles.
Educational, research and information services
also provided.
****

Rowland Institute For Science
100 Edwin H. Land Blvd.
City: Cambridge
State/Province: MA
Postal Code: 02142
Country: United States of America
Voice Phone: 1 617 497 4657
Contact # 1: Jean-Marc Fournier

Business Description: Scientific holography research. NDT, Lippman photography
****

Royal Holographic Art Gallery
122 Market Square
560 Johnson Street
City: Victoria
Postal Code: V8W 3C6
Country: Canada
Voice Phone: 1 604 384 0123
Fax Phone: 1 604 384 0123
email: royal@islandnet.com
Web Address: www.islandnet.com/-roylal/
index.htm
Contact # I: Derek Galon
Business Description: Gallery offers holograph-
ic art, gifts , limited editions, custom work, se-
lection from best studios, and fashion accesso-
ries. We ship worldwide. Also some film, plates,
developer, lasers and holographic kits. Russian
Holograms.
****

Royal Institute Of Technology
Dept Of Industrial Metrology
City: Stockholm
Postal Code: 10044
Country: Sweden
Voice Phone: 46 08 790 7823
Fax Phone: 4608 790 8219
Contact # 1: Lennart Svennson
Business Description: Holography education.
Holographic Non-Destructive testing.
****

Ruey-Tung, Miss. Hung
A 202
Chigasati-Coat Nango 6-7-12
City: Chigasaki-Shi
State/Province: Kanagawa
Postal Code: 253
Country: Japan
Voice Phone: 81 0467 857 750
Contact #1: Hung Ruey-Tung
Business Description: Fine Artist.
****

Rutherford & Appleton Labs
Chilton
City: Didcot
State/Province: England
Postal Code: OXII OQX
Country: United Kingdom
Voice Phone: 44 123 582 1900
Fax Phone: 44 123 544 6733
email: robert.sekulin@rl.ac.uk
Contact # I: Robert Sekulin
Business Description: Particle measurements;
Holographic non-destructive testing.
****

Saab-Scania
S-581
City: Linkoping
Postal Code: 88
Country: Sweden
Voice Phone: 46 013 129 020
Contact #1: Sven Malmqvist
Business Description: Holographic non-
destructive testing; scientific holography re-
search.
****

Saginaw Valley State University
2250 Pierce Road
City: University Center
State/Province: MI
Postal Code : 48710-0001
Country: United States of America

Voice Phone: 1 517 790 4000
Fax Phone: 1 517 7902717
Contact #1: Hsuan Chen
Business Description: Course instruction on holography; research includes HOEs, multiplex and rainbow holography.
****

Saint Mary's College
Art Department
City: Notre Dame
State/Province: IN
Postal Code: 46556
Country: United States of America
Voice Phone: 1 219 284 4000
Contact # 1: Doug Tyler
Business Description: Holographic fine artist. Extensive portfolio. Holography Instructor. Call for class schedule.
****

SAM Museum
3-27-3 Isoji , Minato-ku
City: Osaka
Postal Code: 552
Country: Japan
Voice Phone: 81 6 572 0036
Fax Phone: 81 65748136
Contact # 1: Akinobu Fukuda
Business Description: Museum that exhibits holograms.
****

San Jose State University
Physics Dept. and Inst. for Modem Optics
One Washington Square
City: San Jose
State/Province: CA
Postal Code: 95192-0 1 06
Country: United States of America
Voice Phone: 1 408 924 5245
Fax Phone: 1 408 924 2917
Web Address: http: //fire.sjsu.edu
Contact # I: Ramen Bahuguna
Business Description: Research and development work on 1) holographic fingerprint sensor, 2) holographic fingerprint verification, 3) display holography on DCG, and 4) holographic optical elements.
****

Sandia National Laboratories
P.O. Box 5800
City: Albuquerque
State/Province: NM
Postal Code: 87185
Country: United States of America
Voice Phone: 1 505 845 0011
Web Address: -irn. sand ia.gov/
Business Description: Sandia National Laboratories is able to do research in all phases of holography.
****

Scharr Industries
40 East Newberry Road
City: Bloomfield
State/Province: CT
Postal Code: 06002
Country: United States of America
Voice Phone: 1 860 243 0343
Voice Phone: 1 800 284 7286
Fax Phone: 1 860 242 7499
Contact # 1: Peg Horne
Business Description: Scharr holographic e bossing: wide web, on polyester, polypropylene, polyethylene, PVC, and nylon. We coat, laminate, metalize standard and custom patterns. Products include film to paper, board, transfer film, PSA and static cling.
****

School Of Holography
Museum Of Holography/Chicago
1134 W. Washington Blvd.
City: Chicago
State/Province: IL
Postal Code: 60607
Country: United States of America
Voice Phone: 1 312 226 1007
Fax Phone: 1 312 829 9636
Contact # I: Loren Billings
Business Description: Founded in 1978, the oldest continuous school of holographic instruction in the world. Basic courses in holography have been taught to thousands of students. In addition there are special workshops and tutorials for advanced study.
****

Science Kit & Boreal Labs
777 East Park Drive
City: Tonawanda
State/Province: NY
Postal Code: 14150-6784
Country: United States of America
Voice Phone: 1 716 874 6020
Fax Phone: 1 716 874 9572
Business Description: Suppliers of educational science materials especially suitable for junior and senior high school coursework. Comprehensivemail order catalog includes holography kits, holggraphy books, related optical components and mDre.
****

Semicon Austria
Morellenfeldgasse 41
City: Graz
Postal Code: A-8010
Country: Austria
Voice Phone: 43 0316 38 25 41
Fax Phone: 43 0316 38 24 03
Business Description: Collection of Russian art holograms for sale.
****

SETEC Oy
PO Box 31
City: SF-0671 Vant~
Country: Finland .
Voice Phone: 358 0 89 411
Fax Phone: 358 891 887
****

Shandong Academy of Sciences
Keyuan Road
City: Jinan Shan dong
Postal Code: 250014
Country: People's Rep. of China
Voice Phone: 86 615 615102 316
Contact # I: Prof. Zhu De Shun
Business Description: Laser & holography exhibit.
****

Sharon McCormack Holography
P.O. Box 38
City: White Salmon
State/ Province: WA
Postal Code: 98672
Country: United States of America
Voice Phone: 1 509 493 4850
Voice Phone: 1 509 493 1334
Fax Phone: 1 509 493 4830
email : sharon@gorge.net
Web Address: gorge. net/business/holography
Contact # 1: Sharon McCormack
Business Description: Holographic Fine Artist. Complete holographic stereogram production, including 360 degree viewab le. Filming, animation and computer graphics services offered. Also exhibit, design, and consultation services. Extensive commercial portfolio of work for major corporate clients.
****

Sharp Corp.
Tokyo Research Laboratories
Research Dept2 ; 271 , Kashiwa
City: Kashiwa
Postal Code: 227
Country: Japan
Voice Phone: 81 0471 346 166
Fax Phone: 81 0471 346 119
Contact # I: Shunichi Sato
Business Desc ription: Optics research
****

Shipley Chemical Co.
455 Forrest Street
City: Marlboro
State/Province: MA
Postal Code: 01752
Country: United States of America
Voice Phone: 1 800 345 3100
Voice Phone: 1 508 481 7950
Fax Phone: 1 508 485 9113
Contact # 1: Stu Price
Business Description: Primary manufacturer of photoresist material for coating onto substrates. Sold wholesale by quarts and gallons. For precoated plates, see listing for Towne Technologies and Process Technology.
****

Silhouette Technology Inc.
10 Wilmot Street
City: Morristown
State/Province: NJ
Postal Code: 07962-1479
Country: United States of America
Voice Phone: 1 201 539 2110
Fax Phone: 1 201 539 5797
Contact # I: Toicia Murphay
Business Description: Produces custom HOEs under contract. HOP maker. Heads-up display. DOE & DOE printers.
****

Silicon Graphics
2011 North Shoreline blvd.
City: Mountain View
State/Province: CA
Postal Code: 94043
Country: United States of America
Voice Phone: 1 415 960 1940
Fax Phone: 1 415 960 1737
Business Description: Silicon Graphics produces high end computer graphics stations ideal for rendering and modeling holographic stereograms.
****

Sillcocks Plastics International
310 Snyder Avenue
City: Berkeley Heights
State/Province: NJ
Postal Code: 07922
Country: United States of America
Voice Phone: 1 908 665 0300
Fax Phone: 1 908 665 9254
email: spisales@sillcocks.com
Web Address: sillcocks.com
Business Description: Producer of flat plastic products, printed or unprinted, which can feature hot-stamped holograms and laminated holograms. Products include credit cards, promotional cards and other custom specialties and POP products.

CORPORATE LISTINGS

****

Silver Dragon Holography
2540 Professional Road
City: Richmond
State/Province: VA
Postal Code: 23235
Country: United States of America
Voice Phone: 1 804 272 9284
Fax Phone: 1 8043202100
Contact # 1: Jenny Garrett
Business Description: International technology transfer for embossing equipment.

****

Silverbridge Group
Box 489
City: Powassan
Postal Code: POH IZ0
Country: Canada
Voice Phone: 1 705 724 6164
Fax Phone: 1 705 724 6249
Contact # 1: James Hepburn
Business Description: Limited edition DCG holograms in large format size.

****

Simian Co.
298 Harvey West
City: Santa Cmz
State/Province: CA
Postal Code: 95060
Country: United States of America
Voice Phone: 1 408 457 9052
Fax Phone: 1 408 457 9051
Contact # 1: Debbie Haines
Business Description: Manufacturer of high quality masters for embossed holography. Originations can be 2D/3D, 3D, animation and motion, or any combination. High production capacity with quick turnaround. SEE HOLOGRAM ON FRONT COVER AND ADVERTISEMENT

****

Sinclair Optics, Inc.
6780 Palmyra Road
City: Fairport
State/Province: NY
Postal Code: 14450
Country: United States of America
Voice Phone: 1 716 425 4380
Fax Phone: 1 716 425 4382
Contact #1: Bmce Capron
Business Description: Software for Computer Generated Holograms.

****

Slavich Joint Stock Company
2 Mendeleeva Square
City: Pereslavl-Zalessky
Postal Code: 152140
Country: Russia
Business Description: Makers of ultra fine grain silver halide emulsion for use in holography. USA distributors include 3Deep Hologram and Holicon.

****

Smith & McKay Printing Co. Inc.
96 North Almaden Boulevard
City: San Jose
State/Province: CA
Postal Code: 95110-2490
Country: United States of America
Voice Phone: 1 408 292 8901
email: smithmckay@aol.com
Contact #1: Dave McKay
Business Description: Expe11 hot-stampers of foil holograms onto paper products. Dimension-

al printing and fine lithography. Parent company of Holographic Impressions.

****

Societa Olografica !talia (Soi)
Via Degli Eugenii 23
City: Roma
Postal Code: 00178
Country: Italy
Voice Phone: 39 6 718 0976
Fax Phone: 39 6 718 5172
Contact # 1: Luigi Attardi
Business Description: Produces artistic/commercial holograms of any type.

****

Sonoma State University
Physics Dept.
1801 E. Cotati Ave.
City: Rohnert Park
State/Province: CA
Postal Code: 94928
Country: United States of America
Voice Phone: 1 707 664 2119
Contact #1: Steve Anderson
Business Description: Holography workshops offered. Call for class schedule.

****

Sophia University
Faculry Of Science & Technology
7-1 , Kioi-Cho Chiy04a-Ku
City: Tokyo
Postal Code: 102
Country: Japan
Fax Phone: 81 03 3238 3341
Contact # 1: Kazue Ishikawa
Business Description: Holography research.

****

Sopra
26 & 28 me Pierre Joingnequx
City: Bois-Colombes
Postal Code : 92270
Country: France
Voice Phone: 33 1 47 81 09 49
Fax Phone: 33 1 42 42 29 34
Contact #1: Robert · Stehle
Business Descripticm: Laser equipment, holographic kit and camera for interferometry.

****

Southern Indiana Holographies
6841 Newburgh Rd
City: Evansville
State/Province: IN
Postal Code: 47715
Country: United States of America
Voice Phone: 1 8124740604
Fax Phone: 1 812473 0981
email: Uohann@msn.com
Contact # 1: Larry Johann
Business Description: Holographic Fine Artist.

****

Spatial Holodynamics (India) Pvt. Ltd.
1041105 Shah & Nahar Estate
Off. Dr. E. Moses Road
City: Worli, Bombay
Postal Code: 400 018
Country: India
Voice Phone: 91 22 493 0975
Voice Phone: 91 22492 1069
Fax Phone: 91 22 495 0585
Contact # 1: Yogesh Desai
Business Description: Holographic embossing using the latest DI-HO System. Total service from designing of holograms up to holographic shims. Alternately, if desired, glass Photo Resist

Masters.

****

Spatial Imaging Ltd.
6 Marlborough Rd.
City: Richmond
State/Province: England
Postal Code: TWI06JR
Country: United Kingdom
Voice Phone: 44 (0) 181 332 1948
Voice Phone: 44 1932 564 899
Fax Phone: 44 (0) 181 332 1948
email: spatial@dircon.co.uk
Web Address: dircon.co.uklspatiall
Contact # I: Rob Munday
Business Description: Full design, origination and shimming service. We specialize in DIHO, digital, full colour, animated embossed holograms from video or computer images. Also available - conventional 2D/3D, 3D, dot matrix, photopo1 ymer and multi -colour silver halide holograms including pulsed laser portraits. We also produce and distribute a wide range of holographic retail products.

****

Spectra-Physics Lasers Inc.
1330 Terra Bella Ave.
City: Mountain View
State/Province: CA
Postal Code: 94039-7013
Country: United States of America
Voice Phone: 1 800 775 5273
Voice Phone: 1 415 966 5596
Fax Phone: 1 415 964 3584
email: splaser@ix.netcom.com
Web Address: www.splasers.com
Contact #1: Alfred Feitisch
Business Description: World's largest supplier of CW and pulsed gas and solid state laser systems, including a comprehensive optical accessories line and a worldwide customer service network. SEE OUR ADVERTISEMENT

****

Spectratek Inc.
5405 Jandy Place
City: Los Angeles
State/Province: CA
Postal Code: 90066
Country: United States of America
Voice Phone: 1 310 822 2400
Fax Phone: 1 310 822 2660
Contact # 1: Marty Kelem
Business Description: Spectratek manufacturers the highest quality diffraction patterns which are the only ones available without seams or pattern breaks. These patterned films are available in a variety of fonnats , including adhesive backed for labels, laminated to card stock, or films for packaging and other applications.

****

Spectrogon Ab
(Company Headquarters)
Box 2076
City: Taby
Postal Code: S-18302
Country: Sweden
Business Description: Manufactures & designs interference filters for IR, visible & UV spectral regions; narrow and bandpass, long/shortwavepass, isolation-line, & ND filters; atmospheric windows; diffraction gratings; AR & metallic coatings.

****

Spectrum Corporation

608 Sangita Complex
City: Ahmedabad
Postal Code: 380 006
Country: India
Voice Phone: 91 79 419 080
Fax Phone: 91 79 436 457
Business Description: Distributor of James River Products embossing machines.
****

SPIE
The International Society for Optical Engineering (Society of Photo-Optical Instrumentation Engineers)
P.O. Box 10
City: Bellingham
State/Province: WA
Postal Code: 98227
Country: United States of America
Voice Phone: 1 360 676 3290
Voice Phone: 1 800 483 9034
Fax Phone: 1 360 647 1445
email: info-optolink-request@mom.spie.org (use help)
Business Description: SPIE is a nonprofit educational society dedicated to advancing engineering and scientific applications of optical, electro-optical, and optoelectronic instlUmentation, systems, and technologies.
SEE OUR ADVERTISEMENT
****

SPIE's Holography Working Grp. Newsletter
P.O. Box 10
City: Bellingham
State/Province: WA
Postal Code: 98227-0010
Country: United States of America
Voice Phone: 1 360 676 3290
Fax Phone: 1 360 647 1445
email: info-holo-request@spie.org
Contact # 1: Sunny Bains
Business Description: The Holography Working Group newsletter is published semiannually by SPIE for its International Technical Working Groups on Holography.
****

Spindler & Hoyer GmbH & Co.
Koenigsallee 23
Postfach 33 53
City: Goettingen
Postal Code: 37070
Country: Germany
Voice Phone: +49 (0)551 6935-0
Voice Phone: +49 (0)5 6935-971
Fax Phone: +49 (0)551 6935-166
Contact # I: Mr. Keilholz
Business Description: Manufacturer of precision optics, mechanics and laser technology.
****

Springer-Verlag New York
175 Fifth Ave.
City: NY
State/Province: NY
Postal Code: 100 1 0
Country: United States of America
Voice Phone: 1 212 460 1500
Voice Phone: 1 800 777 4643
Fax Phone: 1 201 348 4505
email: orders@springer-ny.com
Web Address: www.springer-ny.com
Contact # I: Ken Quinn
Business Description: Publishers of an Optical Sciences series of books, including Silver Halide Recording Materials for Holography, by H. Bjelkhagen.
****

Stanford University
Mechanical Engineering Dept.

Mai 1 Code 4021
City: Stanford
State/Province: CA
Postal Code: 94305-4021
Country: United States of America
Voice Phone: 1 415 723 2123
Fax Phone: 1 415 723 3521
email: dnelson@leland.stanford.edu
Contact # 1 : Drew Nelson
Business Description: Use of holographic interferometry (with rapid thelmoplastic recording of holograms) for measurements of small defonnations and for residual stresses in materials via stress release technique.
****

Star Magic
275 Amsterdam St.
City: NY
State/Province: NY
Postal Code: 10023
Country: United States of America
Voice Phone: 1 212 7692020
Web Address: www.starmagic.com
Business Description: Retail store featuring Space Age gifts, holograms, novelties, etc.
****

Star Magic
745 Broadway
City: New York
State/Province: NY.
Postal Code: 10003
Country: United States of America
Voice Phone: 1 212 228 7770
Web Address: www.starmagic.com
Business Description: Retail store featuring Space Age gifts, holograms, novelties, etc.
****

Star Magic
1256 Lexington Ave. (85th St.)
City: New York
State/Province: NY
Postal Code: 10028
Country: United States of America
Voice Phone: 1 212 988 0300
Business Description: Retail store featuring Space Age gifts, holograms, novelties, etc.
****

Star Magic
4026 24th Street
City: San Francisco
State/Province: CA
Postal Code: 94114
Country: United States of America
Voice Phone: 1 415 641 8626
Business Description: Retail store featuring Space Age gifts, holograms, novelties, etc.
****

Starcke, Ky.
Ratastie 6
City: Kokemaki
Postal Code: 32800
Country: Finland
Voice Phone: 358 39 5460 700
Fax Phone: 358 39 5467 230
Contact #1: Ari-Veli Starcke
Business Description: Starcke KY is the leading company selling holograms in Scandinavia. Hologram Hot Stamping.
****

Steinbichler Optotechnik GmbH
Am Bauhof 4
City: Neubeuern
Postal Code: D-83115
Country: Germany

Voice Phone: 49 8035 87040
Fax Phone: 49 8035 10 1 0
Contact #1: H. Steinbichler
Business Description: Development and sales of optical measuring and test systems, e.g. holographic interferometer, ESPI (electronic speckle Pattern Interferometer), contour measurement systems, non-destmctive inspection (shearography), image analysis software.
****

Stephens, Anait
1685 Fernald Point Lane
City: Santa Barbara
State/Province: CA
Postal Code: 93108
Country: United States of America
Voice Phone: 1 805 969 5666
Fax Phone: 1 805 969 5666
Contact #1: Anait Amtunoff Stephens
Business Description: Pioneer Holographic Fine Artist. Extensive portfolio includes reflection and transmission holograms, pulsed works and true color reflection.
****

Steuer KG GmbH & Co.
Ernst-Mey-Strasse 7
City: Leinfelden-Echterdingen
Postal Code: 70771
Country: Germany
Voice Phone: +49 (0)711 16068 0
Fax Phone: +49 (0)711 16068 63
Contact #1: Mr. Seitz
Business Description: Manufacturer of holographic hot-stamping machines.
****

STI
56 Cheny Street
City: Bridgeport
State/Province: CT
Postal Code: 06605
Country: United States of America
Voice Phone: 1 203 333 5503
Voice Phone: 1 800 538 5170
Fax Phone: 1 203 336 8570
email: simian1laol.com
Web Address: stiovd.com
Contact #1: Richard Roule
Business Description: STI is a manufacturer of complex holographic products. We offer a wide variety of products for security, promotional, collectable, and packaging applications. The company specializes in the development of security devices for anti counterfeiting and identification applications.
SEE HOLOGRAM ON FRONT COVER AND OUR ADVERTISEMENT
****

STI - Europe
Huttons Yard
City: Mapledurwell
State/Province: England
Postal Code: RG25 2LP
Country: United Kingdom
Voice Phone: 44 1256 346208
Fax Phone: 44 1256 329238
Business Description: STI is a manufacturer of complex holographic products. We offer a wide variety of products for security, promotional, collectable, and packaging applications. The company specializes in the development of security devices for anti cOllnterfeiting and identification applications.
SEE OUR ADVERTISEMENT
****

Stichting Voor Holographie En Laseropiek

CORPORATE LISTINGS

Prinsengracht 675
City: 1017 1T Amsterdam
Country: Netherlands

****

Stoltz Ag
Tafernstrasse 15
City: Baden Dattwil
Postal Code: CH-5405
Country: Switzerland
Voice Phone: 41 056 840 lSI
Contact # 1: Beat Ineichen
Business Description: Holographic non-destructive testing.

****

Studio Fuer Holographie
Waldfriedenweg 10
City: Eichenau
Postal Code: 82223
Country: Germany
Voice Phone: +49 (0)8 14 70831
Fax Phone: +49 (0)814 70831
Contact # 1: Carlo Dr. Schmelzer
Business Description: Products: mastering and copy services (rainbow/reflection), production of mass-run embossed holograms, open stock images, art-pieces.

****

Studio Weil-Alvaron
Ostra Tullgatan 8
S-211 28
City: Malmo
Country: Sweden
Voice Phone: 46 08141 70831
Contact #1: Lektor H. Herman Wei!
Business Description: Hans Weil's inventions were made in the period 1933-193 7; while Gabor invented holography in 1948, the laser was invented 1962 and the first laser-illuminated hologram was exposed as late as 1964.

****

Superbin Co. Ltd
3F-339
Section 2 Ho Ping E Road
City: Taipei
Postal Code: 10662
Country: Taiwan
Voice Phone: 886 02 701 3626
Fax Phone: 886 02 70 1 3531
Contact # 1: Edward Hwang
Business Description: Exclusive Chinese representative of Coherent (Argon, Krypton Laser, Dye Laser); Continuum (ruby Laser, Nd: YAG la ser); Newport (optical components) .Also supply embossed hologram- manufacturing equipment/material and consulting service.

****

Swede Holoprint
Duvhoksgatan 6A
City: Malmo
Postal Code: 21460
Country: Sweden
Voice Phone: 46 040 898 21
Contact # 1: Bjorn Wahlberg
Business Description: Artistic holography; Artistic marketing consultant.

****

Swift Instruments
1190 North 4th St.
City: San Jose
State/Province: CA
Postal Code: 95112
Country: United States of America

Voice Phone: 1 408 293 2380
Business Description: Microscope objectives and related optics.

****

Swiss Federal Inst Of Technology
Laboratory Of Photoelasticity
Ramistrasse 10 I
City: Zurich
Postal Code: CH-8092
Country: Switzerland
Voice Phone: 41 0380 246 000
Contact # 1: Walter Schumann
Business Description: Holographic non-destructive testing; Scientific and industrial research.

****

Synchron Pty Ltd.
Chempet
P.O. Box 36921
City: Capetown
Postal Code: 7442
Country: South Africa
Voice Phone: 27 21 551 1790
Fax Phone: 27 21 52 5291
Contact # 1: Sean Kritzinger
Business Description: Agent and distributor of holographic foils and labels. Also in house hot stamping of holographic images.

****

Synchronicity Holog'"r ams
Box 4235, Route 1
City: Lincolnville
State/Province: ME
Postal Code: 04849
Country: United States of America
Voice Phone: 1 207 763 3182
email: holo@midcoast.com
Contact # 1: Arlene Jurewicz
Business Description: Synchronicity Holograms provides outreach educational presentations on all aspects of holography for primary grades through high school. Available for workshops and presentations. Also conducts research on the educational aspects of holography.

****

Syracuse University
Department of Chemistry
City: Syracuse
State/Province: NY
Postal Code: 13244-4100
Country: United States of America
Voice Phone: 1 315 443 4880
Fax Phone: 1 315 443 4070
email: sponsler@syr.edu
Web Address: -che.syr.edu/
Contact #1: Michael B. Sponsler
Business Description: Holographic materials research, liquid crystalline photopolymers for switchable gratings.

****

Tair Hologram Company
Behterevsky, 8
City: Kiev
State/Province: 252053
Country: Ukraine
Voice Phone: 38 044 269 13 77
email: root@alextech.kiev.ua
Contact # 1: Alexander Monchak
Business Description: Fine art holograms.

****

Tama Art University
Department Of Physics
1723 Yarimizu Hachiouji-Shi

City: Tokyo
Country: Japan
Voice Phone: 81 0426 768 611
Fax Phone: 81 0426 762 935
Contact # 1: Hidetoshi Katsuma
Business Description: Research on Holographic TV, Holography Movie

****

Tamarack Storage Devices
12112 Technology Blvd., Suite 101
City: Austin
State/Province: TX
Postal Code: 78727
Country: United States of America
Voice Phone: 1 512 250 3100
Contact # 1: John Stockton
Business Description: R&D on optical memory for computers.

****

Technical Marketing Services
925 Park Ave.
City: Laguna Beach
State/Province: CA
Postal Code: 92651
Country: United States of America
Voice Phone: 1 714 497 1659
Fax Phone: 1 714 497 5331
email: 75021.3617@compuserve.com
Contact # I: David Kuntz
Business Description: TMS Combines technical strength with extensive experience in marketing in the electro-optics industry. Offers support in all areas of marketing, including PR and ad agency services, marketing planning and technical writing. Specializes in low cost, applications oriented marketing programs.

****

Technical University @ Eindhoven
Faculty Of Architecture
Calibre Institute PO Box 513
City: Eindhoven
Postal Code: NL-5600MB
Country: Netherlands
Contact # I: Geert T. A. Smelzer
Business Description: Academic research, computer generated holograms

****

Technical University Of Budapest
Institute Of Precision Mechanics
Applied Biophysics Laboratory
City: Budapest
Postal Code: H-1621
Country: Hungary
Business Description: Medical holography research.

****

Technical University Of Wroclaw.
Institute Of Physics
Wybrvzeze Wyspianskiego 27
City: Wroclaw
Postal Code: PL-50-370
Country: Poland
Contact # 1: Henryk Kasprzak
Business Description: Academic, scientific research

****

Technical University Zvolen
Faculty of Wood Technology
Dept. of Physics and Applied Mechanics
City: Zvolen
Postal Code: SK-960 53
Country: Slovakia

Voice Phone: 42 855 635
Fax Phone: 42 855 321 8 11
email: stano@tuzvo.sk
Contact # I: Stanislav Urgela
Business Description: Holographic interferometry. Measurement of temperature fields, deformations and vibrations applied to wood technology, wooden plates, musical instruments and material quality control.
****

Technische Fachhochschu le Berlin
FB 2 / Labor Fuer Laseranwendungen
Seestrasse 64
City: Berlin
Postal Code: 13347
Country: Germany
Voice Phone: +49 (0)30 4504 39 17
Voice Phone: +49 (0)3 4504 3918
Fax Phone: +49 (0)30 4504 3959
emai l: eichler@tfh-berlin .de
Contact #1: Juergen Prof. Dr. Eich ler
Business Description: Holographic Interferometry, Display Holography, Medical Applications.
****

Technische Universitaet Berlin
Optisches Institut, Sekr. P II
Strasse des 17. Juni 135
City: Berlin
Postal Code: 10623
Country: Germany
Voice Phone: +49 (0)30 314 22498
Voice Phone: +49 (0)3 3 14 22097
Fax Phone: +49 (0)30 314 26888
email: eichler@physik.tu-berlin.de
Web Address: http://www.physik .tu-berl in .de
Contact #1: Hans Joachim Prof. Dr. Eichler
Business Description: Scientific holography research: Optical holographic data storage, realtime holography, material research, semiconductors, liquid crystals, new lasers.
****

Technoexan Ltd
Polytechnicheskaya, 26
City: St. Petersburg
Postal Code: 194021
Country: Russia
Voice Phone: 7 812 247 9383
Voice Phone: 7 812 247 5273
Fax Phone: 7 812 247 5333
Contact # I: Igor Lovygin
Business Description: Power semiconductors, Lasers, optoelectronics devices (IR range), many channel 1- and p- diodes, equipment for high format art and picture hologram, holographic registers, school packages for showing optic effects.
****

Technolas Laser Technik Gmbh
Lochhamer Schlag 19
City: Graefelfing
Postal Code: 8032
Country: Germany
Voice Phone: +49 (0)89 854 5040
Fax Phone: +49 (0)89 854 561
Business Description: Lasers, medical
****

Textile Graphics, In c.
(see Holography Presses On)
State/Province: MI
Country: United States of America
SEE OUR ADVERTISEMENT
****

The Foreign Dimension
The Peak Gall eria
Level 2, Shops 29 & 42, The Peak
City: Hong Kong
Country: Hong Kong
Voice Phone: 852 2 849 6361
Business Description: Holography shop/showroom offering all varieties of holograms for sale to the public. We also offer holograms for sale wholesale to other businesses. See listing for Foreign Dimension and ad.
****

The Hologram Company #1
900 North Point Street
City: San Francisco
State/Provi nce: CA
Postal Code: 94109
Country: United States of America
Voice Phone: 1 41 5 775 9356
Business Desc ription: Located in hi storic Ghiradelli Square on the waterfront, selling all types of holographic wall art, jewelry, souvenirs and related gift items.
****

The Hologram Company #2
Barefoot Land ing Suite 4722A
Highway 17 South
City: North Myrtle Beach
State/Province: SC
Postal Code: 29582
Country: United States of America
Voice Phone: 1 803 272 3583
Business Description: Located in the golfing capital of the US, se lling all types of holographic wall art, jewelry and gift items.
****

The Hologram Company #3
The Florida Mall Shopping Center Room 344
8001 South Orange Blossom Trail
City: Orlando
State/Province: FL
Postal Code: 32809
Country: United States of America
Voice Phone: 1 407 856 9072
Business Desc ription: Central Florida vacation land location, selling all types of holographic wall art, jewelry and gift items.
****

The Hologram Company #4
370 Horton Plaza
City: San Diego
State/Provi nce: CA
Postal Code: 92101
Country: United States of America
Voice Phone: 1 6195578371
Business Description: Sunny Southern California location, selling all types of holographic wall art, jewelry and gift items.
****

The Hologram Company #5
Downtown Plaza Suite B-87
547 L Street
City: Sacramento
State/Province: CA
Postal Code: 95814
Country: United States of America
Voice Phone: 1 9 16 498 0305
Business Description: California Capital location, selling all types of holographic wall art, jewelry and gift items.
****

The Hologram Company #6
New Orleans Center - Room 498

City: New Orleans
State/Province: LA
Postal Code: 70112
Country: United States of America
Voice Phone: 1 504 529 5700
Business Description: Located in the mall adjoining the Superdome, se lling all types of holographic wall art, jewelry and gift items.
****

The Hologram Company #7
Broadway at the Beach
1215 Celebrity Circle MI47
City: Myrtle Beach
State/Province: SC
Postal Code: 29577
Country: United States of America
Voice Phone: 1 803 444 3583
Business Description: Located in the brand new, exciting Myrtle Beach Mall , selling all types of holographic wall art, jewelry and gift items.
****

The Hologram Store, Ltd.
#2673, 8770 - 170 St.
City: Edmonton
Postal Code: T5T 412
Country: Canada
Voice Phone: 1 403 444 3333
Fax Phone: 1 403 444 4455
Business Description: Retail stores located in Canada and U.S. specializing in holographic and sc ience products. Wholesale and mail order available.
****

The Holography, Laser & Photonics Center
2018 R Street, N.W.
City: Washington
State/Province: DC
Postal Code: 20009
Country: United States of America
Voice Phone: 1 202 667-6322
Fax Phone: 1 202 265-8563
Contact # 1: Maria-Florica Busuioc
Business Description: Features a permanent international show: Hologram, Laser and Photonic Image of the Future. The collection includes 75 masterpieces from: Europe, Asia and North America. It is the most informative and entertaining Center for the technologies of the furure. The museum is visited by guided tour only, and reservations are required.
****

The HaLOS Corporation
Route 12 North
City: Fitzwilliam
State/Province: NH
Postal Code: 03447
Country: United States of America
Voice Phone:--i ~03 585 3400
Fax Phone: 1 60'3 585 6936
email: mwitt@holoscorp.com
Web Address: holoscorp.comlholos
Contact #1: Martin Witt
Business Description: The HOLOS Corporation was formed to market and manufacture a broad spectrum of holographic technologies based on patents and innovations developed by its founding sc ienti sts. HOLOS will become the worldwide leader in manufacturing monochrome and full color image di splay holograms on silver halide coated glass plates and photopol ymer.
SEE OUR ADVERTISEMENT
****

The Institute of Applied Optics
National Academy of Sciences of Ukraine

CORPORATE LISTINGS

10-G Kudryavaskaya St.
City: Kiev
Postal Code: 254053
Country: Ukraine
Voice Phone: 380 44 212 21 58
Fax Phone: 380 44 212 48 12
email: vmarkov@iao.freenet.kiev.ua
Contact # 1: Vladimir B. Markov
Business Description: R&D in the areas of: display holography (reflection, transmission, mastering) and colour holography for museum items reproduction; holographic interferometry and its application in industry and cultural heritage protection; diffraction on the volume grating; laser physics, including properties of the cavity, tunable lasers, etc.; nonlinear optical effects and multi-beam interaction
****

The London Holographic Image Studio
9 Warple Mews
Warple Way
City: London
State/Province: England
Postal Code: W3 ORF
Country: United Kingdom
Voice Phone: 44 181 740 5322
Fax Phone: 44 181 740 1733
email: 101630.2754@compuserve.com
Contact #1: Martin Richardson
Business Description: Commissioned holograms up to 1 x 2 meters, pulse-portraiture, movie stereograms and mass production of silver halide holograms. Catalogues on request.
****

The Regal Press Inc.
Holographics Di vis ion
129 Gui ld Street
City: Norwood
State/Province: MA
Postal Code: 02062
Country: United States of America
Voice Phone: 1 617 769 3900
Voice Phone: 1 800 447 3425
Fax Phone: 1 617 55 1 0466
Contact # 1: William Duffey
Business Desc ription: The Regal Press, Inc. has expertise in all areas of print production, including engraving, lithography, thermography, embossing, foil -stamping, and holography. We hold a worldwide patent for REGAL MARQUE, a simulated private watermarking process , and we provide our customers with REGAL EXPRESS guaranteed overnight delivery and 24-hour rush Business Cards.
SEE OUR ADVERTISEMENT
****

Third Dimension Arts Inc.
1241 Andersen Drive, Suites C
City: San Rafael
State/Province: CA
Postal Code: 9490 I
Country: United States of America
Voice Phone: 1 415 485 1730
Voice Phone: 1 800 622 4656
Fax Phone: 1 415 485 0435
Contact # 1: Dara Haskell
Business Desc ription: Third Dimension Arts Inc. manufacturers of dichromate jewelry, gifts, and hologram watches. Suppliers to: the gift, jewelry, museum, and entertainment industry. Custom designs welcome!
****

Thorlabs Inc .
435 Route 206
City: Newton

State/Province: NJ
Postal Code: 07860
Country: United States of America
Voice Phone: 1 201 579 7227
Fax Phone: 1 20 1 383 8406
Web Address: thorlabs.com
Business Description: High quality equipment for optics and photonics research including first surface mirrors, optical component mounts, power meters, etc.
****

Three-D Light Gallery
109-A The Commons
City: Ithaca
State/Province: NY
Postal Code: 14850
Country: United States of America
Voice Phone: 1 607 273 11 87
Contact # I: Jonathan Pargh
Business Description: Artistic holography; holography gallery.
****

Tjing Ling Industrial Research.
130 Keelung Road
Section Iii
City: Taipei
Country: Taiwan
Voice Phone: 886 86 207 041 856
Business Description'" Fine art originals.
****

TNO Institute of Applied Physics
Department Of Optics
PO Box 155
City: Ad Delft
Postal Code: NL-2600
Country: Netherlands
Contact #1: Ruud L. Van Renesse
Business Description: Academic and industrial applications.
****

Tokai University
Department of Electro Photo Optics
1117 Kitakanam~
City: Hiratsuka City
Postal Code: 259-12
Country: Japan
Voice Phone: 81 0463 58 121 I
Fax Phone: 81 0463 59 2594
Contact # 1: Hideshi Yokota
Business Description: Holography research - artistic holography.
****

Tokyo Institute Of Technology
Imaging Science And Engineering
4259 agatsuda Midori-Ku
City: Yokohama
Postal Code: 227
Country: Japan
Voice Phone: 81 045 922 11 II
Fax Phone: 81 045921 1492
Contact #1: Masahiro Yamaguchi
Business Description: Holographic Display, 3-D Imaging Science. Also Dr. Toshio Honda.
****

Tokyo Institute of Technology
Imaging Science & Engineering Lab.
4529 Nagatsuta Midori-ku
City: Yokohama
Postal Code: 226
Country: Japan
Voice Phone: 81 45 921 5183
Fax Phone: 81 45 921 1492
email: guchi@ho.isl.titech.ac.jp

Contact # 1: Mashahiro Yamaguchi
****

topac GmbH, Department Holography
Carl - Miele Str. 202-204
City: Guetersloh
Postal Code: D3 331 1
Country: Germany
Voice Phone: 49 05241 803302
Fax Phone: 49 05241 8060870
email: http ://members.aol.com/tophol!
hhome_e.htm
Web Address: tophol@aol.com
Contact # I: Uwe Sarda
Business Description: topac GmbH is a full service hologram producer with complete production line for all hologram processes. Speciality is embossed hologram production for security and authenticity devices.
****

Topcon Inc.
75 -1
Hasunuma-Machi Ttabasi-Ku
City: Tokyo
Postal Code: 174
Country: Japan
Vo ice Phone: 81 0339663141
Fax Phone: 81 03 3966 2140
Contact # 1: Reiji Hashimoto
Business Description: Hologram manufacturer.
****

Toppan Printing Co., Ltd.
Image Technology Laboratory
1-3-3 Suido Bunkyo-Ku
City: Tokyo
Country: Japan
Voice Phone: 81 03 3817 2873
Fax Phone: 81 03 5684 7600
Contact # 1: Teiichi Nishioka
Business Description: Research in Holography, 3-D TV, Computer Graphics
****

Toppan Printing Co., Ltd.
4-2-3 Takanodai-Minami
Sugito-Machi, Saitama-ken, Kita-Katsuskika-Gun
Postal Code: 345
Country: Japan
Voice Phone: 81 0480 33 9079
Fax Phone: 81 0480 33 9022
email: fiwata@tri.toppan.co.jp
Contact # 1: Fujio Iwata
Business Description: Holographic Display, manufacturer of embossed holograms, Lippman, and Multiplex. Also Mr. Susumu Takahashi and Mr. Toshiki Toda at this address.
****

Toppan Printing Co., Ltd.
Tech. Research Inst. Tsukuba Research Lab.
4-2-3 Takanodai - Minami
City: Saitama
Postal Code: 345
Country: Japan
Voice Phone: 81 480 339079
Fax Phone: 81 480 339022
email: fiwata @tri.toppan.co.jp
Contact # 1: Fujio Iwata
****

Total Register Inc.
3 Quarry Road
City: Brookfield
State/Province: CT
Postal Code: 06804
Country: United States of America

Voice Phone: 1 203 740 0199
Fax Phone: 1 203 740 0177
Contact # I: John Gallagher
Business Description: Manufacturer of registration devices for hot-stamping presses. Manufacturer of registered rotary die cutting equipment. Registered hologram sheeting and die cutting services.

****

Towne Technologies
6-10 Bell Ave.
City: Somerville
State/Province: NJ
Postal Code: 08876-0460
Country: United States of America
Voice Phone: 1 908 722 9500
Fax Phone: 1 908 722 8394
email: sales@townetech.com
Web Address: townetech.com
Contact # I: Sal LoSardo
Business Description: Towne Technologies is a producer of fine quality holographic photoresist plates with or without a sub- layer of Iron-Oxide. These plates are spin-coated with striation free photoresist in sizes up to 15 x15. SEE OUR ADVERTISEMENT

****

Toyama National College Of Marit
1-2 Ebie-Neriya
City: Shinminato
Postal Code: 933 02
Country: Japan
Voice Phone: 8 1 0766 860 511
Contact #1: Dr. Kenji Kinoshita
Business Description: Holographic Stereogram

****

Transfer Print Foils, Inc.
(a division of Holopak Technologies, Inc. Company)
21B Cotters Lane - P.O. Box 538
City: East Brunswick
State/Province: NJ
Postal Code: 088 1 6
Country: United States of America
Voice Phone: 1 908 238 1800
Voice Phone: 1 800 235 3645 x445
Fax Phone: 1 908 65 1 1660
Contact # I: Marc O. Wootner
Business Description: T.P.F. provides holographic products for a variety of end uses. From security images (as transfer coatings or laminated patches) which enhance the security of credit cards, licenses, bank documents, tickets and gate passes; to value added decorative finishes for greeting cards, book jackets, trophies and picture frames.

****

Trend
Miramarska 85
City: Zagreb
Postal Code: 41000
Country: Yugoslavia
Voice Phone: 381 041 511 42
Contact # 1: Dalibor Vukicevic
Business Description: Gallery.

****

Triple-D Laser Imaging
Bergselaan 13-B
City: Rotterdam
Postal Code: 3037 BA
Country: Netherlands
Voice Phone: 31 10 465 6331
Fax Phone: 31 10 465 6331
email: aca@wirehub.nl

Web Address: wirehub.nl?-aca/
Contact # 1: A.C. Akveld
Business Description: New business specializing in the sale of holograms and related optical equipment.

****

Turing Institute
77 - 81 Dumbarton Road
City: Glasgow
State/Province: Scotland
Postal Code: G 1 I 6PP
Country: United Kingdom
Voice Phone: 44 141 3376410
Fax Phone: 44 141 3390796
email: P.Mowforth@turing.gla.ac . uk
Web Address: turing.gla.ac.uk
Contact # 1: Peter Mowforth
Business Description: Three dimensional imaging systems based on multiple cameras, digital processing and proprietary software. Designed especially for medical and clinical applications where precise 3D modeling and accurate measurement are required. The Turing Institute is the trading name of Greenagate Ltd. (Scotland).

****

Tyler Group
218 Linden Avenue
City: Moorestown
State/Province: NJ
Postal Code: 08057
Country: United States of America
Voice Phone: 1 609' 234 1'800
Fax Phone: 1 609 866 0351
Business Description: Holographic image security consulting and application services.

****

U.K. Gold Purchasers, Inc.
DBA Holograms Unlimited
110 Central Park Mall
City: San Antonio
State/Province: TX
Postal Code: 78216
Country: United States of America
Voice Phone: 1 210 530 0045
Voice Phone: 1 800 722 7590
Fax Phone: 1 210 530 0048
Web Address: www.e~n.com /-mainlinklart/ ral /Index.htm .
Contact #1: Marvin Uram
Business Description: Full line distributor of hologram and related products for specialty retailers - representing more than 80 firms. One stop shopping at competitive prices.

****

Uk Optical Supplies
84 Wimborne Road West
City: Wimborne
State/Provi nce: England
Postal Code: BH21 2DP
Country: United Kingdom
Voice Phone: 44 1202 886 831
Fax Phone: 44 1202 886 831
Contact # I: Ralph Cullen
Business Description: Supplying probably the world's largest selection of Holographic/Optical components which are best quality, best value. Designed by experienced holographers. Component selection and laboratory/studio set-up advice freely available. Also buys and sells a large selection of used and surplus equipment.

****

Ultra-Res Corporation
1395 Greg St. - Suite 107
City: Reno

State/Province: NV
Postal Code: 89431
Country: United States of America
Voice Phone: 1 702 355 1 J 77
Fax Phone: 1 702 359 6273
email: alex@acds.com
Contact #1: Alex Chaihorsky
Business Description: Manufacturers of the Ultra-Res Instant Holographic camera system. Capable of making small transmission and reflection holograms in stantly, without darkroom processing, on slides or film rolls. Especially useful for interferometry, setup veri fication, HOEs for phase filters and proofing for embossed runs. SEE OUR ADVERTISEMENT

****

Uniphase Lasers
(a di vision of Uniphase Corp.)
163 Baypoint Parkway
City: San Jose
State/Province: CA
Postal Code: 95134
Country: United States of America
Voice Phone: 1 408 434 1800
Voice Phone: 1 800 644 8674
Fax Phone: 1 408 433 3838
Contact # 1: Tanis Mofchetti
Business Description: Manufacturer of HeNe lasers suitable for holography.

****

Unistay
(Innovative Frame Products)
1627 5 Monterey Road
City: Morgan Hill
State/Province: CA
Postal Code: 95037
Country: United States of America
Voice Phone: 1 408 776 7829
Fax Phone: 1 408 779 3600
Contact #1: Bob Manrubia
Business Description: Unistay is the developer and manufacturer of the MICROBRIGHT hologram light. Suitable for the display of embossed and photopolymer holograms up to 8x 1 0 inches. Perfect for point of purchase or retail sales. Low voltage, light weight.

****

United Association Manufacturer's Representatives
34071 La Plaza, Suite 120
P.O. Box 986
City: Dana Point
State/Province: CA
Postal Code: 92629
Country: United States of America
Voice Phone: 1 714 240-4966
Fax Phone: 1 714240-7001
Contact #1: Karen Mazzola
Business Description: Provide va luable services that help manufacturers and reps come together for mutual benefit.

****

Univ. de Liege
Sart Tilman
Hololab, Physique B5
City: Liege
Postal Code: B 4000
Country: Belgium
Voice Phone: 32 4 166 3626
Fax Phone: 32 4 166 2355
email: lion@gw.unipc.ulg.ac.be
Contact # 1: Yves F. Lion

****

Universidade Do Porto

CORPORATE LISTINGS

Laboratorio De Fisica
Praca Gomes Teixeira
City: Porto
Postal Code: P-4000
Country: Portugal
Contact # 1: Oliverio Soares
Business Description: Holographic non-destructive testing; Academic holography research.
****

Universita Di Roma
La Sapienza Dipt Di Fisica
Piazzale Aldo Moro 2
City: Rome
Postal Code: 1-001 85
Country: Italy
Contact # I: Paolo De San tis
Business Description: Scientific research
****

Universite De Neuchatel
Institut De Microtechnique
2, Rue A L Breguet
City: Neuchatel
Postal Code: CH-2000
Country: Switzerland
Voice Phone: 41 038 246 000
Contact #1: Rene Dandliker
Business Description: Industrial research.
****

Universite Laval
Dept Physique - e.O.P.L
Pavi lion Vachon
City: University City
Postal Code: G1K 7P4
Country: Canada
Voice Phone: 1 4186562131
Voice Phone: 1 418 656 3436
Contact # I: Roger A. Lessard
Business Description: Holography education. Research in holographic recording materials (photopolymer) for optical data storage, CGH and diffractive optics.
****

University Of Tsukuba
Institute Of Art & Design
City: 1 -I , Tennodai
State/Province: Tsukuba
Postal Code: 305
Country: Japan
Voice Phone: 81 298 53 2883
Fax Phone: 81 298 53 6508
Contact # 1: Shunsuke Mitamura
Business Description: Artistic holography, holography education.
****

University Of Alabama
University At Huntsville
Center For Applied Optics
City: Huntsville
State/Province: AL
Postal Code: 35899
Country: United States of America
Voice Phone: 1 205 890 6030
Fax Phone: 1 205 895-6618
Contact # 1: Chandra Vikram
Business Description: Scientific holography research, NDT.
****

University Of Alicante
Applied Physics/Cent De Holograf
Facultad De Ciencias
City: Alicante Apdo
Postal Code: 99
Country: Spain

Voice Phone: 34 566 1200
Contact #1: A. Fimia.
Business Description: Artistic holography; HOEs; workshops.
****

University Of Arizona
Optical Science Center
City: Tucson
State/Province: AZ
Postal Code: 85721
Country: United States of America
Voice Phone: 1 602 621 6997
Fax Phone: 1 602 6219 x6 13
Contact # I: Dick Powell
Business Description: Industrial and scientific holography research; Holographic interferometry; Holographic non-destructive testing.
****

University Of Bologna
Via Fiumazzo 347
City: Belricetto
Postal Code: 1-48010
Country: Italy
Contact # I: Pier Luigi Capucci
Business Description: Artistic holography research & education.
****

University Of Dayton
Research Institute "
300 College Park
City: Dayton
State/Province: OH
Postal Code: 45469-0 102
Country: United States of America
Voice Phone: 1 513 2293515
Fax Phone: 1 513 229 3433
email: huff@udri.udayton.edu
Web Address: udri.udayton.edu
Contact #1: Lloyd Huff, Ph.D.
Business Description: Scientific research, Industrial research; courses.
****

University of Erlanger
Physics Institute, Dept. of Optics
Staudtstrasse 7 -
City: Erlangen
Postal Code: D-91058
Country: Germany
Voice Phone: 49 9131 858395
Fax Phone: 49 9131 13508
email: schwider@move.physik.unierlangen.de
Contact # I: J. Schwider
Business Description: We offer design and manufacturing of micro lenses and diffractive optical elements (lenses, beamsplitters, beam shaping elements). We can test micro optical elements by means of interferometers.
****

University Of Michigan
Dept. of Electrical Engineering
Room 1108 EECS Bui lding
City: Ann Arbor
State/Province: MI
Postal Code: 48109-2122
Country: United States of America
Voice Phone: 1 313 764 9545
Fax Phone: 1 3 13 763 1503
Contact # 1: Emmet Leith
Business Description: Scientific holography research. Design HOEs. Courses on holography.
****

University of Muenster
Laboratory of Biophysics
Robert-Koch-Str. 45
City: Muenster

Postal Code: 48129
Country: Germany
Voice Phone: +49 (0)251 83 6888
Fax Phone: +49 (0)251 83 8536
email: biophys@gabor.uni-muenster.de
Contact # I: Gert von Bally
Business Description: holography and Interferometry in medicine. Environmental research and cultural heritage protection.
****

University of Munich
Inst itute Of Medical Optics
Barbarastrasse 16
City: Munich
Posta l Code: 80797
Country: Germany
Voice Phone: +49 (0)89 2105 3000
Fax Phone: +49 (0)89 12406301
Contact # 1: Mr. Zurek
Business Description: Medical Holography, Scientific holographic research.
****

University Of Oxford
Holography Group
Department of Engineering Science
City: Parks Road
State/Province: England
Postal Code: OXI 3PJ
Country: United Kingdom
Voice Phone: 44 186 527 3099
Fax Phone: 44 186 527 3905
Contact # 1: G. Yang
Business Description: Scientific holography research. Graduate and undergraduate courses
****

University Of Rochester
The Institute Of Optics
City: Rochester
State/Province: NY
Postal Code: 14627
Country: United States of America
Voice Phone: 1 7162754722
Fax Phone: 1 716 244 4936
Web Address: optics.rochester.edu:80801
Contact #1: Beverley Holloway
Business Description: Scientific and industrial holography research; interferometry; particle testing & measurement.
****

University Of Southern California
Department Of Physics
University Park
City: Los Angeles
State/Province: CA
Postal Code: 90089-0484
Country: United States of America
Voice Phone: 1 213 740 1134
Contact # 1: Jack Feinberg
Business Description: Scientific holography research; interferometry.
****

University of Stuttgart
Institute Of Applied Optics
Pfaffenwaldring 9
City: Stuttgart
Postal Code: 70569
Country: Germany
Voice Phone: +49 (0)711 685 6075
Fax Phone: +49 (0)711 685 6586
Contact # I: Hans Tiziani
Business Description: Scientific holography research; interferometry, NDT, HOE's
****

University Of Tokyo
Faculty Of Engineering
Hongo 7-3-1 Bunkyo-Ku
City: Toyko
Country: Japan
Contact # I: T. Uyemura
Business Description: Scientific and Medical holography research; Interferometry.
****

University Of Wisconsin/Madison
Dept. Of Engineering - Professional Development
432 North Lake Street
City: Madison
State/Province: WI
Postal Code: 53706
Country: United States of America
Voice Phone: 1 608 262 8708
Fax Phone: 1 608 263 3160
Contact # I: Elaine Bower
Business Description: Continuing Education courses on laser system design and application. Call or write for catalog.
****

Unterseher & Associates
709 112 West Glen Oaks Blvd.
City: Glendale
State/Province: CA
Postal Code: 91202
Country: United States of America
Voice Phone: 1 818 549 0534
Contact # I: Fred Unterseher
Business Description: Artistic holography and holography education. Originates masters for mass production holograms in embossed, DCG and photopolymer materials. Both pulsed and CW lasers available.
****

Uvex Safety Inc.
10 Thurber Blvd.
City: Smithfield
State/Province: RI
Postal Code: 02917
Country: United States of America
Voice Phone: 1 40 1 232 1200
Business Description: Manufacturer of industrial, medical and laser safety eyewear, as well as disposable and reusable respirators.
****

Van Leer Metallized Products
24 Forge Park
City: Franklin
State/Province: MA
Postal Code: 02038
Country: United States of America
Voice Phone: 1 508 541 7700
Voice Phone: 1 800 343 6977
Fax Phone: 1 508 541 7788
Contact #1: Harry Mann
Business Description: Manufacturer of Holo-PRISM holographic and VALVAC conventional metallized papers. Van Leer produces diffraction patterns or multi-dimensional holographic images on direct metallized papers. Existing patterns or customized images are available.
****

Vincennes University
1002 North First Street
City: Vincennes
State/Province: IN
Postal Code: 47591
Country: United States of America
Voice Phone: 1 812 888 8888

Contact # I: Richard Duesterberg
Business Description: Offering holography workshops for high school teachers, & college level courses in holography. We have 4 research-grade optical tables, as well as argon and krypton lasers. Call for more details.
****

Virtual Image (a division of Printpack, Inc.)
1050 Northfield Court
City: Roswell
State/Province: GA
Postal Code: 30076
Country: United States of America
Voice Phone: 1 770 751 0704
Fax Phone: 1 770 751 0806
email: jlecompt@printpack.com
Contact #1: Joe LeCompte
Business Description: Manufacturer of high quality, embossed BOPP holographic film, primarily for the packaging industry. Specialty is large runs in wide web. SEE OUR ADVERTISEMENT
****

Visual Visionaries
2011 Clement St., Suite 4
City: San Francisco
State/Province: CA
Postal Code: 94121
Country: United States of America
Voice Phone: 1 415 666 0779
Business Description: Consulting and marketing firm. Exhibitions and educational services. 12 years experience with holographic production, display and sales. Professional and reliable.
****

Volkswagen AG
Forschung und Entwicklung
Brieffach 1785
City: Wolfsburg
Postal Code: 38436
Country: Germany
Voice Phone: +49 (0)536 925 824
Fax Phone: +49 (0)536 972 444
Contact # 1: M.-A. Dr. Beeck
Business Description: industrial research, Interferometry; Holographic non -destructive testing, Laser combustion diagnostics.
****

Volvo-Flygmotor
S-461
City: Trollhattan
Postal Code: 81
Country: Sweden
Voice Phone: 46 0520 9447 1
Contact # I: Robert Frankmark.
Business Description: Holographic non-destructive testing.
****

Voxel
26081 Merit Circle - Suite 117
City: Laguna Hills
State/Province: CA
Postal Code: 92653
Country: United States of America
Voice Phone: 1 714 348 3207
Fax Phone: 1 714 348 8665
email: dlee@voxel.com
Contact # I: David Lee
Business Description: Medical imaging research. Company is developing a 3D visual display for non-invasive imaging techniques to be used as a diagnostic tool.
****

Waseda University

Dept Of Applied Physics
School Of Science & Engineering
City: Tokyo
Postal Code: 160
Country: Japan
Voice Phone: 81 03 209 321
Business Description: Medical holography research.
****

Wave Mechanics
450 North Leavitt
City: Chicago
State/Province: IL
Postal Code: 60612
Country: United States of America
Voice Phone: 1 312829 WAVE
Fax Phone: 1 312 829 8557
Contact # I: Deni Drinkwater-Welch
Business Description: Artistic holographer; silver halide transmission and reflection; consultant.
****

Wavefront Research, Inc.
616 West Broad Street
City: Bethlehem
State/Province: PA
• Postal Code: 1801 8-5221
Country: United States of America
Voice Phone: 1 610 974 8977
Fax Phone: 1 610 974 9896
email: tws@wavres.com
Contact #1: Thomas Stone
Business Description: Basic and applied research and development in holographic optical elements, applications of holography, non linear optics, and novel optical systems.
****

Wavefront Technology
15149 Garfield Ave.
City: Paramount
State/Province: CA
Postal Code: 90723
Country: United States of America
Voice Phone: 1 310 634 0434
Fax Phone: 1 310 634 0434
Contact #1: Joel Petersen
Business Description: Specialists primarily serving the embossing industry. Embossed hologram maste ring, recombining and ganging. Prototype, short run embossing. Rigid sheet embossing up to 4 x 8 foot in transmission or reflection.
****

Wesley, Ed
2124 West Irving Park Road
City: Chicago ...
State/Province: IE
Postal Code: 60618-3924
Country: United States of America
Voice Phone: 1 312 539 3672
Contact #1: Ed Wesly
Business Description: Holographic fine artist, author, and researcher. Instructor at the School of the Art Institute of Chicago. Consulting services offered. Pulsed laser specialist.
****

Whiley Foils Limited
Firth Road
Houston Industrial Estate
City: Livingston
State/Province: Scotland
Postal Code: EH54 5DJ
Country: United Kingdom

CORPORATE LISTINGS

Voice Phone: 44 150 643 8611
Fax Phone: 44 150 643 8262.
Contact # I: M ick Barker
Business Description: Whi ley Foils Limited is a long-established manufacturer of stamping foils. We have developed special base materi als for Holographic embossing and market these and other Holographic fo ils worldwide.
****

Wild Style Entertainment
1201 Park Ave. Suite 203A
City: Emeryville
State/Province: CA
Postal Code: 94608
Country: United States of America
Voice Phone: 1 510 654 8395
Fax Phone: 1 510 654 8396
Business Description: Computer imaging and animation for holography.
SEE OUR ADVERTISEMENT
****

Witchcraft Tape Products, Inc.
P.O. Box 937
City: Coloma
State/Province: MI
Postal Code: 49038
Country: United States of America
Voice Phone: 1 616 468 3399
Voice Phone: 1 800 521 0731
Fax Phone: 1 616 468 3391
Contact # I: Ronald Warczynski
Business Description: Full service manufacturer of quality embossed holograms, preci sion registered die cutting hot stamping, specialty lamination, assembly and packaging. This is our 24th year of furnishing quality products to the industry. SEE OUR ADVERTISEMENT
****

Wonders of Holography Gallery
PO Box 1244
City: Jeddah
Postal Code: 21431
Country: Saudi Arabia
Voice Phone: 966 2 652 0052
Fax Phone: 966 2 651 1325
Contact # 1: A.M. Baghdadi
Business Description: Retail holograms of all kinds. We are the first and only holography gallery in the Gulf countries. We resell all types of holograms and we produce laser shows.
****

Worcester Polytechnic Institute
Mechanical Engineering Department
100 Institute Road
City : Worcester
State/Province: MA
Postal Code: 01609-2280
Country: United States of America
Voice Phone: 1 508 831 5536
Fax Phone: 1 508 831 57 13
Contact # I: Ryszard Pryputniewicz
Business Description: Scientific, medical & industrial holography research ; Interferometry; Holographic non-destructive testing.
****

Wyko Corporation
2650 East Elvira Road
City: Tucson
State/ Province: AZ
Postal Code: 85706-7123
Country: United States of America
Voice Phone: 1 520741 1044
Fax Phone: 1 520 294 1799
email: sales@wyko.com

Web Address: wyko.com
Contact # 1: James Wyant
Business Description: Scientific holography research ; Interferometry and analysis.
****

X-IAL
Les Algorithmes
Parc D'lnnovation
City: Illkirch
Postal Code: F-67400
Coun try: France
Voice Phone: 33 88 67 44 90
Fax Phone: 33 88 67 80 06
Contact # I: Dr. Christian D. Liegeois
Business Description: Stereograms for embossing; HOEs designed and manufactured.
****

Yarovoy, Leonid
P.O.Box 164
Ci ty: Kiev
Postal Code: 252 191
Country: Ukraine
email: eeoc@gluk.apc.org
Contact # 1: Leonid Yarovoy
Business Description: Fiber-based holographic systems, holography correlators. Image holography, portraits.
****

Zero Gravity "
Pier 39 - KI05
City: San Francisco
State/Province: CA
Postal Code: 94133
Country: United States of America
Voice Phone: 1 415 989 5277
Business Description: Retail gift store offering holograms and other unique products as you wander through enchanting theme settings. Includes a huge hologram gallery of pictures, jewelry and novelty items. Also see Galaxies Unlimited listing.
****

Zero Gravity
West Edmonton Mall
16308770 - 170 SI.
City: Edmonton
Postal Code: T5T4M2
Country: Canada
Vo ice Phone: 1 403 413 70 10
Business Description: Retail gift store offering holograms and other unique products as you wander through enchanting theme settings. Includes a huge hologram gallery of pictures, jewelry and novelty items. Also see Galaxies Unlimited listing.
****

Zero Gravity
Changi Airport - Terminal 2
Departure Tra nsit North
Postal Code: 0817
Country: Singapore
Business Description: Retail gift store offering holograms and other unique products as you wander through enchanting theme settings. Includes a huge hologram gallery of pictures, jewelry and no velty items.
****

Zero Gravity
3A River Valley Road
1-4 Clark Quay
Postal Code: 0817
Country: Singapore
Business Description: Retail gift store offering

holograms and other unique products as you wander through enchanting th eme settings. Includes a huge hologram gall ery of pictures, jewelry and novelty items.
****

Zero Gravity
Forum Shops at Ceasar's
3500 Las Vegas Blvd. S.
City: Las Vegas
State/Province: NV
Postal Code: 89109
Country: United States of America
Voice Phone: 1 702 731 3565
Business Description: Retail gift store offering holograms and other unique products as you wander th rough enchanting theme settings. Includes a huge hologram gall ery of pictures, jewelry and novelty items. Also see Galaxies Unlimited listing.
****

Zero Gravity
Aloha Tower Marketp lace
101 Ali Moana Blvd.
City : Honolulu
State/Prov ince: HW
Postal Code: 96813
Country: United States of America
Voice Phone: 1 808 545 2355
Business Desc ription: Retail gift store offering holograms and other unique products as you wander through enchanting theme settings. In cludes a huge hol ogram gallery of pictures, jewelry and novelty items. Also see Galaxies Unlimited listing.
****

Zone Holografix Studios
5338 B Vineland Ave.
City: North Hollywood
State/Province: CA
Postal Code: 91601
Country: United States of America
Voice Phone: 1 8 18 985 8477
Fax Phone: 1 8 18 549 0534
Contact # I: Fred Unterseher
Business Description: Originates masters for mass production holograms in embossed, DCG and photopolymer materials. Both pulsed and CW lasers avai lab le. In struction. SEE OUR ADVERTISEMENT

REGAL *HOLOGRAPHICS*
brings a new dimension to printing

Corporations report dramatic increase in sales with use of Holograms in marketing and advertising programs

It's called the most powerful tool in promotion and marketing since the photograph. Holograms reach out and grab attention. Marketing surveys confirm their effectiveness. **Available now at Regal Press.** Now your company can take advantage of the 3-dimensional world of Holographics. Regal Press is the only printer in New England to produce and hot stamp custom and stock holograms. **Free samples.** Explore the exciting possibilities of holography and how it can benefit your company's communications. Write (on your letterhead) to Regal Holographics for an informative brochure, complete with samples. The time for a new dimension in printing has come to New England. And it's at Regal. Today.

The Regal Press Inc.

129 Guild St., Norwood, MA 02062 (617) 769-3900

This ad was printed on new Champion Kromekote® 2000, 1S Cover/.012, which is ideal for hot stamping holograms.

MARKETING
WITH

POLAROID
MIRAGE
HOLOGRAMS

COUNTERFEITERS BEWARE!
A Mirage hologram is the perfect device for product security and authentication. It is nearly impossible to duplicate and offers the highest level of security available at a relatively low cost. This Super Bowl XXV ticket was copy-proof and became an impressive collectors item.

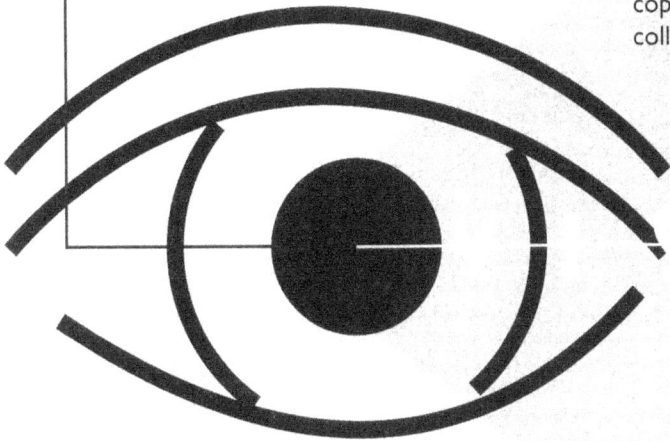

PHENOMENAL ATTRACTION!
Holographic stickers – promotional or premium use. Available in a variety of themes – Nature, Dinosaurs, Space and licensed images like The Mask™, Marvel Super Heroes™ or customize your own.

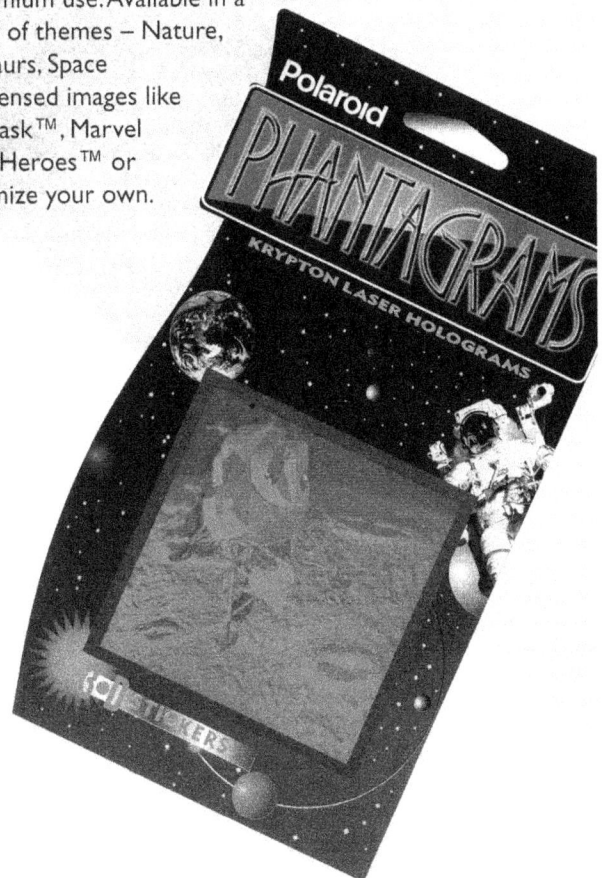

THE EYES HAVE IT!
Holographic sunglasses now available with Marvel Super Hero™ characters, college team logos, or Americana Images. They can be customized for any need with corporate, sports and event logos or images.

For more information call
800-237-5519 Fax: (617) 386-8857

MORE THAN A
MIRAGE

There is no stronger, more effective way for a company to boost its image, improve overall customer retention and increase product impact than by using Polaroid Mirage holograms. A consumer scanning the shelf or standing at a counter will notice and remember a package with the kind of visual depth and beauty that a Polaroid hologram offers.

Polaroid Mirage holograms take advantage of photopolymer technology, a process that allows light to reflect through several layers of film to create a sense of depth that is easily visible under a wide range of lighting conditions. By comparison, conventional embossed holography provides only a surface image, resulting in limited three-dimensionality visible only under a narrow range of lighting conditions.

Coca-Cola achieved a 100% sell-through when they updated an historic Christmas ad campaign with hologram trading cards.

Polaroid holograms not only fascinate the eye they also provide security for documents and products. Their unique appearance makes them difficult to counterfeit and their physical design makes them hard to tamper with. The images add real value by securing the credibility and authenticity of a company's product. The applications for Mirage holograms are wide ranging.

Polaroid's 3D sunglasses also add another means for a company to advertise its logo or products. Customized sunglasses can be created for an event, sports image, corporate image or logo.

Polaroid also supplies a line of cost effective holographic elements called Imagix™ Holographic Reflectors that dramatically improve the brightness, contrast, and thus readability of reflective and transflective LCDs. The Imagix™ Holographic Reflector opens up a new range of design parameters for the display engineer.

From product promotion and authentication to premium incentives, a Mirage hologram does more than look spectacular it gets your product or company noticed by consumers.

Casinos can now authenticate their chips with holograms that are virtually tamper-proof. Counterfeits can now be detected on sight, streamlining security and decreasing losses from fraud.

Mirage three-dimentional images have achieved tremendous recognition in packaging. Suzanne Vega's Days of Open Hand, the recording industry's first compact disc to feature a custom designed hologram, was honored with a Grammy Award for best album package. A&M Records noted that Polaroid's photopolymer technology "is superior to previous technologies. The colors pop out, have more depth, and create a more sophisticated product".

POLAROID

MIRAGE

HOLOGRAMS

For more information call
800-237-5519 Fax: (617) 386-8857

Holograms are a hot campus collectable, a real departure from sweatshirts and other ordinary mementos.
They can get serious results when used for student fund raising

Businesses Listed by Country

Argentina - Centro de Arte Holografico
Argentina - Centro De Investifaciones
 Opticas
Australia - 3D Optical Illusions
Australia - Australian Holographics
Australia - Clemenger Perth Fry Ltd
Australia - Holograms Fantastic and Optical
 Illusions
Australia - Lazart Holographics
Australia - Moonbeamers
Australia - New Dimension Holographics
Australia - Parallax Gallery
Austria - Holography Center of Austria
Austria - Semicon Austria
Belgium - Agfa - Gevaert N.V
Belgium - Free University Of Brussels.
Belgium - Laboratory Vinckiner
Belgium - Multifacet
Belgium - Univ. de Liege
Canada - Capilano College
Canada - Centre d' Art Holographique et
 Photonique
Canada - Claudette Abrams
Canada - Concordia University
Canada - Deep Space Holographics
Canada - Dimension 3
Canada - Fringe Research Holographics
Canada - General Holographics - Inc.
Canada - Harvard Apparatus - Canada
Canada - Holocrafts
Canada - Hologram Development Corp.
Canada - Holography and Media In stitute of
 Quebec
Canada - Laser Inspeck
Canada - Lasiris Inc.
Canada - Les Productions Hololab!
Canada - Meli ssa Crenshaw Holography
 Studio Internati onal
Canada - Mu's Laser Works
Canada - Ontario College Of Art
Canada - Ontario Science Centre
Canada - Photon League Of Holographers
 Ontario
Canada - Royal Holographic Art Gallery
Canada - Silverbridge Group
Canada - The Hologram Store - Ltd.
Canada - Universite Lava l
Canada - Zero Gravity
Columbia - Imagenes Holograficas De
 Columbia
Czech Repub lic - Czechoslovak Academy
 Of Science
Denmark - Bbt Instrumenter Aps.
Denmark - Ibsen Micro Strucmres N S
Finland - Ad-Holograms Oy
Finland - Heptagon Oy
Finland - SETEC Oy
Finland - Starcke - Ky.
France - Aerospatiale
France - Atelier Holographique De Paris
France - H.Ologrammi
France - Holo 3
France - Holo-Laser
France - Hologram Industries
France - Holomedia France
France - Laser Movement
France - Magic Laser
France - P.S.A Peugeot Citroen
France - Photonics Systems Laboratory
France - Quantel
France - Sopra
France - X-IAL
Germany - 3D Vision
Germany - AB Rueck Holoart
Germany - Academy of Media Arts Cologne
Germany - Adlas G.M.B.H. & Co Kg.
Germany - AHT 3D-Medien
Germany - AKS Holographie-Galerie

GmbH
Germany - AKS Holographie-Gallerie
 GmbH
Germany - Andreas Wappelt - Photonics
 Direct
Germany - Arbeitskreis Holografie B. V
Germany - Armin Klix Holographie
Germany - Artbridge Light Studios
Germany - Baier Praegepressen
Germany - BIAS
Germany - CHIRON Technolas GmbH
Germany - Coherent Luebeck GmbH
Germany - Curt Abramzik
Germany - Daimler Benz Aerospace
Germany - Deutsche Gesellschaft fur
 Holografie
Germany - Die Dritte Dimension
Germany - Dietmar Oehlmann
Germany - Dornier Medizintechnik GmbH
Germany - ETA-Optik Gmbh
Germany - Fielmann-Verwaltung KG
Germany - Gresser - E. - KG
Germany - HOL 3 - Galerie fur
 Holographie GmbH
Germany - Holo GmbH
Germany - Hologram Company RAKO
 GmbH
Germany - Holographic Systems Muenchen
 GmbH
Germany - Holographie Anubis
Germany - Holographie Fachstudio Bad
 Rothenfe lde
Germany - Hologr-aphie _Labor
Germany - Holoptics
Germany - Holopublic Unbehaun
Germany - HoloVis ion
Germany - HRT Holographic Recording
 Technologies GmbH
Germany - Institut fur Angewandte Physik
Germany - Kolbe-Druck mit
 Tochtergesellschaften
Germany - Krystal Holographics Vertriebs-
 GmbH
Germany - Laserfilm Eckard Knuth
Germany - Lauk & Partner GmbH
Germany - Leonhard Kurz GmbH
Germany - Leseberg - Dr. Detlef
Germany - Magick Signs Holografie
Germany - Magick signs Holografie
Germany - Melles ~ot GmbH
Germany - Metrologic Instruments GmbH
Germany - Meulien Odile
Germany - Newport Gmbh
Germany - Optical Coating Laboratory
 GmbH
Germany - Optical Test Equipment
Gennany - OWIS Gmbh
Germany - Phantastica
Germany - Physik lnstmmente (PI) GmbH
 & Co.
Germany - Rofin-Sinar Laser GmbH
Germany - Spindler & Hoyer GmbH & Co.
Germany - Steinbichler Optotechnik GmbH
Germany - Steuer KG GmbH & Co.
Germany - Studio Fuer Holographie
Germany - Technische Fachhochschule
 Berlin
Germany - Technische Universitaet Berlin
Germany - Technolas Laser Technik Gmbh
Germany - topac GmbH - Department
 Holography
Germany - University of Erlanger
Germany - University of Muenster
Germany - Uni versity of Munich
Germany - University of Stuttgart
Germany - Volkswagen AG
Greece - Cavomit
Greece - Hellenic Institute Of Holography
Hong Kong - Far East Holographics
Hong Kong - Foreign Dimension

Hong Kong - The Foreign Dimension
Hungary - Ap Holografika Studio
Hungary - Hol ogram Varga Miklos
Hungary - Optopol Panoramic Metrology
 Consulting
Hungary - Technical Univers ity Of
 Budapest
India - Advance Photonics
India - Better Labels Mnfg Co
India - Coated Specialties
India - Hi-Glo Holo Images Pvt. Ltd.
India - Holostik India Pvt. Ltd.
India - I.S. Gill
India - Ojasmit Holographics
India - Print-M-Boss
India - Spatial Holodynamics (India) Pvt.
 Ltd.
India - Spectrum Corporation
Indones ia - Holography Di vision
Indonesia - P. T. Pura Barutama
Indonesia - Pt. Pura Nusapersada
Israel - Glaser - Technical Consulting
Israel - Holo-Or Ltd
Israel - Holography Israel
Italy - Autoadesivi Sri
Italy - Cise Spa Technologie Innovative
Italy - CISE Tecnologie Innovative SpA
Italy - Conte M. Vulcano
Italy - Db Electronic Instruments S.R.L.
Ital y - Diaures S.A. Holography Division
Italy - Diavy sri
Italy - Holofar Lab (Sri)
Italy - Holographic Service
Italy - Societa Olografica ltalia (Soi)
Italy - Universita Di Roma
Italy - Uni vers ity Of Bologna
Japan - Asahi Glass Co.
Japan - Brainet Corporation
Japan - Canon Inc. R&D Headquarters
Japan - Central Glass Co. - Ltd.
Japan - Chiba University
Japan - Dai Nippon Printing Co. - Ltd.
Japan - Fuji Electric Co. Ltd
Japan - Fuj itsu Laboratories Ltd.
Japan - HODIC Holographic Display Al1ists
 & Engineers Club
Japan - Hyogo Prefectual Museum of
 Modern Art
Japan - Ishii - Ms. Setsuko
Japan - Japan Communication Arts Co.
Japan - Keio University
Japan - Kimmon Electric Co. - Ltd.
Japan - Laboratories of Image Information
Japan - Light Dimension - Inc.
Japan - Marubun Corporation
Japan - Mazda Motor Corp.
Japan - Ministry Of International Trade
Japan - Mitsubishi Heavy Industries Ltd.
Japan - Newport (Asian Office)
Japan - Nihon University
Japan - Nippon Polaroid K.K.
Japan - Nippondenso Co. Ltd.
Japan - Nippondenso Co. - Ltd.
Japan - Nissan Motor.
Japan - Numazu College Of Technology
Japan - Ruey-Tung - Miss. Hung
Japan - SAM Museum
Japan - Sharp Corp.
Japan - Sophia University
Japan - Tama Art Umversity
Japan - Tokai Univers ity
Japan - Tokyo Institute of Technology
Japan - Tokyo Inst itute Of Technology
Japan - Topcon Inc.
Japan - Toppan Printing Co. Ltd.
Japan - Toppan Printing Co. - Ltd.
Japan - Toppan Printing Co. Ltd.
Japan - Toyama National College Of Marit
Japan - University Of Tsukuba
Japan - University Of Tokyo

Japan - Waseda University
Latvia - Physics Institute. Latvian
Lithuania - Geola
Mexico - Centro de Investigaciones en
Optica - A.c.
Mexico - Hologramas - S.A. de c.Y.
Mexico - Jimenez-Ceniceros - Antonio
Netherlands - Dutch Holographic Laboratory
BV
Netherlands - Foundation Ideecentrum.
Netherlands - Optilas B.Y.
Netherlands - Optische Fenomenen
Netherlands - Stichting Voor Holographie
En Laseropiek
Netherlands - Technical University @
Eindho ven
Netherlands - TNO Institute of Applied
Physics
Netherlands - Triple-D Lase. Imaging
Norway - Interferens Holografi D.A.
Norway - Norges Tekniske Hogskole.
Pakistan - AK Relunan Traders
Pakistan - Dimensions
Panama - Holographic Dimensions
People 's Rep. of China - Beijing Fantastic
Hologram Product Corp
People 's Rep. of China - Beijing Hologram
Printing Technology Co
People's Rep. of China - Beijing Inst. of
Posts
People's Rep. of China - Beijing Institute Of
Posts
People's Rep. of China - Beijing Nonnal
University.
People 's Rep. of China - Beijing Sanyou
Laser Images Co
People 's Rep. of China - China Ann Arbor
Holographical Institute
People 's Rep. of China - Morning Light
Holograms
People 's Rep. of China - Shandong
Academy of Sciences
Poland - Holografia Polska
Poland - Holographic Dimensions - Poland
S.A.
Poland - Hololand S.c.
Poland - Institute Of Plasma Physics
Poland - Technical University Of Wroclaw.
Portugal - 3D Holographics
Portugal - INET! - Institute of Infonnation
Technologies
Portugal - Universidade Do Porto
Russia - Denisyuk - Yuri N.
Russia - Siavich Joint Stock Company
Russia - Technoexan Ltd
Saudi Arabia - Wonders of Holography
Gallery
Singapore - Zero Gravity
Singapore - Zero Gravity
Slovakia - Technical University Zvolen
South Africa - Synchron Pty Ltd.
South Korea - Dan Han Optics
Spain - Holosco - Ernest Barnes
Spain - Karas Studios S.L.
Spain - Lasing S.A. -
Spain - Museu D' Holografia
Spain - University Of Alicante
Sweden - Dialectica Ab
Sweden - High Tech Network
Sweden - HoloMedia Ab/Hologram
Museum.
Sweden - Holovision AB
Sweden - Karolinska Institutet
Sweden - KTH
Sweden - Lulea University Of Technology
Sweden - Lund Institute Of Tech.
Sweden - Martinsson Elektronik Ab.
Sweden - Royal Institute Of Technology
Sweden - Saab-Scania
Sweden - Spectrogon Ab

Sweden - Studio Weil-Alvaron
Sweden - Swede Holoprint
Sweden - Volvo-Flygmotor
Switzerland - Galerie IIIusoria
Switzerland - Holo-Service
Switzerland - Holo-Service.Fries
Switzerland - Holodesign Studies
Switzerland - Hologramm Werkstatt &
Galerie
Switzerland - Holos Art Galerie
Switzerland - Stoltz Ag
Switzerland - Swiss Federal Inst Of
Technology
Switzerland - Universite De Neuchatel
Taiwan - Fong Teng Technology
Taiwan - Hiat Image Technology Group -
Inc.
Taiwan - Holo Images Tech Co. - Ltd.
Taiwan - Holo Impressions
Taiwan - Holo Impressions Inc
Taiwan - Industrial Technology Research
Ins!.
Taiwan - Infox Corporation
Taiwan - Institute Of Optical Science
Taiwan - Superb in Co. Ltd
Taiwan - Tjing Ling Industrial Research.
Thailand - Electro-Optics Lab - NECTEC
Thailand - New Horizons (Thai land) - Ltd.
Ukraine - Environmental Education and
Information Ctr.
Ukraine - Feofaniya L,td.
Ukraine - Tair Hologr~m Company
Ukraine - The Institute of Applied Optics
Ukraine - Yarovoy - Leonid
United Arab Emerates - Hololaser Gallery
UK - 3D Images
UK - 3D-4D Holographics
UK - A.H. Prismatic - Ltd.
UK - Action Tapes
UK - Advanced Holographic Laboratories
UK - Ag Electro-Optics Ltd.
UK - Amazing World Of Holograms.
UK - Applied Holographics - Pic.
UK - Astor Universal Ltd.
UK - Barr & Stroud - Ltd.
UK - Beddis Kenley (Machinery) Ltd.
UK - Bemrose Production Products
UK - Boyd - Patrick
UK - British Aerospace Pic.
UK - BTG pic
UK - Checkpoint
UK - Courtauld Acetate
UK - Creative Holography Index - The
UK - Customer Service Instrumentation
UK - Datasights Ltd.
UK - De La Rue Holographics Ltd.
UK - Dimuken
UK - Electro Optics Developments Ltd.
UK - Embossing Technology Ltd
UK - Expanded Optics Limited
UK - Focal Image Ltd.
UK - Galvoptics Ltd
UK - Global Images
UK - Graham Saxby
UK - Holocrafts Europe Limited.
UK - Holograms 3D
UK - Holographics (Uk) Ltd.
UK - Holomex Ltd.
UK - iC Holographics
UK - Imperial College Of Science
UK - International Data Ltd.
UK - International Hologram Manufacturers
Association
UK - Jayco Holographics
UK - K.C. Brown Holographics
UK - Kendall Hyde Ltd.
UK - Laser International
UK - Laza Holograms Ltd.
UK - Light Impressions International - Ltd.
UK - LOT Oriel

UK - Loughborough Univ. Of Tech.
UK - Lumonics Ltd.
UK - Margaret Benyon Holography Studio
UK - National Physical Laboratory
UK - Op-Graphics (Holography) Ltd.
UK - OpSec - England
UK - Optical Works Ltd.
UK - Oxford Holographics
UK - Pepper - Andrew
UK - Pilkington Optronics
UK - Ralph Cullen Holographics
UK - Reconnaissance International Ltd.
UK - Richmond Holographic Studios
UK - Rolls-Royce Pic
UK - Rutherford & Appleton Labs
UK - Spatial Imaging Ltd.
UK - ST! - Europe
UK - The London Holographic Image
Studio
UK - Turing Institute
UK - Uk Optical Supplies
UK - University Of Oxford
UK - Whiley Foils Limited
USA - 21 st Century Finishing Inc.
USA - 3 Deep Hologram Company
USA - 3-D Systems
USA - 3-D Worldwide Holograms - Inc .
USA - 3D Holograms Inc.
USA - 3M - Safety and Security Systems
USA - A.D. Tech (Advanced Deposition
Technologies)
USA - A.H. Prismatic - Inc.
USA - Acme Holography
USA - AD 2000 - Inc.
USA - Advanced Holographic Laboratories
USA - Advanced Optics - Inc.
USA - Advanced Technology Program
USA - Aerotech Inc.
USA - Agfa - a division of Bayer Corp.
USA - Alabama A&M University
USA - Amagic Technologies Inc.
USA - American Bank Note Holographics
USA - American Holographic Inc.
USA - American Laser Corporation
USA - American Paper Optics Inc.
USA - American Propylaea Corporation
USA - Ana MacArthur
USA - Another Dimension
USA - APA Optics Inc.
USA - Applied Optics
USA - Art Institute Of Chicago (The School
of the ...)
USA - Art Lab
USA - Art - Science & Technology Institute
(AST!)
USA - Automated Holographic Systems
USA - Avant-Garde Studio
USA - Barilleaux - Rene Paul
USA - Batelle Pacific Northwest National
Laboratory
USA - Bellini - Victor
USA - Berkhout - Rudie
USA - Blue Ridge Holographics - Inc.
USA - Bobst Group
USA - Booth - Roberta
USA - Brandtjen & Kluge - Inc. -
USA - Bridgestone Graphic Technologies -
Inc.
USA - Broadbent Consulting Services
USA - Burleigh Instruments - Inc.
USA - Cambridge Laser Labs
USA - Carl M. Rodia And Associates
USA - Casdin-Silver Holography
USA - CFC Applied Holographics
USA - Cherry Optical Holography
USA - Chromagem Inc.
USA - Chronomotion
USA - Cifelli - Dan
USA - City Chemical
USA - Coburn Corporation

USA - Coherent - Inc. - Laser Group
USA - Continental Optical
USA - Control Module Inc.
USA - Control Optics
USA - Corion Corp.
USA - Creative Label
USA - Crown Roll Leaf - Inc.
USA - CVI Laser Corporation
USA - Datacard Corporation
USA - Deem - Rebecca
USA - DeFreitas - Frank
USA - Dell Optics Company - Inc.
USA - Diamond Images - Inc.
USA - Diffraction Ltd.
USA - Dimensional Arts
USA - Dimensional Cinematography Co.
USA - Dimensional Foods Co.
USA - Direct Holographics
USA - Doris Vila Holographics
USA - DuPont (see E.!. DuPont De Nemours & Co.)
USA - E.!. DuPont De Nemours & Co.
USA - Ealing Electro-Optics Inc .
USA - Eastman Kodak Company
USA - Edmund Scientific Company
USA - EI Don Engineering
USA - Electro Optical Industries - Inc.
USA - Elusi ve Image
USA - Engineering Animation - Inc.
USA - Evolution Design - Inc.
USA - Excitek Inc.
USA - Fantastic Holograms
USA - Fast Light Inc.
USA - Feroe - James
USA - Fisher Scientific
USA - FLEXcon
USA - Flight Dynamics
USA - Foil Stamping and Embossing Association
USA - FoilMark Holographic Images
USA - Fornari - Arthur David
USA - Forth Dimension Holographics
USA - Frank DeFreitas Holography Studio
USA - Fresnel Technologies Inc.
USA - G.M. Vacuum Coating Lab - Inc.
USA - Galaxies Unlimited - Inc.
USA - General Design
USA - Glass Mountain Optics
USA - Gorglione - Nancy
USA - Hallmark Capital Corp.
USA - Holage
USA - Holart Consultants
USA - Holicon Corporation.
USA - Holo Sciences - LLC
USA - Holo-Spectra
USA - Holo/Source Corporation
USA - HoloCom
USA - Holografica
USA - Hologram Fantastic
USA - Hologram Land
USA - Hologram Research - Inc.
USA - Hologram World - Inc.
USA - Holograms and Lasers International
USA - Holograms and Lasers International
USA - Holograms International
USA - Holographic Applications
USA - Holographic Design Systems
USA - Holographic Dimensions
USA - Holographic Images Inc.
USA - Holographic Impressions
USA - Holographic Industries - Inc.
USA - Holographic Label Converting (HLC)
USA - Holographic Optics Inc.
USA - Holographic Products
USA - Holographic Studios
USA - Holographics Inc.
USA - Holographics North Inc.
USA - Holography Institute of San Francisco

USA - Holography Marketplace
USA - Holography Presses On (HPO)
USA - Holophile - Inc.
USA - Holotek
USA - Holovision Systems Inc.
USA - HoloWebs - Inc.
USA - Honeywell Technology Center
USA - Hughes Power Products - Inc.
USA - IBM Almaden Research Center
USA - ICI Polyester
USA - Illinois Institute Of Technology
USA - Illuminations
USA - Imagen Holography - Inc .
USA - Images Company
USA - Imagination Pl antation
USA - 1m Edge Technology
USA - Industrial Technology Institute
USA - Infinity Laser Laboratories
USA - Infrared Optical Products - Inc.
USA - Innovative Technology Associates
USA - Inrad - Inc.
USA - Inside Finishing Magazine
USA - Integraf
USA - Interactive Industries Inc.
USA - Intrepid World Communications
USA - Ion Laser Technology - Inc.
USA - James River Products
USA - Jeffery Murray Custom Holography
USA - Jodon Inc.
USA - JR Holographics
USA - Kaiser Optical Systems - Inc.
USA - Kan - Mike
USA - Kauffman - ·· John _
USA - Keystone Scientific Co.
USA - Kinetic Systems - Inc.
USA - Kreischer Optics - Ltd.
USA - Krystal Holographics International Inc .
USA - Krystal Holographics Intern ational Inc.
USA - L.A.S.E.R. News
USA - Laboratory for Optical Data Processing
USA - Lake Forest College
USA - Larry Lieberman Holography
USA - Lasart Ltd.
USA - Laser Affiliates
USA - Laser and Motion Development Company
USA - Laser Arts SocSety For Education and Research
USA - Laser Drive Inc.
USA - Laser Focus World
USA - Laser Holography Workshop
USA - Laser Images
USA - Laser Innovations
USA - Laser Institute Of America
USA - Laser Las Vegas
USA - Laser Light Designs
USA - Laser Light Ltd.
USA - Laser Media - Inc.
USA - Laser Optics - Inc.
USA - Laser Reflections
USA - Laser Resale Inc.
USA - Laser Technical Services
USA - Laser Technology - Inc.
USA - Lasermetrics - Inc.
USA - Lasersmith - Inc.(The)
USA - Laserworks
USA - Lawrence Berkeley Laboratory
USA - Lazer Wizardry
USA - Lenox Laser
USA - Letterhead Press - In c.
USA - Lexel Laser - Inc.
USA - LiCONiX
USA - Light Impressions International - Ltd.
USA - Light Wave Gallery
USA - Lightrix - Inc.
USA - Linda Law Holographics

USA - Lone Star Illusions
USA - Lopez 's Gallery International
USA - Louis Paul Jonas Studios - Inc.
USA - Lumenx Technologies - Inc.
USA - Luminer Printing and Converting
USA - M.LT. (Massachusetts Institute of Technology)
USA - M.!.T. Museum
USA - M.O.M. Inc.
USA - MacShane Holography
USA - Man/Environment - Inc.
USA - Marks - Gerald
USA - MasterPrint Holography - Inc.
USA - McCain Marketing & Graphic Design
USA - McMahan Electro-Optic
USA - Media Interface - Ltd.
USA - Melles Griot
USA - Meredith Instruments
USA - Mesmerized Holographic Marketing
USA - MetroLaser
USA - Metrologic Instruments - Inc.
USA - MGM Converters Inc.
USA - Midwest Laser Products
USA - Mitutoyo Measuring In struments (MTI Corp.)
USA - Multiplex Moving Holograms
USA - Museum Of Holography/Chicago
USA - MWK Industries
USA - Navidec Inc. (formerly ACI Systems - Inc.)
USA - NeoVision Productions
USA - New Light Industries
USA - New York Hall Of Science
USA - New York Holographic Laboratories
USA - Newport Corporation
USA - Nimbus Manufacturing - Inc.
USA - Norland Products - Inc.
USA - Northern Illinois University
USA - OIE Research
USA - Odhner Holographics
USA - Omnichrome Corp.
USA - OpSec - Corporate Headquarters
USA - OpSec - USA
USA - Optical Corporation Of America
USA - Optical Research Services
USA - Optical Society of America (OSA)
USA - Optics Plus Inc.
USA - Optimation
USA - Optimation Holographics
USA - Optineering
USA - Optitek
USA - Oregon Institute of Technology
USA - Oregon Laser Consultants
USA - Oriel Instruments
USA - Pacific Holographics Inc.
USA - Panatron Inc.
USA - Pasco Scientific
USA - Peacock Laboratories - Inc.
USA - Pennsylvania Pulp & Paper Co.
USA - Photon Cantina Ltd.
USA - PhotoRic!> Spectra
USA - Phys ical Optics Corporation.
USA - Pink - Patty
USA - Planet 3-D
USA - Point Source Productions
USA - Polaroid Corporation
USA - Polymer Image
USA - Potomac Photonics - Inc.
USA - Process Technologies
USA - PullTime 3-D Laboratories
USA - Rainbow Symphony Inc.
USA - Ralcon
USA - Real Image
USA - Reconnaissance International Ltd.
USA - Red Beam - Inc.
USA - Reva 's Holographic Illusions
USA - Reynolds Metals Co.
USA - Rice Systems
USA - Richard Bruck Holography

USA - Richardson Grating Laboratory
USA - Richmond Development Group
USA - Robert Sherwood Holographic
    Design
USA - Rochester Inst. Of Technology
USA - Rochester Photonics Corporation
USA - Rolyn Optics
USA - Ross Books
USA - Rowland Inst itute For Science
USA - Saginaw Valley State University
USA - Saint Mary's College
USA - San Jose State University
USA - Sandia National Laboratories
USA - Scharr Industries
USA - School Of Holography
USA - Science Kit & Boreal Labs
USA - Sharon McCormack Holography
USA - Shipley Chemical Co.
USA - Silhouette Technology Inc.
USA - Silicon Graphics ~ .-
USA - Sillcocks Plastics International
USA - Silver Dragon Holography
USA - Simian Co.
USA - Sinclair Optics - Inc.
USA - Smith & McKay Printing Co. Inc.
USA - Sonoma State University
USA - Southern Indiana Holographics
USA - Spectra-Physics Lasers Inc.
USA - Spectratek Inc.
USA - SPIE
USA - SPIE's Holography Working Group
    Newsletter
USA - Springer-Verlag New York
USA - Stanford University
USA - Star Magic
USA - Star Magic
USA - Star Magic
USA - Star Magic
USA - Stephens - Anait
USA - STI
USA - Swift Instruments
USA - Synchronicity Holograms
USA - Syracuse University
USA - Tamarack Storage Devices
USA - Technical Marketing Services
USA - Textile Graphics - Inc.
USA - The Hologram Company #1
USA - The Hologram Company #2
USA - The Hologram Company #3
USA - The Hologram Company #4
USA - The Hologram Company #5
USA - The Hologram Company #6
USA - The Hologram Company #7
USA - The Holography - Laser &
    Photonics Center
USA - The HOLOS Corporation
USA - The Regal Press Inc.
USA - Third Dimension Arts Inc.
USA - Thorlabs Inc.
USA - Three-D Light Gallery
USA - Total Register Inc.
USA - Towne Technologies
USA - Transfer Print Foils - Inc.
USA - Tyler Group
USA - U.K. Gold Purchasers - Inc.
USA - Ultra-Res Corporation
USA - Uniphase Lasers
USA - Unistay
USA - United Association Manufacturer's
    Representatives
USA - University Of Alabama
USA - University Of Arizona
USA - University Of Dayton
USA - University Of Michigan
USA - Un iversity Of Rochester
USA - University Of Southern California
USA - University Of Wisconsin/Madison
USA - Unterseher & Associates
USA - Uvex Safety Inc.
USA - Van Leer Metallized Products

USA - Vincennes University
USA - Virtual Image (a division of
    Printpack - In c.)
USA - Visual Visionaries
USA - Voxel
USA - Wave Mechanics
USA - Wavefront Research - Inc.
USA - Wavefront Technology
USA - Wesley - Ed
USA - Wild Style Entertainment
USA - Witchcraft Tape Products - Inc.
USA - Worcester Polytechnic Institute
USA - Wyko Corporation
USA - Zero Gravity
USA - Zero Gravity
USA - Zero Gravity
USA - Zero Gravity
USA - Zone Holografix
Yugoslavia - Trend

Industry People

Last Name, First Name, Business, Country

Abendroth, Detlev, AKS HolographieGalerie GmbH, Germany

Abouchar, Natalalie, Foreign Dimension, Hong Kong

Abrams, Claudette, Claudette Abrams, Canada

Abrams, Claudette, Photon League Of Holographers Ontario, Canada

Abramson, Nils, KTH, Sweden

Abramzik, Curt, Curt Abramzik, Germany

Agehall. , Christer, High Tech Network, Sweden

Akveld, A.C., Triple-D Laser Imaging, Netherlands

Albrecht, Gerd M., Phantastica, Germany

Albright, Steve, Optimation Holographics, USA

Alten, Susanne, Spindler & Hoyer GmbH & Co., Germany

Andersen, Chad, Meredith Instruments, USA

Anderson, Mike, Holomex Ltd., UK

Anderson, Steve, Sonoma State University, USA

Ando, Hiroshi, Nippondenso Co. , Ltd ., Japan

Andrade, Ana Alexandra, INETJ - Institute of Information Technologies , Portugal

Andrews, Mathew, 30-40 Holographi cs, UK

Anoff, Mark, Another Dimension, USA

Aprile, Silvio, New Horizons (Thai land), Ltd., Thailand

Aranguren, Mrs. Cintia, Centro de Arte Holografico, Argentina

Arkin, Bill, Holo-Spectra, USA

Attardi, Luigi, Societa Olografica Italia (Soi), Ita ly

Aymar, Bill, Laser Las Vegas, USA

Baghdadi, A.M. , Wonders of Holography Gallery, Saudi Arabia

Baghdadi, Abdul Wahab, Hololaser Gallery, United Arab Emerates

Bagley, Sheila, A.H. Prismatic, Inc. , USA

Bahuguna, Ramen, San Jose State University, USA

Bains, Sunny, SPIE's Holography Working Group Newsletter, USA

Baker, Marion, Krystal Holographics International Inc ., USA

Balogh, Tibor, Ap Holografika Studio, Hungary

Bar, Edgar, Holo-Service, Switzerland

Barefoot, Paul D. , Holophile, Inc. , USA

Barilleaux, Rene Paul, Barilleaux, Rene Paul, USA

Barker, Mick, Whiley Foils Limited, UK

Barre, Pascal, Holos Art Galerie, Switzerland

Bassin, Barry, Infrared Optical Products, Inc., USA

Bazargan, Kaveh, Focal Image Ltd., UK

Bear, Sol , Hologram World, Inc. , USA

Beauregard, Alain, Lasiris Inc., Canada

Beeching, Dave, CFC Applied Holographics, USA

Begleiter, Erich, Dimensional Foods Co., USA

Behrmann, Greg, Potomac Photonics, Inc., USA

Bellini, Victor, Bellini, Victor, USA

Benito, Ramon, Karas Studios S.L. , Spain

Bentley, John, Dimuken, UK

Benton, Stephen, M.l. T. (Massachusetts Institute of Technology), USA

Benyon, Margaret, Margaret Benyon Holography Studio, UK

Berkhout, Rudie, Berkhout, Rudie, USA

Bianchi, Herman-Josef, Arbeitskreis Holografie B.V., Germany

Billeri, Ralph, Control Module Inc. , USA

Billings, Loren, Museum Of Holographyl Chicago, USA

Billings, Loren, School Of Holography, USA

Billings, Robert, Holographic Design Systems, USA

Billo, Andreas, Institut fur Angewandte Physik, Germany

Bjelkhagen, Hans, American Propylaea Corporation, USA

Blosvern, Moss, Kinetic Systems, Inc. , USA

Bobeck, Paula, E.!. DuPont De Nemours & Co., USA

Bohan, Brian, Cambridge Laser Labs, USA

Bolognini, Nestor, Centro De Investifaciones Opticas, Argentina

Boone, Pierre, Laboratory Vinckiner, Belgium

Booth, Roberta, Booth, Roberta, USA

Bosco, Eric, Centre d'Art Holographique et Photonique, Canada

Botos, Steve A. , Aerotech Inc., USA

Bourque, Nina, Hologram Fantastic, USA

Bower, Elaine, University Of Wisconsinl Madison, USA

Boyd, Patrick, Boyd, Patrick, UK

Bradshaw, Roy, Real Image, USA

Brill, Louis, Illuminations, USA

Broadbent, Donald C., Broadbent Consulting Services, USA

Broeders, Jan M., Optische · Fenomenen, Netherlands

Brown, David, Optical Research Services, USA

Brown, George, Barr & Stroud, Ltd., UK

Brown, John, Light Impressions International, Ltd., UK

Brown, Kevin, Holographic Dimensions, USA

Brown, Kevin, K.C. Brown Holographics, UK

Bruck, Richard, Holicon Corporation. , USA

Bruck, Richard, Ri chard Bruck Holography, USA

Bruegmann, Machteld, Hologram Company RAKO GmbH, Ger~any

Buell, Richard, James River Products, USA

Bunkenburg, Jo, Rochester Photonics Corporation , USA

Burder, David, 3D Images, UK

Burgmer, Brigitte, Deutsche Gesellschaft fur Holografie, Germany

Burke, Ed, Hologram Development Corp. , Canada

Burney, Michael, Chronomotion, USA

Burns, Joseph, Hologram Research, Inc. , USA

Busano, Albertus, P.T. Pura Barutama, Indonesia

Bussard, Jan, Holography Presses On (HPO), USA

Bussard, Jan, Textile Graphics, Inc., USA

Bussaut, Laurent, Art, Science & Technology Institute (ASTJ), USA

Busuioc, Maria-Florica, The Holography, Laser & Photonics Center, USA

Butteriss, Tony, New Dimension Holographics, Australia

Butteriss, Tony, Parallax Gallery, Australia

Cantos, Brad, Holage, USA

Capron, Bruce, Sinclair Optics, Inc. , USA

Capucci, Pier Luigi, Uni versity Of Bologna, Italy

Carlsson, Torgny, KTH, Sweden

Casasent, David, Laboratory for Optical Data Processing, USA

Casdin-Sil ver, Harriet, Casdin-Silver Holography, USA

Cason, Thad, Infinity Laser Laboratories, USA

Castagna, Luigi , Holomedia France, France

Caulfield, John , Alabama A&M University, USA

Chaihorsky, Alex, Ultra-Res Corporation, USA

Chang, Long, Amagic Technologies Inc. , USA

Chantler, Sylvia, National Physical Laboratory, UK

Cheimets, Alex, 3 Deep Hologram Company, USA

Chen, Alex c.T., Infox Corporation, Taiwan

Chen, Hsuan, Saginaw Valley State University, USA

Cherry, Greg, Cherry Optical Holography, USA

Chiang, Mark, Fong Teng Technology, Taiwan

Chiarot, Roy, Photon Cantina Ltd., USA

Chiou, Billy, Holo Impressions, Taiwan

Chiou, Craig, Holo Images Tech Co., Ltd., Taiwan

Chou, Billy, Hiat Image Technology Group , Inc ., Taiwan

Christakis, Anne-Marie, Magic Laser, France

Cifelli, Dan, Cifelli , Dan, USA

Clarke, Walter, Global Images, UK

Claudius, Peter, Multiplex Moving Holograms, USA

Claytor, Linda H. , Fresnel Technologies Inc., USA

Conklin, Don, Glass Mountain Optics, USA

Connors, Betsy, Acme Holography, USA

Cooper, Nick, Oxford Holographics, UK

Cope, Jonathan, Laza Holograms Ltd. , UK

Cordner-Guled, Valeska, HOL 3, Galerie fur Holographie GmbH, Germany

Cossa, Rich, Planet 3-D, USA

Cossette, Marie-Andree, Holography and Media Institute of Quebec, Canada

Cote, Paul, FoilMark Holographic Images, USA

Coursen, Dan, G.M. Vacuum Coating Lab, Inc., USA

Cox, Dr. J. Allen, Honeywell Technology Center, USA

Creath, Kathy, Optineering, USA

Crenshaw, Melissa, Melissa Crenshaw Holography Studio International, Canada

Crenshaw, Milessa, Capilano College, Canada

Cross, Lloyd, 3-D Systems, USA

Cubberly, George, Excitek Inc. , USA

Cullen, Karoline, Holocrafts, Canada

Cullen, Ralph, Ralph Cullen Holographics, UK

Cullen, Ralph, Dk·.Optical Supplies, UK

Curiel, Yoram, OpSec - Corporate Headquarters, USA

Cvetkovich, Thomas J., Chromagem Inc., USA

D' Entremont, Joseph P., Lenox Laser, USA

Da-Hsiung, Hsu, Beijing Institute Of Posts, People's Rep. of China

Dahsiung, Hsu, Beij ing In st. of Posts, Peo ple 's Rep. of China

Damer, Cynthi a, AD 2000, Inc. , USA

Dandliker, Rene, Universite De Neuchatel, Switzerland

Dausmann, Gunther, Holographic Systems Muenchen GmbH, Germany

Davis, Ernie, MWK Industries, USA

Davis, Gene, Fast Light Inc. , USA

Dayus, Ian, A.H. Prismatic, Ltd. , UK

de Roos, Marcus, Deep Space Holographics, Canada
DeBerry, Larry, Galaxies Unlimited, Inc. , USA
Deem, Rebecca, Deem, Rebecca, USA
Deem, Rebecca, Zone Holografix, USA
DeFillipo, Elmer, Lazer Wizardry, USA
DeFreitas, Frank, DeFreitas, Frank, USA
DeFreitas, Frank, Frank DeFreitas Holography Studio, USA
del-Prete, Sandro, Galerie Illusoria, Switzerland
Delvo, Pierino, CISE Tecnologie Innovative SpA, Italy
Denisyuk, Yuri N., Denisyuk, Yuri N., Russia
Desai, Yogesh, Spatial Holodynamics (India) Pvt. Ltd., India
Deutschman, Bill, Oregon Laser Consultants, USA
Deutschmann, Gunter, AHT 3D-Medien, Germany
Deutschmann, Gunter, Holographie Fachstudio Bad Rothenfelde, Germany
Diamond, Mark, 3-D Worldwide Holograms, Inc. , USA
Diamond, Mark, Diamond Images, Inc ., USA
Dietrich, Edward, OpSec - Corporate Headquarters, USA
Dion, David, FoilMark Holographic Images, USA
Dondi, Alesandro, Diavy srl, Italy
Dowley, Mark, LiCONiX, USA
Dr. Windeln, Wilbert, ETA-Optik Gmbh, Germany
Dr. Beeck, M.-A., Volkswagen AG, Germany
Dr. Birenheide, Richard, HRT Holographic Recording Technologies GmbH, Germany
Dr. Leseberg, Detlef, Leseberg, Dr. Detlef, Germany
Dr. Schmelzer, Carlo, Studio Fuer Holographie, Germany
Dr. Spanner, Karl, Physik Instrumente (PI) GmbH & Co. , Gennany
Dreelaw, Dawn, International Data Ltd., UK
Drinkwater-Welch, Deni, Wave Mechanics, USA
Duesterberg, Richard, Vincennes University, USA
Duffey, William, The Regal Press Inc., USA
Duignan, Michael , Potomac Photonics, Inc. , USA
Dumra, Sumant, APA Optics Inc . , USA
Duplica, John, 21 st Century Finishing Inc., USA
Dutton, Keith, Laser International, UK
Easterlang, Lund, Imagen Holography, Inc., USA
Edgar, John, Brandtjen & Kluge, Inc." USA
Edhouse, Simon, Australian Holographics, Australia
Erickson, Ronald R., Media Interface, Ltd., USA
Evan, Alan, M.O.M. Inc . , USA
Faddis, Terry, Optimation Holographics, USA
Farina, Joseph A., Laser Holography Workshop, USA
Fattal, Isaac, Krystal Holographics International Inc., USA
Fee, Renee, Holografica, USA
Feinberg, Jack, University Of Southern California, USA
Feingold, M., P.S.A Peugeot Citroen, France
Feitisch, Alfred, Spectra-Physics Lasers Inc., USA
Felix, Patricia, Holograms and Lasers Inter-

national, USA
Felix, Patricia, Holograms and Lasers International, USA
Felix, Perry, Holograms and Lasers International, USA
Felix, Perry, Holograms and Lasers International , USA
Fernandez, Nancy, Oriel Instruments, USA
Feroe, James, Feroe, James, USA
Ferris, Lorraine, 3-D Worldwide Holograms, Inc ., USA
Fimia., A., University Of Alicante, Spain
Fischer, Julian, HoloVision, Germany
Fischler, Ben, Imagination Plantation, USA
Fisher, Gary, Man/Environment, Inc ., USA
Fitzpatrick, Colleen, Rice Systems, USA
Florence, Mike, James River Products, USA
Ford, Richard, Courtauld Acetate, UK
Formosa, Joe, Van Leer Metallized Products, USA
Fornari, Arthur David, Fornari, Arthur David, USA
Forsberg, Mona, HoloMedia Ab/Hologram Museum. , Sweden
Fournier, Jean-Marc, Rowland Institute For Science, USA
Francois, Laurent, H.Ologrammi, France
Frankmark. , Robert, Volvo-Flygmotor, Sweden
Frieb, M.T. , Holographie Anubis, Germany
Fries, Urs, Holo-Service.fries, Switzerland
Frisk, E.O., Optical Works Ltd., UK
Fukuda, Akinobu, SAM Museum, Japan
Fukuma, Mineko, Japan Communication Arts Co. , Japan
Fuller, Mary, Foil Stamping and Embossing Association, USA
Gabrielson, Dan, Pennsylvania Pulp & Paper Co. , USA
Gallagher, Dan, Total Register Inc., USA
Gallagher, John, Total Register Inc. , USA
Galon, Derek, Royal Holographic Art Gallery, Canada
Garcia, Diego, M.I.T. Museum, USA
Garrett, Jenny, Silver Dragon Holography, USA
Garrett, Steve, Midwest Laser Products, USA
Gauchet, Pascal, Atelier H~lographique De Paris, France
(}aynor, Joseph , Innovative Technology Associates, USA
Gibb, Don, Ion Laser Technology, Inc., USA
Gibb, Jim, Automated Holographic Systems, USA
Gibson, J.A. , Ag Electro-Optics Ltd., UK
Gillespie, Don, EI Don Engineering, USA
Gillespie, Mike, Jodon Inc . , USA
Ginouves, Paul, Coherent, Inc. - Laser Group, USA
Glaser, Shelly, Glaser - Technical Consulting, Israel
Glazer, Stewart, Crown Roll Leaf, Inc." USA
Gnatovskii , Alexander, Environmental Education and Information Ctr., Ukraine
Goldstein, Robert, Lasermetrics, Inc., USA
Golen, VP, Grace, Holographic Dimensions, Poland S.A. , Poland
Gorglione, Nancy, Cherry Optical Holography, USA
Gorglione, Nancy, Gorglione, Nancy, USA
Gorglione, Nancy, Laser Affiliates, USA
Gougeon, Pierre, Dimension 3, Canada
Graham, Ben, Lexel Laser, Inc. , USA
Green, John, LOT Oriel, UK
Greguss, Pal, Optopol Panoramic Metrology Consulting, Hungary
Grichine, Mike, Geola, Lithuania

Gunther, John E., Hughes Power Products, Inc. , USA
Gustafsson. , Jonny, Holovision AB, Sweden
Gutekunst, Horst, Hologramm Werkstatt & Galerie, Switzerland
Guy, Desmet, Multifacet, Belgium
Haines, Debbie, Simian Co., USA
Halkes, Adrian J., Far East Holographics, Hong Kong
Hall, Patricia M. , Hallmark Capital Corp., USA
Halliwell, N., Loughborough Univ. Of Tech., UK
Haney, Lorri, Nimbus Manufacturing, Inc. , USA
Hankin, Alan, Bridgestone Graphic Technologies, Inc ., USA
Harden, Jim, Light Wave Gallery, USA
Harding, Kevin, Industrial Technology Institute, USA
Hardy, Nick, Op-Graphics (Holography) Ltd., UK
Harris, Ken, Dimensional Arts, USA
Harris, Ken, HoloCom, USA
Harrison, Ann Marie, Intrepid World Communications, USA
Hartman, John, Batelle Pacific Northwest National Laboratory, USA
Hashimoto, Chikara, Central Glass Co., Ltd., Japan
Hashimoto, Reiji, Topcon Inc ., Japan
Haskell, Dara, Third Dimension Arts Inc. , USA
Hassen, Chuck, Holo Sciences, LLC, USA
Hatton, Keith, Checkpoint, UK
Heck, David, Spectra-Physics Lasers Inc., USA
Heil, Wendy, Advanced Optics, Inc. , USA
Hein, Elke, Die Dritte Dimension, Germany
Hell, Mikael, Martinsson Elektronik Ab. , Sweden
Hennigan, Jill, Omnichrome Corp. , USA
Hepburn, James, Silverbridge Group, Canada
Herman Weil, Lektor H., Studio Weil- Alvaron, Sweden
Herr, Doug, Bobst Group, USA
Hess, Bob, Point Source Productions, USA
Hill, Dean, OpSec - USA, USA
Hillard, Bill, NeoVision Productions, USA
Hinz, Daniel, Melles Griot GmbH, Germany
Hoefer, Dan, American Laser Corporation, USA
Hoffstadt-Braeutigam, Irmhild, topac GmbH, Department Holography, Germany
Holden, Laurence, Astor Universal Ltd., UK
Hollinsworth, T.R., Expanded Optics Limited, UK
Holloway, Beverley, University Of Rochester, USA
Horn, Rolf, HoloMedia Ab/Hologram Museum., Sweden
Horne, Peg, Scharr Industries, USA
Horvath, Josef, Czechoslovak Academy Of Science, Czech Republic
Hoskins, Gregory, Robert Sherwood Holographic Design, USA
Hsu, Jonathan, Holo Impressions Inc, Taiwan
Hudson, Phillip M.G., De La Rue Holographics Ltd. , UK
Huff, Marilyn, Amagic Technologies Inc., USA
Huff, Ph.D., Lloyd, University Of Dayton, USA
Hurst, Andrew, Pilkington Optronics, UK
Hurwitz, Noah, Imagination Plantation, USA

Hwang, Edward, Superbin Co. Ltd, Taiwan

Hyttinen, Ikka, Ad-Holograms Oy, Finland

ide, Makoto, Nippon Polaroid K.K. , Japan

Inagaki, Takehumi, Fujitsu Laboratories Ltd., Japan

Ineichen, Beat, Stoltz Ag, Switzerland

Infantes, Mrs., Karas Studios S.L. , Spain

Inoue, Yutaka, Brainet Corporation, Japan

Iovine, John, Art Lab, USA

Ishihara, Dr. Satoshi, Ministry Of International Trade, Japan

Ishii, Setsuko, Ishii, Ms. Setsuko, Japan

Ishikawa, Kazue, Sophia University, Japan

Iverson, Mark, Datacard Corporation, USA

Iwata, Fujio, Toppan Printing Co. , Ltd. , Japan

Iwata, Fujio, Toppan Printing Co., Ltd. , Japan

Jain, Hemant, Coated Specialties, India

James, Randy, Pacific Holographics Inc., USA

Jamison, Pamela, Light Impressions International, Ltd. , USA

Jeong, T.H. , Integraf, USA

Jeong., T.H. , Lake Forest College, USA

Jerit, John, American Paper Optics Inc., USA

Jiang

Jiang, Prof. Yaguang, China Ann Arbor Holographical Institute, People 's Rep. of China

Jimenez-Ceniceros, Antonio, JimenezCeniceros, Antonio, Mexico

Johann, Larry, Southern Indiana Holographics, USA

Jorgensen, Dean, Optimation, USA

Jung, Dieter, Academy of Media Arts Cologne, Germany

Junger, Mr. , CHIRON Technolas GmbH, Germany

Jurewicz, Arlene, Synchronicity Holograms, USA

Kan, Mike, Kan, Mike, USA

Kane, Brian, General Design, USA

Karaganova, Svetlana, Australian Holographics, Australia

Kasprzak, Henryk, Technical University Of Wroclaw. , Pol and

Kassover, Kathy, Crown Roll Leaf, Inc." USA

Katsuma, Hidetoshi, Tama Art Umversity, Japan

Kauffman, John, Kauffman, John , USA

Keilholz, Mr. , Spindler & Hoyer GmbH & Co., Germany

Kelem, Marty, Spectratek Inc. , USA

Keller, Manuela, Steuer KG GmbH & Co. , Germany

Kelly, Ed, Keystone Scientific Co. , USA

Kendall, Christen, Metrologic Instruments, Inc., USA

Kendall, M. , Kendall Hyde Ltd. , UK

Kenny, Mike, MWK Industries, USA

Kerpen, Uta, Fielmann-Verwaltung KG, Germany

Kettel, Klaus, Krystal Holographics Vertriebs-GmbH, Germany

Kilpatrick, Jack, Laser Resale Inc., USA

King, R. Eric, Laser Innovations, USA

Kinoshita, Dr. Kenji, Toyama National College Of Marit, Japan

Kirk, Ronald L. , Holovision Systems Inc., USA

Klix, Armin, Armin Klix Holographie, Germany

Knuth, Eckard, Laserfilm Eckard Knuth, Germany

Koch, Mr. , Optical Coating Laboratory GmbH, Germany

Koizumi, Fumihiko, Asahi Glass Co., Japan

Kontnik, Lewis, Reconnaissance International Ltd. , USA

Kooi, A. , Optilas BV, Netherlands

Koril, Gary, Creative Label, USA

Koril, Jerry, Creative Label, USA

Kornienko, Sergey, Feofaniya Ltd ., Ukraine

Korradai, Giorgio, Autoadesivi Sri, Italy

Kottova, Alena, Ontario Sci ence Centre, Canada

Kreischer, Cody, Kreischer Optics, Ltd., USA

Krick, Reva, Reva 's Holographic Illusions, USA

Krick, Robert, Reva 's Holographic Illusions, USA

Kritzinger, Sean, Synchron Pty Ltd. , South Africa

Krueger, Dave, Holograms International, USA

Krueger, Jean, Holograms International, USA

Kuntz, David, Technical Marketing Services, USA

Kuwayama, Tetsuro, Canon Inc . R&D Headquarters, Japan

Labelle, Scott, Holographic Label Converting (HLC), USA

Lacey, Lee, Holo/Source Corporation, USA

Lancaster, Ian, Reconnaissance International Ltd. , UK

Langer, W. , Dornier Medizintechnik GmbH~ Germany

Lansing, Joseph, Electro Optical Industries, Inc. , USA

Larim, Jim, Laser Optics; Inc., USA

Larson, Ann, Laser Images, USA

Larson, Steve, Laser Images, USA

Lauder, Dea, American Propylaea Corporation, USA

Lauk, Mathias, Lauk & Partner GmbH, Germany

Law, Linda, Linda Law Holographics, USA

LeCompte, Joe, Virtual Image (a di vision of Printpack, Inc.), USA

Lee, David, Voxel, USA

Lefloc 'H, c., Aerospatiale, France

Leith, Emmet, University Of Michigan, USA

Lekki, Walt, Optical Corporation Of America, USA

Lembessis, Alkis, Hellenic Institute Of Holography, Greece

Lembessus, Alkis, Cavomit, Greece

Lessard, Roger A. , Universite Laval, Canada

Lev, Steven, Chromagem Inc., USA

Levine, Chris, iC Holographics, UK

Levine, Jeff Jeffrey, Mesmerized Holographic Marketing, USA

Levy, Rob, Holo/Source Corporation, USA

Levy, Uri, Holo-Or Ltd, Israel

Liberato, Pablo, Evolution Design , Inc., USA

Lieberman, Dan, Hol ogramas, S.A. de C.Y., Mexico

Lieberman, Dan, HoloWebs, Inc ., USA

Lieberman, Larry, Larry Lieberman Holography, USA

Liegeois, Dr. Christian D. , X-IAL, France

Lifshen, Alan, Lone Star Illusions, USA

Lind, Michael, Batelle Pacific Northwest National Laboratory, USA

Lion, Yves F., Univ. de Liege, Belgium

Lissack, Selwyn, Laserworks, USA

Liu, Michael , MasterPrint Holography, Inc., USA

Liu, Wai-Min, Control Optics, USA

Lkegami, Dr. Koji, Numazu College Of

Techno logy, Japan

Long, Mike, Pacific Holographics Inc. , USA

Lopez, Argelia, Elusive Image, USA

Lopez, Jesus, Lopez's Gallery International, USA

LoSardo, Sal, Towne Technologies, USA

Love, Valerie, Op-Graphics (Holography) Ltd., UK

Lovygin, Igor, Technoexan Ltd, Russia

Lucy, Thomas, Holo GmbH, Germany

Luton, Chris, Holocrafts Europe Limited ., UK

MacArthur, Ana, Ana MacArthur, USA

MacShane, Jim, MacShane Holography, USA

Madsen, Henrik, Ibsen Micro Structures N S, Denmark

Magarinos, Jose R. , Holographic Optics Inc. , USA

Malmqvist, Sven, Saab-Scania, Sweden

Malott, Michael, Laser Light Designs, USA

Mann, Harry, Van Leer Metallized Products, USA

Manrubia, Bob, Uni stay, USA

Margolis, Mark, Rainbow Symphony Inc., USA

Markov, Vladimir B., The Institute of Applied Optics, Ukraine

Marks, Gerald, Marks, Gerald, USA

Marks, Gerald, PullTime 3-D Laboratories, USA

Martinez, Guillermo, Evolution Design, Inc., USA

Masamori, Ichiro, Mazda Motor Corp., Japan

Maslenkov, Michael, Technoexan Ltd, Russia

Massuda, Elle, Harvard Apparatus, Canada, Canada

Mathieu, Marie-Christiane, Les Productions Hololab , Canada

Mazzero, Francois, H.Ologrammi, France

Mazzola, Karen, United Association Manufacturer's Representatives, USA

McCain, Richard, McCain Marketing & Graphic Design, USA

McCarthy, Kevin, Laser Media, Inc., USA

McCormack, Sharon, Sharon McCormack Holography, USA

McGarry, Dan, Rochester Photonics Corporation, USA

McGaw, Trevor, Holograms Fantastic and Optical Illusions, Australia

McGrew, Steve, New Light Industries, USA

McKay, Dave, Holographic Impressions, USA

McKay, Dave, Smith & McKay Printing Co. Inc. , USA

McLeod, Don, Corion Corp., USA

McMahan, Robert, McMahan Electro-Optic, USA

Medford, Amy E., Avant-Garde Studio, USA

Mendoza, Ph.B. , Fernando, Centro de Investigaciones en Optica, A.C., Mexico

Merritt, Dave, Louis Paul Jonas Studios, Inc., USA

Metz, Michael , ImEdge Technology, USA

Meulien, Odile, Artbridge Light Studios, Germany

Meulien, Odile, Meulien Odile, Germany

Meuse, Ron, Mu's Laser Works, Canada

Meyer, Steve, MGM Converters Inc., USA

Meyrueis, P., Photonics Systems Laboratory, France

Mikes, Thomas, American Holographic Inc., USA

Miller, Doug, Krystal Holographics International Inc., USA

Mistry, Rohit, Jayco Holographics, UK

Mitamura, Shunsuke, University Of

Tsukuba, Japan
Mitchell, Astrid, Applied Holographics, PIc ., UK
Mizuno, Toru, Nippondenso Co., Ltd., Japan
Mofchetti, Tanis, Uniphase Lasers, USA
Molin, Nils-Erik, Lulea University Of Technology, Sweden
Monaghan, Brian, Pennsylvania Pulp & Paper Co. , USA
Monberg, Ed, Laser and Motion Development Company, USA
Monchak, Alexander, Tair Hologram Company, Ukraine
Moore, Lon, Red Beam, Inc., USA
Moree, Sam, New York Holographic Laboratories, USA
Morrison, Dan, Laser Technical Services, USA
Morterud, Alan P. , Imagen holography, Inc., USA
Mortier, Frank, Agfa - Gevaert N.V. , Belgium
Mouroulis, P. , Rochester In s!. Of Technology, USA
Mowforth, Peter, Turing In stitute, UK
Mueller, Joachim, Gresser, E. , KG, Germany
Mulvaney, Mark, Letterhead Press, Inc. , USA
Munday, Rob, Spatial Imaging Ltd., UK
Munzer, Hubert, OWlS Gmbh, Germany
Murata, M., Mitsubishi Heavy Industries Ltd., Japan
Murphay, Toicia, Silhouette Technology Inc. , USA
Murray, Jeffery, Jeffery Murray Custom Holography, USA
Murray, Jeffrey, Holography Institute of San Francisco, USA
Murray, Jeffrey, Laser Arts Society For Education and Research, USA
Murray, Maria, Inrad, Inc. , USA
Muth, August, Lasart Ltd. , USA
Naaman, Bill, Mitutoyo Measuring Instruments (MT! Corp.), USA
Naeve, Ambjorn, Dialectica Ab, Sweden
Nakajima, Dr. Masato, Keio University, Japan
Neister, Ed, Lumenx Technologies, Inc., USA
Nelson, Drew, Stanford University, USA
Neu, Martha, Polaroid Corporation, USA
Nevin, Karen, Diffraction Ltd. , USA
Newman, John, Laser Technology, Inc ., USA
Ninomiya, Shinji, Kimmon Electric Co., Ltd. , Japan
Nishioka, Teiichi , Toppan Printing Co., Ltd. , Japan
Noble, Marcus, Technical Marketing Services, USA
Noems, Benny, Metrologic Instruments GmbH, Germany
Norman, Kenneth, Control Module Inc. , USA
Norton, Dan, Polymer Image, USA
O'Brien, Roger, Holotek, USA
O'Connor, Tom, Lasersm ith, Inc.(The), USA
Odhner, Jefferson E. , Odhner Holographics, USA
Oehlmann, Dietmar, Di etmar Oeh lmann, Germany
Oishi, Mariko, Light Dimension, Inc ., Japan
Okada, Dr. Katsuyaki, HOmC Holographic Display Artists & Engineers Club, Japan
Olmo, Anthony, 21 st Century Finishing Inc. , USA

Olson, Bernadette, Laser Reflections, USA
Olson, Ron, Laser Refl ec tions, USA
Orr, Edwina, Richmond Holographic Studios, UK
Orszag, Alain, Quantel, France
Osada, RB, Fantastic Holograms, USA
Oteri, Lance, Holographic Label Converting (HLC), USA
Owen, Harry, Kaiser Optical Systems, Inc., USA
Page, Michael , Ontario College Of Art, Canada
Pahnke, Roland, Kolbe-Druck mit Tochtergese llschaften, Germany
Paletz, Jim, Hologram World, Inc. , USA
Palmer, Christopher, Richardson Grating Laboratory, USA
Pargh , Jonathan. Three-D Light Gallery, USA
Parker, Bill, Diffraction Ltd. , USA
Parker, Dr. Steve, British Aerospace Pic., UK
Parker, Ric, Roll s-Royce PIc, UK
Patterson, Rich, Reynolds Meta ls Co., USA
Patters son, Sven-Goran, Lund Institute Of Tech. , Sweden
Paxton, Chuck, Photon Cantina Ltd., USA
Payne, Patty, Burleigh Instruments, Inc., USA
Pecheux, Patrice, Laser Movement, France
Pepper, Andrew, Creative Holography Index, The, UK "
Pepper, Andrew, Pepper, Andrew, UK
Perry, John, Holographics North Inc ., USA
Petersen, Joel, Wavefront Technology, USA
Peterson, Jeff, Inside Finishing Magazine, USA
Pfiel, Larry, Dimensional Arts, USA
Phillips, Alan J., Action Tapes, UK
Phillips, Jacque, Direct Holographics, USA
Phillips, Ronald, Interact ive Industries Inc. , USA
Pierce, Robert, Oregon Institute of Technology, USA
Pink, Patty, Pink, Patty, USA
Plotnick, Harvey, Laser Media, Inc ., USA
Poe, Nelson, Blue Ridge Holographics, Inc ., USA
Powell , Dick, University .of Arizona, USA
Price, Stu, Shipley Chemical Co., USA
Pricone, Robert, Holographic Industries, Inc ., USA
Prof. Dr. Eichler, Hans Joachim, Technische Universitaet Berlin, Gern1any
Prof. Dr. Eichler, Juergen, Techni sche Fachhochschule Berlin, Germany
Prof. Juptner, Werner, BIAS, Germany
Provence, Steve, Blue Ridge Holographics, Inc. , USA
Pryputniewicz, Ryszard, Worcester Polytechnic Institute, USA
Quanhong, Wang, Beijing Hologram Printing Technology Co, People 's Rep. of China
Quinn, Ken, Springer-Verlag New York, USA
Radi, Filippo, Autoadesivi Sri , Italy
Rallison, Richard, Ralcon, USA
Randazzo, Dean, Holographic Images Inc ., USA
Rankin , Kevin, Omnichrome Corp. , USA
Rayfie ld, Dave, Krystal Holographics International Inc ., USA
Redzikowski , Mark, Agfa - a divi sion of Bayer Corp., USA
Reichert, Uwe, 3D Vision, Germany
Reinhart, Werner, Leonhard Kurz GmbH, Germany
Rezny, Abe, Laser Light Ltd. , USA
Rhody, Alan , Holography Marketplace,

USA
Rhody, Alan, Ross Books, USA
Rich, Chri s, Wavefront Technology, USA
Richardson, Martin, The London Holographic Image Studio, UK
Rickert, Sue, Hologram Land, USA
Rincon, Angelika, Hologramas, S.A. de c.v., Mexico
Rizzi, M. Luciana, Cise Spa Technologie Innovati ve, Italy
Robb, Jeffery, Spatial Imaging Ltd., UK
Roberts, Judy, JR Holographics, USA
Roberts, Terry, 3D Optical Illusions, Australia
Robinson, David, National Physical Laboratory, UK
Robinson, Deborah, Lightrix, Inc., USA
Robinson, George, Hologram Land, USA
Robur, Lubomir, Feofaniya Ltd., Ukraine
Rodia, Carl M. , Carl M. Rodia And Associates, USA
Roedianto, Rendy, Holography Division, Indonesia
Ross, Franz, Holography Marketplace, USA
Ross, Franz, Ross Books, USA
Ross, Jonathan, Holograms 3D, UK
Ross, Michael, IBM Almaden Research Center, USA
Rossing, Thomas, Northern Illinois University, USA
Roule, Richard, ST!, USA
Rueck, A. B., AB Rueck Holoart, Germany
Ruey-Tung, Hnng, Ruey-Tung, Miss. Hung, Japan
Ryden, Hans, Karolinska Institutet, Sweden
Saarinen, Jyrki, Heptagon Oy, Finland
Sakai, Miss Tomoko, HODIC Holographic Display Artists & Engineers Club, Japan
Santis, Paolo De, Universita Di Roma, Italy
Sapan, Jason, Holographic Studios, USA
Sarda, Uwe, topac GmbH, Department Holography, Germany
Sato, Shunichi, Sharp Corp. , Japan
Saxby, Graham, Graham Saxby, UK
Schaefer, Rick, Clemenger Perth Fry Ltd, Australia
Scheir, Peter, AD 2000, Inc., USA
Scher, Marcy, 3D Holograms Inc ., USA
Scher, Phil, 3D Holograms Inc ., USA
Schipper, Wilfried, Hologram Company RAKO GmbH, Germany
Schlewitt, Carsten, Optical Test Equipment, Germany
Schrieber, Matthew, Holographic Images Inc., USA
Schumann, Walter, Swiss Federal Inst Of Technology, Switzerland
Schwartzman, Frederic, Foreign Dimension, Hong Kong
Schweer, Joerg, Holoptics, Germany
Schweitzer, Dan, New York Holographic Laboratories, USA
Schwider, J., University of Erlanger, Germany
Sciammarella, Cesar, Illinois In stitute Of Technology, USA
Seitz, Mr. , Steuer KG GmbH & Co., Germany
Sekulin, Robert, Rutherford & Appleton Labs, UK
Shafer, Brad, Engineering Animation, Inc. , USA
Shah, Kailesh, Ojasmit Holographics, India
Shahjahan, Mr., Dimensions, Pakistan
Sharma, Govind, Holostik India Pvt. Ltd. , India
Sharpe, Frank, Datasights Ltd., UK
Sherwood, Robert, Robert Sherwood Holographic Design, USA
Shie, Rick, Physical Optics Corporation. ,

USA

Shimon, Hameiri, Holography Israel, Israel

Shun, Prof. Zhu De, Shandong Academy of Sciences, People's Rep. of China

Shvarts, Dr. Kurt, Physics Institute. Latvian, Latvia

Sikorsky, Zbigniew, Institute Of Plasma Physics, Poland

Simson, Bernd, General Holographics, Inc. , Canada

Simson, Paula, General Holographics, Inc., Canada

Singh, Ravinder, Print-M-Boss, India

Siveriver, Leonid, Avant-Garde Studio, USA

Sivy, George, Richmond Development Group, USA

Skipnes, Olav, Interferens Holografi D.A., Norway

Smelzer, Geert T. A., Technical University @ Eindhoven, Netherlands

Smith, Carol, Laser Drive Inc. , USA

Smith, S.D., Beddis Kenley (Machinery) Ltd., UK

Smith, Steven L. , Lasersmith, Inc.(The), USA

Soales, Bob, CVI Laser Corporation, USA

Soares, Oliverio, Universidade Do Porto, Portugal

Soewandi, Edy, Pt. Pura Nusapersada, Indonesia

Song, Chung, Dan Han Optics, South Korea

Song, Li, Laser Inspeck, Canada

Sott, Gudrun, AKS Holographie-Galerie GmbH, Germany

Souparis, Hughes, Hologram Industries, France

Sowdon, Michael, Fringe Research Holographics, Canada

Spiegel, Gary, Newport Corporation , USA

Spierings, Walter, Dutch Holographic Laboratory BV, Netherlands

Spina, Thomas, Luminer Printing and Converting, USA

Sponsler, Michael B., Syracuse University, USA

St. Cyr, Suzanne, Holographic Applications, USA

Starcke, Ari-Veli, Starcke, Ky. , Finland

Stehle, Robert, Sopra, France

Steinbichler, H., Steinbichler Optotechnik GmbH, Germany

Steinfeld, Belle, Dell Optics Company, Inc., USA

Stelter, Manfred, Process Technologies, USA

Stephens, Anait Arutunoff, Stephens, Anait, USA

Stepien, Pawel, Hololand S.C., Poland

Stich, Boguslaw, Holografia Polska, Poland

Stockier, Len, OIE Research, USA

Stockton, John, Tamarack Storage Devices, USA

Stone, Thomas, Wavefront Research, Inc., USA

Stooss, Richard, Krystal Holographics Vertriebs-GmbH, Germany

Styns, Erik, Free University Of Brussels., Belgium

Su, Dr.].1. , Industrial Technology Research Inst. , Taiwan

Sugarman, Stephen, Holographic Products , USA

Surana, Rajendra, Ojasmit Holographies, India

Sutaria, D.K., Better Labels Mnfg Co, India

Svensson, Lennart, Royal Institute Of Technology, Sweden

Sweeney, Dr. Eugene, BTG pic, UK

Swinehart, Patricia, CFC Applied

Holographics, USA

Synowiec, George, Lumonics Ltd., UK

Taylor, Rob, Forth Dimension Holographies, USA

Taylor, Tom, Direct Holographies, USA

Thiemon, Ms., Daimler Benz Aerospace, Germany

Tholen, Maureen, 3M - Safety and Security Systems, USA

Thoma, John, Advanced Holographic Laboratories, USA

Thomas, Jackie, Laser Institute Of America, USA

Thompson, Bridget, iC Holographics, UK

Thuston - Lighty, Cathy, M.LT. Museum, USA

Thwaites, Hal M., Concordia University, Canada

Tidmarsh, David, Applied Holographics, Pic., UK

Tiemon, M., Dornier Medizintechnik GmbH, Germany

Titizian, Lia, Optical Research Services, USA

Tiziani, Hans, University of Stuttgart, Germany

Tobin, John, Moonbeamers, Australia

Toland, Lee, Meredith Instruments, USA

Tolia, Dr. Arun, Spatial Holodynamics (India) Pvt. Ltd., India

Tong, Chen Guo, Morning Light Holograms, People 's Rep. of China

Townsend, Patrick" Navidec Inc . (formerly ACI Systems, Inc.), QSA

Trayner, David, Richmond Holographic Studios, UK

Tribillon, Dr.Jean Louis, Holo-Laser, France

Trolinger, James D. , MetroLaser, USA

Tsujiuchi. , Jumpei, Chiba University, Japan

Tuffy, Francis, Advanced Holographic Laboratories, UK

Tunnadine, Graham, 3D-4D Holographics, UK

Turnage, Mark, OpSec - England, UK

Tyler, Doug, Saint Mary's College, USA

Tzong, Tang Yaw, Institute Of Optical Science, Taiwan

Unbehaun, Klaus, Holopublic Unbehaun, Germany

Unterseher, Fred, Unt~rs eher & Associates, USA .

Unterseher, Fred, Zone Holografix, USA

Upatnieks, Juris, Applied Optics, USA

Uram, Marvin, U.K. Gold Purchasers, Inc. , USA

Urgela, Stanislav, Technical University Zvolen, Slovakia

Uwe, Saurda, Holographie Labor, Germany

Uyemura, T. , University Of Tokyo, Japan

Valdivia, Allison, Optics Plus Inc., USA

Van Renesse, Ruud L., TNO Institute of Applied Physics, Netherlands

Varga, Miklos, Hologram Varga Miklos, Hungary

Varney, Chris, Electro Optics Developments Ltd. , UK

Venkateswaran, Sagar, Peacock Laboratories, Inc., USA

Vikram, Chandra, Unive rsity Of Alabama, USA

Vila, Doris, Doris Vila Holographics, USA

Vogel, Jon, Holographics (Uk) Ltd. , UK

von Bally, Gert, University of Muenster, Germany

Vukicevic , Dalibor, Trend, Yugoslavia

Vulcano, Marino, Conte M. Vulcano, Italy

Wada, Takashi, Dai Nippon Printing Co., Ltd. , Japan

Wahlberg, Bjorn, Swede Holoprint, Sweden

Wale, R. D. , Galvoptics Ltd. , UK

Walters, Glenn J., A.D. Tech (Advanced Deposition Technologies), USA

Wanlass, Mike Mike, Spectratek Inc. , USA

Wanyun, Huang, Beijing Normal University., People 's Rep. of China

WappeJt, Andreas, Andreas Wappelt - Photonics Direct, Germany

Wappelt, Andreas, Technische Universitaet Berlin, Germany

Warczynski, Ronald, Witchcraft Tape Products, Inc., USA

Warneski , John, Jodon Inc. , USA

Wegeler, Marc, Lauk & Partner GmbH, Germany

Weil, Jeffrey, Holographic Dimensions, USA

Weinstein, Beth, New York Hall Of Science, USA

Wen, Pei, Beijing Sanyou Laser Images Co, People's Rep. of China

Wesly, Ed, Art Institute Of Chicago (The School of the ...), USA

Wesly, Ed, USA

White, John, Coburn Corporation, USA

White, Steve, Electro Optical Industries, Inc., USA

Wilbur, Fred, Elusive Image, USA

Williams, Sareth, San Jose State University, USA

Wilson, Brett, Lazart Holographics, Australia

Witt, Martin, The HOLOS Corporation, USA

Wober, Irmfried, Holography Center of Austria, Austria

Wollenweber, Andrea s, Magick signs Holografie, Germany

Woodward, Keith, American Bank Note Holographics, USA

Woolford, James, Holographic Dimensions, Panama

Wootner, Marc 0., Transfer Print Foils, Inc., USA

Wyant, James, Wyko Corporation, USA

Yamaguchi, Masahiro, Tokyo Institute Of Technology, Japan

Yamaguchi, Mashahiro, Tokyo Institute of Technology, Japan

Yamazaki, Hitoshi, Hyogo Prefectual Museum of Modern Art, Japan

Yang, G., University Of Oxford, UK

Yarovoy, Leonid, Yarovoy, Leonid, Ukraine

Yokota, Hideshi, Tokai University, Japan

Yoon, Melissa, M.LT. (Massachusetts Institute of Technology), USA

Yoshikawa, Dr. Hiroshi, Nihon University, Japan

Zafar, M. , AK Rehman Traders, Pakistan

Zellerbach, Gary, Holart Consultants, USA

Zhiwei, Liu, Beying Fantastic Hologram Product Corp; People's Rep. of China

Zucker, Richard, Bridgestone Graphic Technologies, Inc., USA

Zurek, Mr. , University of Munich, Germany

To preview all of our holography publications:

Visit Our Website
http://www. holoinfo.com

HOLOGRAPHY MARKETPLACE Editions 1-6
HOLOGRAPHY HANDBOOK - *Making Holograms the Easy Way* [Revised 1996]

Featuring pictures of all the collectable holograms included in our books.
For your convenience, orders may be placed over the Internet.

SPECIAL PACKAGE FOR OUR READERS

a **HOLOGRAPHY HANDBOOK** and a **HOLOGRAPHY MARKETPLACE***

Delivered to your door (shipped UPS or Air Mail):
$65 in Continental USA
$75 Alaska, Hawaii, Canada, Mexico
$85 all other countries

TO ORDER:
Call toll free 1-800-367-0930 in USA or phone 1-510-841-2474.
Or fax 1-510-841-2695, email sales@rossbooks.com, or mail to :
ROSS BOOKS, P.O. Box 4340 Berkeley, CA 94704

Include check, money order, or credit card information (card number, name on card, expiration date),
"Ship to" address and contact phone number.

We will ship as soon as your order is received and processed.   * Editions 2, 3, 4, 5, or 6

ATTENTION BOOK COLLECTORS AND HOLOGRAPHY AFICIONADOS

The last remaining copies of HOLOGRAPHY MARKETPLACE **1st** EDITION (1989) are available for sale in their original sealed package for $100 each. Each book features a 4" x 5" MIRAGE ™ polymer hologram from Polaroid (showing the inside of a camera) as well as historically significant information. Supplies very limited.

10

Cross-Index Tables

Businesses That Sell Holograms						
Country	Business Name	Wholesale	Retail	Catalog	Museum /Gallery	Touring Show
Australia	3D Optical Illusions	X		X		
Australia	Australian Holographics	X		X		
Australia	Lazart Holographics	X	X	X	X	
Australia	New Dimension Holographics	X	X			
Australia	Parallax Gallery		X	•		
Austria	Holography Center of Austria	X		X		X
Canada	Claudette Abrams	X	X			
Canada	Fringe Research Holographics		X		X	X
Canada	General Holographics, Inc.	X		X		
Canada	Holocrafts	X	X	X		
Canada	Melissa Crenshaw Studio	X	X			
Canada	Ontario College Of Art				X	
Canada	Ontario Science Centre				X	
Canada	Royal Holographic Art Gallery		X			
Canada	Silverbridge Group	X				
Canada	The Hologram Store, Ltd.		X			
Canada	Zero Gravity		X			
Germany	3D Vision	X	X	X		
Germany	AB Rueck Holoart	X				
Germany	AKS HolographieGalerie GmbH	X	X	X	X	
Germany	AKS HolographieGallerie GmbH		X			
Germany	Andreas Wappelt Photonics Direct	X		X		
Germany	Armin Klix Holographie	X		X		
Germany	Artbridge Light Studios					X
Germany	Curt Abramzik	X		X		
Germany	Deutsche Gesellschaft fur Holo					X
Germany	Die Dritte Dimension	X	X	X		
Germany	ETAOptik Gmbh	X		X		
Germany	HOL 3. Galerie fur Holographie		X			
Germany	Holographie Anubis	X		X		
Germany	Holoptics	X	X			
Germany	HoloVision	X		X		
Germany	Krystal Holographics	X		X		
Germany	Lauk & Partner GmbH				X	
Germany	Magick Signs Holografie	X	X	X		
Germany	Magick signs Holografic		X			
Germany	Phantastica		X			
Germany	topac GmbH.	X		X		
Hong Kong	Foreign Dimension	X		X		
Hong Kong	The Foreign Dimension		X	X		
Hungary	Hologram Varga Miklos	X				
Netherlands	Dutch Holographic Laboratory	X		X		
Poland	Holografia Polska	X		X		
Singapore	Zero Gravity		X	X		

Country	Business Name	Wholesale	Retail	Catalog	Museum /Gallery	Touring Show
South Africa	Synchron Pty Ltd.	X				
Ukraine	Tair Hologram Company		X			
UK	3D Images	X		X		
UK	A.H. Prismatic, Ltd.	X		X		
UK	Amazing World Of Holograms.	X	X	X		
UK	Boyd, Patrick	X	X			
UK	Holocrafts Europe Limited.	X		X		
UK	Holograms 3D	X	X			X
UK	Laza Holograms Ltd.	X		X		
UK	Margaret Benyon	X	X			X
UK	OpGraphics (Holography) Ltd.	X		X		
UK	Oxford Holographics	X	X	X		
UK	Pepper, Andrew	X	X			
UK	Richmond Holographic Studios	X		X		
UK	Spatial Imaging Ltd.	X		X		
UK	The London Holo Image Studio	X		X		
USA	3D Worldwide Holograms, Inc.	X				
USA	3D Holograms Inc.	X		X		
USA	A.H. Prismatic, Inc.	X		X		
USA	AD 2000	X		X		
USA	American Paper Optics Inc.	X		X		
USA	Ana MacArthur	X	X			
USA	Another Dimension	X		X		
USA	Art, Science & Technology Inst				X	
USA	Barilleau, Rene Paul					X
USA	Bellini, Victor			X		
USA	Berkhout, Rudie	X	X			
USA	Booth, Roberta	X	X			X
USA	CasdinSilver Holography	X	X			X
USA	Cherry Optical Holography	X	X	X		X
USA	Diamond Images, Inc.	X		X		
USA	Dimensional Arts	X				
USA	Dimensional Foods Co.	X				
USA	Direct Holographics	X		X		
USA	Doris Vila Holographics	X -				
USA	Elusive Image		X			X
USA	Fantastic Holograms		X			
USA	Fast Light Inc.	X				
USA	Forth Dimension Holographics		X			
USA	Galaxies Unlimited, Inc.		X			
USA	Gorglione, Nancy	X	X	X		X
USA	Holicon Corporation.	X				
USA	Holografica		X			
USA	Hologram Fantastic		X			
USA	Hologram Land		X			
USA	Hologram Research, Inc.	X		X		X
USA	Hologram World, Inc.	X		X		
USA	Holograms and Lasers Intnl		X	X		
USA	Holograms International	X	X	X		
USA	Holographic Design Systems	X		X		
USA	Holographic Images Inc.	X		X		
USA	Holographic Impressions	X		X		
USA	Holographic Industries, Inc.	X		X		
USA	Holographic Studios	X	X	X		
USA	Holographics North Inc.	X		X		
USA	Holography Presses On (HPO)	X		X		
USA	Holophile, Inc.					X
USA	Holography, Laser & Photonics Ctr				X	
USA	HOLOS Corporation	X		X		

Country	Business Name	Wholesale	Retail	Catalog	Museum /Gallery	Touring Show
USA	Hughes Power Products. Inc.	X		X		
USA	Imagen Holography, Inc.	X		X		
USA	Images Company		X	X		
USA	Integraf	X				
USA	Interactive Industries Inc.	X				
USA	Intrepid World Communications	X		X		
USA	Jeffery Murray Custom Holography	X	X	X		
USA	Kauffman, John	X	X			
USA	Krystal Holographics Intl	X		X		
USA	Larry Lieberman Holography	X	X	X		
USA	Lasart Ltd.	X		X		
USA	Laser Affiliates					X
USA	Laser Images	X		X		
USA	Laser Light Designs	X				
USA	Laser Reflections	X	X	X		
USA	Lasersmith, Inc.(The)	X		X		
USA	Lazer Wizardry	X		X		
USA	Light Impressions International.	X				
USA	Light Wave Gallery		X	X		
USA	Lightri	X		X		
USA	Lone Star Illusions		X			
USA	Lopez's Gallery International	X	X	X		
USA	M.I.T. Museum		X		X	
USA	MacShane Holography	X				
USA	Multiplex Moving Holograms	X	X	X		
USA	Museum Of Holography/Chicago		X		X	
USA	NeoVision Productions					X
USA	New York Hall Of Science				X	
USA	New York Holographic Lab		X			
USA	Photon Cantina Ltd.	X		X		
USA	Pink, Patty	X	X			
USA	Planet 3D			X		
USA	Point Source Productions	X	X			
USA	Polaroid Corporation	X				
USA	Polymer Image	X				
USA	Rainbow Symphony Inc.	X		X		
USA	Ralcon	X				
USA	Real Image	X		X		
USA	Red Beam. Inc.	X		X		
USA	Reva's Holographic Illusions		X			
USA	Richard Bruck Holography	X		X		
USA	School Of Holography				X	
USA	Science Kit & Boreal Labs	X		X		
USA	Sharon McCormack Holography	X	X	X		X
USA	Southern Indiana Holographics	X				
USA	Star Magic		X	X		
USA	Star Magic		X	X		
USA	Star Magic		X	X		
USA	Star Magic		X	X		
USA	Stephens, Anait	X	X			
USA	The Hologram Company #X		X			
USA	The Hologram Company #2		X			
USA	The Hologram Company #3		X			
USA	The Hologram Company #4		X			
USA	The Hologram Company #5		X			
USA	The Hologram Company #6		X			
USA	The Hologram Company #7		X			
USA	Third Dimension Arts Inc.	X		X		
USA	ThreeD Light Gallery		X		X	
USA	U.K. Gold Purchasers. Inc.	X		X		
USA	Wave Mechanics		X			
USA	Zero Gravity		X			
USA	Zero Gravity		X			
USA	Zero Gravity		X			
USA	Zero Gravity		X			

Country	Business Name	Pre-Master: 3-D model	Pre-Master: Digital artwork	Master: Silver Halide	Master: Dichro-mate	Master: Photo-polymer	Emboss	Stereo-gram	Multi-Color	CW Laser	Pulse Laser
Mastering Businesses											
Australia	Australian Holographics			X					X	X	X
Australia	Holograms Fantastic		X	X			X	X	X		
Austria	Holography Ctr Austria		X	X				X	X	X	X
Canada	Deep Space Holo	X			X						
Canada	Dimension 3	X	X	X	X	X	X	X	X	X	
Canada	Holocrafts				X						
Canada	Laser Inspeck	X	X				X				
Canada	Melissa Crenshaw			X					X	X	
Canada	Mu's Laser Works			X							
France	H.Ologrammi		X	X				X	X	X	
Germany	AB Rueck Holoart	X	X	X							
Germany	Arbeitskreis Holo B.V.			X							
Germany	Holo GmbH	X	X				X				
Germany	Hologram Co.AKO						X				
Germany	Holo Sys. Muenchen	X	X	X							
Germany	Holographie Anubis										X
Germany	Holographie Labor	X	X	X							
Germany	Holoptics	X	X	X							
Germany	HoloVision	X	X	X				X	X	X	X
Germany	Krystal Holo Vertriebs	X	X			X					
Germany	Laserfilm Eckard Knuth			X				X		X	
Germany	Magic Signs Holografie	X	X	X			X				
Germany	Studio Fur Holographie	X	X	X							
Germany	Tech. Fachhochschule	X	X	X							
Germany	topac GmbH	X	X				X				
Hungary	Hologram Varga Miklos					X					
Hungary	Optopol Panoramic			X							
India	Ojasmit Holographics			X	X	X	X				
Lithuania	Geola			X							X
Mexico	Holo, S.A. de C.V.		X	X			X	X	X	X	
Netherland	Dutch Holo. Labs		X	X		X	X	X	X	X	
Panama	Holo Dimensions		X				X	X	X	X	
Poland	Holo Dimensions		X				X	X	X	X	
Thailand	Electro-Optics Lab				X	X					
Ukraine	Feofaniya Ltd.				X						X
UK	3D-4D Holographics			X						X	
UK	Advanced Holo Labs	X	X	X			X	X	X	X	
UK	Applied Holographic.	X	X	X			X	X	X	X	
UK	Boyd, Patrick			X				X		X	
UK	De La Rue Holo						X				
UK	Embossing Technology	X	X				X				
UK	Holocrafts Europe Ltd				X						
UK	Holographics (Uk) Ltd.		X	X			X				
UK	iC Holographics		X				X	X	X		
UK	Jayco Holographics		X				X	X			
UK	K.C. Brown Holo			X							X
UK	Op-Graphics			X						X	
UK	OpSec - England		X				X	X	X		
UK	Richmond Holographic			X							X
UK	Spatial Imaging Ltd.	X	X	X		X	X	X	X	X	X
UK	STI - Europe		X				X	X	X	X	
UK	The London Holo Imag			X				X		X	X
USA	3-D Worldwide Holo		X	X			X	X	X	X	
USA	Acme Holography	X	X	X			X	X	X	X	
USA	AD 2000	X	X			X	X	X	X		
USA	Advanced Holo Labs	X	X	X			X	X	X	X	
USA	Amagic Technologies Inc.	X	X				X	X	X	X	

Mastering Businesses

Country	Business Name	Pre-Master: 3-D model	Pre-Master: Digital artwork	Master: Silver Halide	Master: Dichromate	Master: Photopolymer	Emboss	Stereogram	Multi-Color	CW Laser	Pulse Laser
USA	Amagic Technologies Inc.	X	X				X	X	X	X	
USA	American Bank Note	X	X	X			X	X	X	X	
USA	Automated Holo Systems		X	X	X						
USA	Avant-Garde Studio	X	X								
USA	Blue Ridge Holographics		X				X	X	X	X	
USA	Bridgestone Graphic Tec						X				
USA	Broadbent Consulting			X	X	X	X				
USA	Casdin-Silver Holography			X							
USA	CFC Applied Holo	X	X				X	X	X	X	
USA	Cherry Optical Holo	X	X	X		X		X	X	X	
USA	Chromagem Inc.		X				X	X	X	X	
USA	Crown Roll Leaf, Inc..	X	X	X			X	X	X		
USA	Deem, Rebecca	X	X	X	X	X	X	X	X	X	X
USA	Diamond Images, Inc.		X					X			
USA	Dimensional Arts	X	X				X	X	X	X	
USA	Doris Vila Holographics		X	X							
USA	Engineering Animation,		X								
USA	Fast Light Inc.						X				
USA	Feroe, James			X			X	X	X	X	
USA	FLEXcon	X	X								
USA	Forth Dimension Holo.			X							
USA	Frank DeFreitas			X				X		X	
USA	General Design		X								
USA	Gorglione, Nancy			X		X			X	X	X
USA	Holicon Corporation.	X	X	X							X
USA	Holo Sciences, LLC	X	X	X				X	X	X	
USA	Holo-Spectra						X				
USA	Holo/Source Corporation		X	X	X	X	X	X	X	X	
USA	HoloCom		X				X				
USA	Hologram Research, Inc.		X	X			X	X	X	X	
USA	Holograms and Lasers			X						X	X
USA	Holographic Design Sys	X	X	X		X	X	X	X	X	
USA	Holographic Dimensions	X	X	X			X	X	X	X	
USA	Holographic Images Inc.			X					X	X	
USA	Holographic Label	X	X				X			X	
USA	Holographic Studios		X	X				X		X	
USA	Holographics Inc.			X					X		X
USA	Holographics North Inc.		X	X				X	X	X	
USA	Images Company	X	X	X							
USA	Imagination Plantation										
USA	Infinity Laser Laboratories	X	X	X		X			X	X	X
USA	James River Products						X				
USA	Jeffery Murray Holo			X					X		
USA	Krystal Holographics Intl.	X	X		X	X			X		
USA	Larry Lieberman			X		X		X	X	X	
USA	Lasart Ltd.	X	X		X	X					
USA	Laser Holo Workshop	X	X								
USA	Laser Images			X	X	X	X	X	X	X	
USA	Laser Reflections	X	X	X		X					X
USA	Lasersmith, Inc.(The)	X	X	X		X	X	X	X	X	
USA	Light Impressions Intl.		X								
USA	Lightrix, Inc.	X	X	X		X	X	X	X	X	
USA	Linda Law Holographics		X								
USA	Lopez's Gallery Intl							X			
USA	Louis Paul Jonas Studios	X	X								
USA	Marks, Gerald		X								

Mass Production

Country	Business Name	Silver Halide Glass	Silver Halide Film	Dichromate	Photopolymer	Embossed
Australia	Australian Holographics	X	X			
Australia	Holograms Fantastic	X	X			X
Canada	Holocrafts			X		
Canada	Melissa Crenshaw	X	X			
France	H.Ologrammi	X	X			
Germany	Hologram Co. RAKO					X
Germany	HoloVision	X	X			
Germany	Krystal Holo Vertriebs				X	
Germany	topac GmbH					X
Hungary	Hologram Varga Miklos				X	
India	Hi-Glo Holo Images					X
India	Ojasmit Holographics		X	X	X	X
Lithuania	Geola		X			
Mexico	Hologramas, S.A.					X
Panama	Holographic Dim					
Poland	Holographic Dim.					
South Africa	Synchron Pty Ltd.					X
Ukraine	Feofaniya Ltd.					X
UK	3D-4D Holographics	X	X			
UK	Advanced Holo. Lab					X
UK	Applied Holographics,		X			X
UK	De La Rue Holographics					X
UK	Embossing Technology					X
UK	Light Impressions Intl					X
UK	Op-Graphics	X	X			
UK	OpSec - England					X
UK	STI - Europe					X
UK	The London Holo Studio	X	X			
USA	3M - Safety Sys					X
USA	A.D. Tech					X
USA	AD 2000				X	X
USA	Advanced Holo Labs					X
USA	Amagic Technologies					X
USA	American Bank Note Holo					X
USA	Bridgestone Graphic Tech					
USA	CFC Applied Holographics					X
USA	Cherry Optical Holography	X	X		X	
USA	Coburn Corporation					X
USA	Control Module Inc.					X
USA	Crown Roll Leaf, Inc.,					X
USA	Datacard Corporation					
USA	Dimensional Arts					X
USA	FLEXcon					X
USA	FoilMark Holo. Images					X
USA	Frank DeFreitas Holo	X	X			
USA	Holicon Corporation.	X	X			
USA	Holo Sciences, LLC	X				
USA	Holo-Spectra					X
USA	Holo/Source Corporation					X
USA	Hologram Research, Inc.	X	X			X
USA	Holograms and Lasers Intl	X	X			
USA	Holographic Design Sys.		X		X	X
USA	Holographic Dimensions					X
USA	Holographic Images Inc.		X			
USA	Holographic Label Convet					X
USA	Holographic Studios	X	X			
USA	Holographics North Inc.		X			
USA	James River Products					X
USA	Jeffery Murray	X	X			

Mass Production

Country	Business Name	Silver Halide Glass	Silver Halide Film	Dichromate	Photopolymer	Embossed
USA	Krystal Holographics Intl.			X	X	
USA	Lasart Ltd.			X		
USA	Laser Reflections	X				
USA	Light Impressions Intl.					X
USA	Lightrix, Inc.				X	X
USA	Multiplex Moving Holo.		X			
USA	Odhner Holographics	X				
USA	OpSec - Corporate HQ					X
USA	OpSec - USA					X
USA	Optimation Holographics					X
USA	Pennsylvania Pulp & Paper					X
USA	Photon Cantina Ltd.		X			
USA	Polaroid Corporation				X	
USA	Polymer Image				X	
USA	Red Beam, Inc.				X	X
USA	Reynolds Metals Co.					X
USA	Richard Bruck Holography	X	X			
USA	Robert Sherwood Holo.					X
USA	Scharr Industries					X
USA	STI					X
USA	The HOLOS Corporation	X	X		X	
USA	The Regal Press Inc.					X
USA	Transfer Print Foils, Inc.					X
USA	Van Leer Metallized Prod.		-			X
USA	Virtual Image					X
USA	Witchcraft Tape Products					X

Businesses That Supply Recording Materials

Country	Business Name	Silver Halide	Dichromate	Photopolymer	Photoresist
Belgium	Agfa - Gevaert N.V.	X			
Denmark	Ibsen Micro Structures A/S				X
Germany	HRT Holographic Recording Tech.	X			
Russia	Slavich Joint Stock Company	X			
UK	Uk Optical Supplies	X			
USA	3 Deep Hologram Company	X			
USA	Agfa - a division of Bayer Corp.	X			
USA	Dimensional Arts				X
USA	E.I. DuPont De Nemours & Co.			X	
USA	Eastman Kodak Company	X			
USA	Hologram Research, Inc.	X			
USA	Images Company	X			
USA	Integraf	X			
USA	Jodon Inc.	X			
USA	Keystone Scientific Co.	X			
USA	Process Technologies				X
USA	Shipley Chemical Co.				X
USA	Towne Technologies				X

Holographic Optical Printers (HOP)

Country	Business Name	dot matrix	step and repeat	Custom Made HOP	Turn Key Complete System Including HOP
UK	Global Images				X
UK	Spatial Imaging Ltd.	X			X
USA	Dimensional Arts	X	X		X
USA	Holo Sciences, LLC			X	
USA	Man/Environment, Inc.	X	X	X	X
USA	New Light Industries			X	X
USA	Potomac Photonics, Inc.	X	X	X	X
USA	Silhouette Technology Inc.		X	X	
USA	Silver Dragon Holography	X			
USA	Ultra-Res Corporation				X

Laser and Optical Supply Businesses

Country	Business Name	Laser Mfg - Distributor	Used Lasers	Laser Power supplies	Optics
Canada	Harvard Apparatus, Canada				X
Canada	Lasiris Inc.				X
Germany	Adlas G.M.B.H. & Co Kg.	X			
Germany	Andreas Wappelt - Photonics Direct	X			
Germany	Coherent Luebeck GmbH	X			
Germany	Gresser, E., KG	X			
Germany	Melles Griot GmbH				X
Germany	Newport Gmbh	X			X
Germany	Optical Coating Laboratory GmbH				X
Germany	Optical Test Equipment				X
Germany	OWIS Gmbh				X
Germany	Physik Instrumente (PI) GmbH & Co.				X
Germany	Rofin-Sinar Laser GmbH	X			
Germany	Spindler & Hoyer GmbH & Co.				X
Germany	Steinbichler Optotechnik GmbH				X
Germany	University of Erlanger				X
Japan	Kimmon Electric Co., Ltd.	X			
Lithuania	Geola	X			X
Mexico	Centro de Investigaciones en Optica,				X
Ukraine	Environmental Education and Info.				X
Ukraine	Yarovoy, Leonid				X
UK	Ag Electro-Optics Ltd.	X			X
UK	Barr & Stroud, Ltd.				X
UK	Customer Service Instrumentation				X
UK	Datasights Ltd.				X
UK	Electro Optics Developments Ltd.				X
UK	Expanded Optics Limited				X
UK	Galvoptics Ltd.				X
UK	Kendall Hyde Ltd.				X
UK	Lumonics Ltd.	X			
UK	Optical Works Ltd.				X
UK	Ralph Cullen Holographics				X
UK	Richmond Holographic Studios	X			
UK	Uk Optical Supplies		X		X
USA	Advanced Optics, Inc.				X
USA	Aerotech Inc.	X		X	X
USA	American Holographic Inc.				X
USA	American Laser Corporation	X			
USA	American Paper Optics Inc.				X
USA	APA Optics Inc.				X
USA	Burleigh Instruments, Inc.				X
USA	Coherent, Inc. - Laser Group	X			
USA	Continental Optical				X
USA	Control Optics				X
USA	Corion Corp.				X
USA	CVI Laser Corporation				X
USA	Dell Optics Company, Inc.				X
USA	Diffraction Ltd.				X
USA	Ealing Electro-Optics Inc.	X			X
USA	Edmund Scientific Company	X			X
USA	El Don Engineering		X		
USA	Electro Optical Industries, Inc.				X
USA	Fresnel Technologies Inc.				X
USA	G.M. Vacuum Coating Lab, Inc.				X
USA	Glass Mountain Optics				X
USA	Holo-Spectra				X
USA	Holographic Optics Inc.				X
USA	Holotek				X

Laser and Optical Supply Businesses

Country	Business Name	Laser Mfg - Distributor	Used Lasers	Laser Power supplies	Optics
USA	Holovision Systems Inc.				X
USA	Honeywell Technology Center				X
USA	Images Company	X	X	X	X
USA	ImEdge Technology				X
USA	Infrared Optical Products, Inc.				X
USA	Inrad, Inc.				X
USA	Ion Laser Technology, Inc.	X		X	
USA	Jodon Inc.	X		X	X
USA	Kaiser Optical Systems, Inc.				X
USA	Kan, Mike				X
USA	Kinetic Systems, Inc.				X
USA	Kreischer Optics, Ltd.				X
USA	Laser and Motion Development Co.	X	X		X
USA	Laser Drive Inc.			X	
USA	Laser Innovations	X	X		
USA	Laser Optics, Inc.				X
USA	Laser Resale Inc.				X
USA	Laser Technical Services	X			
USA	Lenox Laser				X
USA	Lexel Laser, Inc.	X		X	
USA	LiCONiX	X			
USA	Melles Griot	X			X
USA	Meredith Instruments	X	X		
USA	Metrologic Instruments, Inc.	X			X
USA	Midwest Laser Products	X	X		
USA	Mitutoyo Measuring Inst.				X
USA	MWK Industries	X	X		
USA	Navidec Inc. (formerly ACI Sys, Inc.)	X			
USA	Newport Corporation				X
USA	O/E Research				X
USA	Odhner Holographics				X
USA	Omnichrome Corp.	X		X	
USA	Optical Corporation Of America				X
USA	Optics Plus Inc.				X
USA	Optimation				X
USA	Optineering				X
USA	Oriel Instruments				X
USA	Pasco Scientific	X			X
USA	Peacock Laboratories, Inc.				X
USA	Potomac Photonics, Inc.				X
USA	PullTime 3-D Laboratories				X
USA	Richardson Grating Laboratory				X
USA	Rochester Photonics Corporation				X
USA	Rolyn Optics				X
USA	Science Kit & Boreal Labs	X	X		X
USA	Spectra-Physics Lasers Inc.	X		X	
USA	Swift Instruments				X
USA	Syracuse University				X
USA	Thorlabs Inc.				X
USA	Uniphase Lasers	X			
USA	Wavefront Research, Inc.				X

Holographic Computer Memory/Storage

Country	Business Name	Computer Storage
Germany	Technische Universitaet Berlin	X
USA	IBM Almaden Research Center	X
USA	Innovative Technology Associates	X
USA	Laboratory for Optical Data Processing	X
USA	Optitek	X
USA	Tamarack Storage Devices	X

Mass Replication Equipment

Country	Business Name	silver halide	dichromate glass blanks. etc.	Photo-polymer	embossed: Web Press	embossing Foil
Germany	Baier Praegepressen				X	
Germany	Hologram Company RAKO GmbH				X	
Germany	Holographic Systems Muenchen				X	
Germany	Leonhard Kurz GmbH				X	
UK	Action Tapes					X
UK	Beddis Kenley (Machinery) Ltd.				X	
UK	Dimuken				X	
UK	Global Images				X	X
UK	Light Impressions International, Ltd.				X	X
UK	Whiley Foils Limited					X
USA	21st Century Finishing Inc.					
USA	3M - Safety and Security Systems					X
USA	Crown Roll-Leaf, Inc..					X
USA	Dimensional Arts				X	X
USA	E.I. DuPont			X		
USA	FoilMark Holographic Images					X
USA	Holo-Spectra				X	
USA	ICI Polyester					X
USA	James River Products				X	
USA	Krystal Holographics Intl Inc.			X		
USA	Light Impressions International. Ltd.				X	X
USA	New Light Industries				X	
USA	Norland Products, Inc.	X				
USA	Peacock Laboratories, Inc.				X	
USA	Reynolds Metals Co.					X
USA	Silver Dragon Holography				X	
USA	Transfer Print Foils, Inc.					X

Industrial Holography

Country	Business Name	Acoustical Holography	Medical Displays	NDT Industrial	NDT Bio-Medical	HOE	CGH
Canada	Laser Inspeck			X			
Canada	Lasiris Inc.					X	
Finland	Heptagon Oy					X	
Germany	BIAS			X			
Germany	Daimler Benz Aerospace			X		X	
Germany	Dornier Medizintechnik GmbH			X			
Germany	Institut fur Angewandte Physik			X			
Germany	Leseberg. Dr. Detlef					X	X
Germany	Steinbichler Optotechnik			X			
Germany	Technische Fachhochschule					X	
Germany	Technische Universitaet Berlin						X
Germany	Technolas Laser Technik				X		
Germany	University of Muenster				X		
Germany	University of Munich			X	X		
Germany	University of Stuttgart			X		X	
Germany	Volkswagen AG			X			
Israel	Glaser - Technical Consulting					X	
Mexico	Centro de Investigaciones			X			
Slovakia	Technical University Zvolen			X			
Sweden	KTH			X			
Ukraine	Environmental Education and Info			X		X	
Ukraine	Feofaniya Ltd.			X			
Ukraine	The Institute of Applied Optics			X			
UK	Barr & Stroud, Ltd.					X	
UK	British Aerospace Plc.			X			
UK	Electro Optics Developments Ltd.					X	
UK	Expanded Optics Limited					X	
UK	National Physical Laboratory			X			
UK	Pilkington Optronics					X	

		Industrial Holography					
Country	Business Name	Acoustical Holography	Medical Displays	NDT Industrial	NDT Bio-Medical	HOE	CGH
UK	Richmond Holographic Studios					X	
UK	Rolls-Royce Plc			X			
UK	Rutherford & Appleton Labs			X			
USA	American Propylaea Corporation		X			X	
USA	APA Optics Inc.						X
USA	Applied Optics					X	
USA	Automated Holographic Systems					X	
USA	Batelle Pacific NW National Labs	X	X				
USA	Broadbent Consulting Services					X	
USA	Diffraction Ltd.					X	
USA	Flight Dynamics					X	
USA	Holographic Optics Inc.					X	
USA	Holographics Inc.			X			
USA	Holovision Systems Inc.		X				
USA	Illinois Institute Of Technology			X			
USA	ImEdge Technology					X	
USA	Industrial Technology Institute			X			
USA	Integraf			X		X	
USA	Lab for Optical Data Processing						X
USA	Laser Technology, Inc.			X			
USA	Lawrence Berkeley Laboratory			X			
USA	MasterPrint Holography, Inc.					X	
USA	McMahan Electro-Optic			X			
USA	Media Interface, Ltd.		X				
USA	MetroLaser			X	X		
USA	Northern Illinois University	X			X		
USA	Odhner Holographics					X	
USA	Optical Research Services						X
USA	Optineering			X			
USA	Physical Optics Corporation.					X	
USA	Potomac Photonics, Inc.						X
USA	Ralcon					X	
USA	Rice Systems			X	X	X	
USA	Rochester Inst. Of Technology					X	X
USA	Rochester Photonics Corporation					X	
USA	Rowland Institute For Science			X			
USA	Saginaw Valley State University					X	
USA	San Jose State University					X	
USA	Silhouette Technology Inc.					X	
USA	Sinclair Optics, Inc.						X
USA	Stanford University			X			
USA	University Of Alabama			X			
USA	University Of Arizona			X			
USA	University Of Dayton			X			
USA	University Of Michigan					X	
USA	University Of Rochester			X			
USA	University Of Southern California			X			
USA	Voxel		X				
USA	Wavefront Research, Inc.					X	
USA	Worcester Polytechnic Institute			X			
USA	Wyko Corporation			X	X		

Country	Business Name	Assoc. & Groups	Child Classes	College	Private - Regularly Scheduled	Private Instruction	Literature	Seminars
Canada	Capilano College			X				
Canada	Centre d'Art Holo et Photonique			X				
Canada	Holography & Media Inst-Quebec				X	X		
Canada	Mu's Laser Works					X		
Canada	Ontario College Of Art			X				X
Canada	Ontario Science Centre					X		X
Canada	Photon League Of Holographers				X	X		X
Canada	Universite Laval			X				X
Germany	Academy of Media Arts Cologne				X	X		X
Germany	AHT 3D-Medien	X						
Germany	Deutsche Gesellschaft fur Holo	X						
Germany	Holographie Anubis							X
Germany	Holopublic Unbehaun	X						
Israel	Glaser - Technical Consulting					X		
Sweden	KTH			X				
Thailand	Electro-Optics Lab					X		
UK	Creative Holography Index							
UK	Graham Saxby							
UK	Holograms 3D							X
UK	Imperial College Of Science			X				
UK	Intl Hologram Mfg Association	X						
UK	Loughborough Univ. Of Tech.			X				
UK	Ralph Cullen Holographics							X
UK	Reconnaissance International Ltd.							X
UK	University Of Oxford			X				
USA	Advanced Tech. Program	X						
USA	Alabama A&M University			X				
USA	Art Institute Of Chicago			X				
USA	Art Lab				X			
USA	Berkhout, Rudie					X		
USA	Carl M. Rodia And Associates							X
USA	Dan							X
USA	DeFreitas, Frank						X	
USA	Doris Vila Holographics				X	X		
USA	Foil Stamping & Embossing Ass.	X						
USA	HoloCom							X
USA	Holographic Products							X
USA	Holographic Studios					X		
USA	Holography Inst of San Francisco					X		X
USA	Holography Marketplace							
USA	Holophile							X
USA	Industrial Technology Inst.	X						X
USA	Inside Finishing Magazine							
USA	Integraf							
USA	James River Products							X
USA	Lake Forest College			X	X			X
USA	Laser Affiliates	X						X
USA	Laser Arts Soc For Edu & Research	X						
USA	Laser Focus World							
USA	Laser Institute Of America	X						
USA	M.I.T. (Mass Inst of Technology)			X				
USA	MacShane Holography		X			X		
USA	Museum Of Holography/Chicago		X		X	X		X
USA	NeoVision Productions					X		
USA	New York Holographic Lab					X		X
USA	O/E Research					X		

Education, Associations and Publications								
Country	Business Name	Assoc. and groups	Child Classes	College	Private - Regularly Scheduled Courses	Private Instruction	Literature	Seminars
USA	Optical Society of America (OSA)	X					X	
USA	Oregon Institute of Technology			X				
USA	Oregon Laser Consultants							X
USA	Photonics Spectra						X	
USA	Reconnaissance International Ltd.						X	X
USA	Rochester Inst. Of Technology			X				
USA	Ross Books						X	X
USA	Saginaw Valley State University			X				
USA	Saint Mary's College			X				
USA	School Of Holography				X	X		X
USA	Sonoma State University			X				
USA	SPIE	X						
USA	SPIE's Holo Working Group Newsletter	X					X	
USA	Springer-Verlag New York						X	
USA	Synchronicity Holograms		X			X		X
USA	Technical Marketing Services						X	
USA	University Of Michigan			X				
USA	University Of Rochester			X				
USA	University Of Wisconsin/Madison							
USA	Vincennes University		X	X				
USA	Visual Visionaries							X
USA	Wesley, Ed					X		X
USA	Worcester Polytechnic Institute			X				

Web (Internet) Addresses

Country	Business Name	Web address
Australia	Australian Holographics	camtech.com.au/~austholo
Canada	Royal Holographic Art Gallery	www.islandnet.com/~roylal/index.htm
Finland	Heptagon Oy	heptagon.fi
Germany	AKS Holographie-Galerie GmbH	http://members.aol.com/akshol/hhome_d.htm
Germany	Andreas Wappelt - Photonics Direct	http://home.t-online.de/home/wappelt/
Germany	Institut fur Angewandte Physik	http://www.physik.th-darmstadt.de/andreas
Germany	Optical Test Equipment	moeller-wedel.com
Germany	Technische Universitaet Berlin	http://www.physik.tu-berlin.de
Germany	topac GmbH, Department Holography	tophol@aol.com
Japan	Kimmon Electric Co., Ltd.	kimmon.com
Lithuania	Geola	camtech.com.au/~austholo
Mexico	Hologramas, S.A. de C.V.	www.holomex.com.mx
Netherlands	Dutch Holographic Laboratory BV	euroweb.com/DHL
Netherlands	Triple-D Laser Imaging	wirehub.nl?~aca/
UK	3D-4D Holographics	hologram.demon
UK	Light Impressions International, Ltd.	euroweb.com/nlcon/b2b/ehol/limpress.htm
UK	National Physical Laboratory	http://www.npl.co.uk
UK	Reconnaissance International Ltd.	hmt.com/holography/hnews/hnhome.htm
UK	Spatial Imaging Ltd.	dircon.co.uk/spatial/
UK	Turing Institute	turing.gla.ac.uk
USA	3 Deep Hologram Company	member.aol.com/the3d/home.htm
USA	3-D Worldwide Holograms, Inc.	www.3dworldwide.com
USA	3M - Safety and Security Systems	www.mmm.com
USA	AD 2000, Inc.	ad2000.com/ad2000/
USA	Advanced Technology Program	atp.nist.gov
USA	Amagic Technologies Inc.	www.thomasregister.com/amagic
USA	Coburn Corporation	coburn.com
USA	Coherent, Inc. - Laser Group	cohr.com
USA	Control Module Inc.	controlmod.com
USA	Control Optics	http://ourworld.compuserv/home page/control optics
USA	DeFreitas, Frank	holoworld.com
USA	Diamond Images, Inc.	DiamondImages.com
USA	Dimensional Arts	www.holo.com
USA	Fisher Scientific	fisheredu.com
USA	FoilMark Holographic Images	foilmark.com
USA	Frank DeFreitas Holography Studio	holoworld.com
USA	General Design	sfo.com/~bk
USA	Holart Consultants	holo.com/gaz/
USA	HoloCom	holo.com
USA	Hologram Research, Inc.	hologramres.com
USA	Holograms and Lasers International	holoshop.com
USA	Holograms and Lasers International	holoshop.com
USA	Holographic Design Systems	http://www.cris.com/~museumh
USA	Holographic Dimensions	shadow.net/~holodi
USA	Holographic Studios	hmt.com/holography/holostudios
USA	Holography Marketplace	holoinfo.com
USA	Holophile, Inc.	www.connix.com/~barefoot
USA	HOLOS Corporation	holoscorp.com/holos
USA	HoloWebs, Inc.	holomwx.com.mx
USA	ICI Polyester	icipolyseter.com
USA	Images Company	he.net/~imagesco
USA	Imagination Plantation	www.iplant.com
USA	Industrial Technology Institute	iti.org
USA	Kreischer Optics, Ltd.	www.kreisher.com
USA	Krystal Holographics International Inc.	www.khiinc.com
USA	Krystal Holographics International Inc.	www.khiinc.com
USA	Laser and Motion Development Co	lasermotion.com
USA	Laser Institute Of America	creol.ucf.edu/~lia/

Web (Internet) Addresses		
Country	Business Name	Web address
USA	Laser Resale Inc.	laserresale.com
USA	Lasersmith, Inc.(The)	www.lasersmity.com
USA	Laserworks	ourworld.compuserve.com;80 homepage/laserworks/
USA	Lightrix, Inc.	ionserve.com lightrix.html
USA	M.I.T. (Mass. Institute of Technology)	media.mit.edu/groups/spi
USA	Man/Environment, Inc.	armchair.com
USA	Marks, Gerald	nttad.com/asci/gmwork.html
USA	Media Interface, Ltd.	bway.net/~ronholog
USA	Melles Griot	www.mellesgriot.com
USA	Meredith Instruments	www.mi-lasers.com
USA	MetroLaser	www.pages.prodgy.com/metrolaser/home.htm
USA	Metrologic Instruments, Inc.	www.metrologic.com
USA	Midwest Laser Products	www.midwest-laser.com lasers
USA	Mitutoyo Measuring Inst (MTI Corp.)	industry.net mitutoyo
USA	MWK Industries	www.pweb.com/mwk/main.htm
USA	New Light Industries	www.iea.com/~nli
USA	Newport Corporation	newport.com
USA	Optical Research Services	www.opticalres.com
USA	Optical Society of America (OSA)	www.osa.org
USA	Oregon Institute of Technology	oit.osshe.edu
USA	Oriel Instruments	www.oriel.com
USA	Pennsylvania Pulp & Paper Co.	holoprism.com
USA	Polaroid Corporation	holoroid.com or www.poloroid.com
USA	Potomac Photonics, Inc.	potomac_laser.com
USA	Process Technologies	www.exepc.com/~pti
USA	PullTime 3-D Laboratories	nttad.com/asci/gmwork.html
USA	Rainbow Symphony Inc.	www.rainbowsymphony.com
USA	Reynolds Metals Co.	rmc.com
USA	Richardson Grating Laboratory	gratinglab.com
USA	Rochester Photonics Corporation	http://www.rphotonics.com
USA	Ross Books	www.holoinfo.com
USA	San Jose State University	http://fire.sjsu.edu
USA	Sandia National Laboratories	-irn.sandia.gov/
USA	Sharon McCormack Holography	gorge.net business/holography
USA	Sillcocks Plastics International	sillcocks.com
USA	Spectra-Physics Lasers Inc.	www.splasers.com
USA	Springer-Verlag New York	www.springer-ny.com
USA	Star Magic	www.starmagic.com
USA	Star Magic	www.starmagic.com
USA	STI	stiovd.com
USA	Syracuse University	-che.syr.edu/
USA	Thorlabs Inc.	thorlabs.com
USA	Towne Technologies	townetech.com
USA	U.K. Gold Purchasers, Inc.	www.eden.com/~mainlink/art/rai/index.htm
USA	University Of Dayton	udri.udayton.edu
USA	University Of Rochester	optics.rochester.edu:8080/
USA	Wyko Corporation	wyko.com

Zip Code, State, Business - USA Listings		
20009	DC	The Holo. Laser & Photonics Ctr.
20036	DC	Optical Society of America (OSA)
20706	MD	Potomac Photonics
20770	MD	Holographic Applications
20899	MD	Advanced Technology Program
21057	MD	Lenox Laser
21152	MD	OpSec - USA
21202	MD	Holografica
21209	MD	M.O.M. Inc.
22902	VA	Blue Ridge Holographics
22906	VA	Nimbus Manufacturing
23230	VA	Reynolds Metals Co.
23235	VA	Silver Dragon Holography
23236	VA	James River Products
29577	SC	The Hologram Company #7
29582	SC	The Hologram Company #2
30076	GA	Virtual Image
32789	FL	McMahan Electro-Optic
32809	FL	The Hologram Company #3
32826	FL	Laser Institute Of America
32856	FL	Odhner Holographics
33029	FL	Evolution Design
33133	FL	Diamond Images
33133	FL	Zero Gravity
33139	FL	Holographic Images Inc.
33139	FL	Larry Lieberman Holography
33166	FL	3-D Worldwide Holograms
33177	FL	Holographic Dimensions
33442	FL	Another Dimension
33442	FL	Galaxies Unlimited
33868	FL	Infinity Laser Laboratories
35762	AL	Alabama A&M University
35899	AL	University Of Alabama
38132	TN	American Paper Optics Inc.
39201	MS	Barilleaux, Rene Paul
44509	OH	Chromagem Inc.
45469	OH	University Of Dayton
45840	OH	Holovision Systems Inc.
46556	IN	Saint Mary's College
47448	IN	Forth Dimension Holographics
47591	IN	Vincennes University
47715	IN	Southern Indiana Holographics
48009	MI	Intrepid World Communications
48009	MI	American Propylaea Corporation
48103	MI	Jodon Inc.
48103	MI	Applied Optics
48105	MI	Industrial Technology Institute
48106	MI	Kaiser Optical Systems
48109	MI	University Of Michigan
48150	MI	Holo/Source Corporation
48408	MI	El Don Engineering
48710	MI	Saginaw Valley State University
48734	MI	Reva's Holographic Illusions
49038	MI	Witchcraft Tape Products
49117	MI	Laser Holography Workshop
49456	MI	Textile Graphics
49456	MI	Holography Presses On (HPO)
50010	IA	Engineering Animation
53097	WI	McCain Marketing & Design
53154	WI	Process Technologies

Zip Code, State, Business - USA Listings		
53186	WI	Letterhead Press
53706	WI	University Of Wisconsin/Madison
54025	WI	Brandtjen & Kluge
55105	MN	Holographic Products
55144	MN	3M - Safety and Security Systems
55418	MN	Honeywell Technology Center
55420	MN	Advanced Optics
55425	MN	Hologram Fantastic
55425	MN	Hologram Land
55439	MN	Holographic Label Converting
55440	MN	Datacard Corporation
55441	MN	Hologram World
55449	MN	APA Optics Inc.
60004	IL	MacShane Holography
60007	IL	Creative Label
60045	IL	Integraf
60045	IL	Lake Forest College
60048	IL	Holographic Industries
60050	IL	Kreischer Optics
60115	IL	Northern Illinois University
60411	IL	CFC Applied Holographics
60423	IL	Midwest Laser Products
60504	IL	Mitutoyo Measuring Instruments
60521	IL	Fisher Scientific
60603	IL	Art Institute Of Chicago
60607	IL	Holographic Design Systems
60607	IL	Lasersmith
60607	IL	Museum Of Holography/Chicago
60607	IL	School Of Holography
60611	IL	Light Wave Gallery
60611	IL	Lopez's Gallery International
60612	IL	Wave Mechanics
60614	IL	Light Impressions International
60616	IL	Illinois Institute Of Technology
60618	IL	Holicon Corporation.
60618	IL	Richard Bruck Holography
60618	IL	Wesley
60647	IL	Robert Sherwood Holographic
66046	KS	Optimation Holographics
66046	KS	Advanced Holographic Labs
66206	KS	Laser Images
66212	KS	Fast Light Inc.
70112	LA	The Hologram Company #6
75202	TX	Elusive Image
76110	TX	Fresnel Technologies Inc.
77010	TX	Holograms and Lasers Intl.
77010	TX	Holograms and Lasers Intl.
78216	TX	U.K. Gold Purchasers
78727	TX	Tamarack Storage Devices
78746	TX	Lone Star Illusions
78758	TX	Glass Mountain Optics
80112	CO	Navidec Inc.
80205	CO	Reconnaissance International Ltd.
80214	CO	Lazer Wizardry
80222	CO	OpSec - Corporate Headquarters
80249	CO	Fantastic Holograms
81621	CO	Imagen Holography
84092	UT	Optimation
84104	UT	American Laser Corporation
84115	UT	Ion Laser Technology

Zip Code, State, Business - USA Listings		
20009	DC	The Holo. Laser & Photonics Ctr.
20036	DC	Optical Society of America (OSA)
20706	MD	Potomac Photonics
20770	MD	Holographic Applications
20899	MD	Advanced Technology Program
21057	MD	Lenox Laser
21152	MD	OpSec - USA
21202	MD	Holografica
21209	MD	M.O.M. Inc.
22902	VA	Blue Ridge Holographics
22906	VA	Nimbus Manufacturing
23230	VA	Reynolds Metals Co.
23235	VA	Silver Dragon Holography
23236	VA	James River Products
29577	SC	The Hologram Company #7
29582	SC	The Hologram Company #2
30076	GA	Virtual Image
32789	FL	McMahan Electro-Optic
32809	FL	The Hologram Company #3
32826	FL	Laser Institute Of America
32856	FL	Odhner Holographics
33029	FL	Evolution Design
33133	FL	Diamond Images
33133	FL	Zero Gravity
33139	FL	Holographic Images Inc.
33139	FL	Larry Lieberman Holography
33166	FL	3-D Worldwide Holograms
33177	FL	Holographic Dimensions
33442	FL	Another Dimension
33442	FL	Galaxies Unlimited
33868	FL	Infinity Laser Laboratories
35762	AL	Alabama A&M University
35899	AL	University Of Alabama
38132	TN	American Paper Optics Inc.
39201	MS	Barilleaux, Rene Paul
44509	OH	Chromagem Inc.
45469	OH	University Of Dayton
45840	OH	Holovision Systems Inc.
46556	IN	Saint Mary's College
47448	IN	Forth Dimension Holographics
47591	IN	Vincennes University
47715	IN	Southern Indiana Holographics
48009	MI	Intrepid World Communications
48009	MI	American Propylaea Corporation
48103	MI	Jodon Inc.
48103	MI	Applied Optics
48105	MI	Industrial Technology Institute
48106	MI	Kaiser Optical Systems
48109	MI	University Of Michigan
48150	MI	Holo/Source Corporation
48408	MI	El Don Engineering
48710	MI	Saginaw Valley State University
48734	MI	Reva's Holographic Illusions
49038	MI	Witchcraft Tape Products
49117	MI	Laser Holography Workshop
49456	MI	Textile Graphics
49456	MI	Holography Presses On (HPO)
50010	IA	Engineering Animation
53097	WI	McCain Marketing & Design
53154	WI	Process Technologies

Zip Code, State, Business - USA Listings		
53186	WI	Letterhead Press
53706	WI	University Of Wisconsin/Madison
54025	WI	Brandtjen & Kluge
55105	MN	Holographic Products
55144	MN	3M - Safety and Security Systems
55418	MN	Honeywell Technology Center
55420	MN	Advanced Optics
55425	MN	Hologram Fantastic
55425	MN	Hologram Land
55439	MN	Holographic Label Converting
55440	MN	Datacard Corporation
55441	MN	Hologram World
55449	MN	APA Optics Inc.
60004	IL	MacShane Holography
60007	IL	Creative Label
60045	IL	Integraf
60045	IL	Lake Forest College
60048	IL	Holographic Industries
60050	IL	Kreischer Optics
60115	IL	Northern Illinois University
60411	IL	CFC Applied Holographics
60423	IL	Midwest Laser Products
60504	IL	Mitutoyo Measuring Instruments
60521	IL	Fisher Scientific
60603	IL	Art Institute Of Chicago
60607	IL	Holographic Design Systems
60607	IL	Lasersmith
60607	IL	Museum Of Holography/Chicago
60607	IL	School Of Holography
60611	IL	Light Wave Gallery
60611	IL	Lopez's Gallery International
60612	IL	Wave Mechanics
60614	IL	Light Impressions International
60616	IL	Illinois Institute Of Technology
60618	IL	Holicon Corporation.
60618	IL	Richard Bruck Holography
60618	IL	Wesley
60647	IL	Robert Sherwood Holographic
66046	KS	Optimation Holographics
66046	KS	Advanced Holographic Labs
66206	KS	Laser Images
66212	KS	Fast Light Inc.
70112	LA	The Hologram Company #6
75202	TX	Elusive Image
76110	TX	Fresnel Technologies Inc.
77010	TX	Holograms and Lasers Intl.
77010	TX	Holograms and Lasers Intl.
78216	TX	U.K. Gold Purchasers
78727	TX	Tamarack Storage Devices
78746	TX	Lone Star Illusions
78758	TX	Glass Mountain Optics
80112	CO	Navidec Inc.
80205	CO	Reconnaissance International Ltd.
80214	CO	Lazer Wizardry
80222	CO	OpSec - Corporate Headquarters
80249	CO	Fantastic Holograms
81621	CO	Imagen Holography
84092	UT	Optimation
84104	UT	American Laser Corporation
84115	UT	Ion Laser Technology

Zip Code, State, Business - USA Listings

84321	UT	Krystal Holographics Intl
84328	UT	Ralcon
84333	UT	Richmond Development Group
85027	AZ	Polymer Image
85301	AZ	Meredith Instruments
85704	AZ	Holo Sciences
85706	AZ	Wyko Corporation
85721	AZ	University Of Arizona
87185	NM	Sandia National Laboratories
87192	NM	CVI Laser Corporation
87501	NM	Lasart Ltd.
87506	NM	Ana MacArthur
88005	NM	Dimensional Arts
88005	NM	HoloCom
89109	NV	Zero Gravity
89129	NV	Laser Las Vegas
89431	NV	Ultra-Res Corporation
90004	CA	NeoVision Productions
90024	CA	JR Holographics
90027	CA	Booth, Roberta
90045	CA	Laser Media
90064	CA	Man/Environment
90066	CA	Spectratek Inc.
90089	CA	University Of Southern California
90245	CA	Hughes Power Products
90402	CA	Chronomotion
90501	CA	Physical Optics Corporation
90703	CA	MGM Converters Inc.
90723	CA	Wavefront Technology
91012	CA	Photon Cantina Ltd.
91107	CA	Optical Research Services
91202	CA	Rebecca
91202	CA	Unterseher & Associates
91335	CA	Rainbow Symphony Inc.
91406	CA	Holo-Spectra
91601	CA	Zone Holografix
91706	CA	Control Optics
91710	CA	Omnichrome Corp.
91720	CA	MWK Industries
91722	CA	Rolyn Optics
91769	CA	Panatron Inc.
92069	CA	Broadbent Consulting Services
92101	CA	HoloWebs
92101	CA	The Hologram Company #4
92124	CA	Jeffery Murray Holography
92606	CA	Newport Corporation
92614	CA	MetroLaser
92614	CA	Rice Systems
92629	CA	United Assoc. of Mfg. Reps
92648	CA	3 Deep Hologram Company
92648	CA	Holograms International
92651	CA	Technical Marketing Services
92653	CA	Voxel
92663	CA	G.M. Vacuum Coating Lab
92705	CA	Optics Plus Inc.
92714	CA	Amagic Technologies Inc.
92714	CA	Melles Griot
92859	CA	Laserworks
93001	CA	Innovative Technology Associates
93021	CA	Laser Innovations
93108	CA	Stephens, Anait
93111	CA	Electro Optical Industries
94039	CA	Optitek
94039	CA	Spectra-Physics Lasers Inc.
94043	CA	Silicon Graphics

Zip Code, State, Business - USA Listings

94044	CA	Real Image
94080	CA	Lightrix
94107	CA	General Design
94109	CA	The Hologram Company #1
94110	CA	Holart Consultants
94110	CA	Imagination Plantation
94110	CA	Multiplex Moving Holograms
94114	CA	Star Magic
94121	CA	Visual Visionaries
94122	CA	Holage
94122	CA	Illuminations
94124	CA	Kan, Mike
94124	CA	L.A.S.E.R. News
94124	CA	Pink, Patty
94124	CA	Holo. Institute of San Francisco
94133	CA	Zero Gravity
94305	CA	Stanford University
94404	CA	A.H. Prismatic
94509	CA	Laser Light Designs
94538	CA	Lexel Laser
94539	CA	Cambridge Laser Labs
94587	CA	Laser and Motion Development Co
94598	CA	Cifelli, Dan
94608	CA	Feroe, James
94608	CA	Wild Style Entertainment
94611	CA	Red Beam
94704	CA	Holography Marketplace
94704	CA	Ross Books
94720	CA	Lawrence Berkeley Laboratory
94901	CA	Third Dimension Arts Inc.
94928	CA	Sonoma State University
94956	CA	Kauffman, John
95006	CA	Point Source Productions
95037	CA	Unistay
95054	CA	Coherent
95054	CA	LiCONiX
95060	CA	Simian Co.
95062	CA	Pacific Holographics Inc.
95110	CA	Holographic Impressions
95110	CA	Smith & McKay Printing Co. Inc.
95112	CA	Swift Instruments
95113	CA	Laser Reflections
95120	CA	IBM Almaden Research Center
95134	CA	Uniphase Lasers
95192	CA	San Jose State University
95468	CA	3-D Systems
95472	CA	Cherry Optical Holography
95472	CA	Gorglione, Nancy
95472	CA	Laser Affiliates
95661	CA	Pasco Scientific
95814	CA	The Hologram Company #5
96813	HW	Zero Gravity
97212	OR	Foil Stamping & Embossing Ass.
97212	OR	Inside Finishing Magazine
97224	OR	Flight Dynamics
97601	OR	Oregon Laser Consultants
97601	OR	Oregon Institute of Technology
98227	WA	SPIE
98672	WA	Sharon McCormack Holography
99204	WA	New Light Industries
99352	WA	Batelle Pacific NW National Labs

Index

DID YOU BORROW THIS COPY?
GET YOUR OWN COPY DELIVERED TO YOUR DOOR!

COST (includes all shipping and handling fees*. Books shipped UPS or US Postal Air Mail):

$ 25 - Continental USA

$ 30 - Alaska, Hawaii, Canada, Mexico

$ 40 - all other countries

Note: Additional copies are **$19.95** each. Any additional shipping charges will be based on weight.

* not responsible for any additional international entry fees that may be imposed.

TO ORDER

Fill out and send us the accompanying order form by fax or by mail. Include a check (US dollars), money order (US dollars) or your credit card information.

Toll free order phone in USA **1-800-367-0930**

Voice phone **1-510-841-2474**

Fax **1-510-841-2695**

email **sales@rossbook.com**

Internet Website address (orders accepted) - http://www.**holoinfo.com**

Mailing Address: **Ross Books P.O. Box 4340 Berkeley, CA 94704**

- -

Payment Information:

Check_____Money Order_____Credit Card_____

American Express_____Mastercard_____ Visa_____

Card Number_____

Expiration Date_____

Name on Card_____

Authorized Signature _____

"Ship to" Information:

Name:_____

Company:_____

Address:_____

City/State/Province:_____

Country:_____

Postal Code:_____

Phone/Fax:_____

_____**Please send me information about reserving advertising space and rates.**

_____**Please contact me when additional or related publications become available.**

_____**Please send me information about getting my company listed in the Business Directory.**